T0156103

Lecture Notes in Computer Science 13985

The series Lecture Notes in Computer Science (LNCS), including its subseries Lecture Notes in Artificial Intelligence (LNAI) and Lecture Notes in Bioinformatics (LNBI), has established itself as a medium for the publication of new developments in computer science and information technology research, teaching, and education.

LNCS enjoys close cooperation with the computer science R & D community, the series counts many renowned academics among its volume editors and paper authors, and collaborates with prestigious societies. Its mission is to serve this international community by providing an invaluable service, mainly focused on the publication of conference and workshop proceedings and postproceedings. LNCS commenced publication in 1973.

Alberto Abelló · Panos Vassiliadis ·
Oscar Romero · Robert Wrembel
Editors

Advances in Databases and Information Systems

27th European Conference, ADBIS 2023
Barcelona, Spain, September 4–7, 2023
Proceedings

Editors
Alberto Abelló
Universitat Politècnica de Catalunya
Barcelona, Spain

Oscar Romero
Universitat Politècnica de Catalunya
Barcelona, Spain

Panos Vassiliadis
University of Ioannina
Ioannina, Greece

Robert Wrembel
Poznan University of Technology
Poznan, Poland

ISSN 0302-9743 ISSN 1611-3349 (electronic)
Lecture Notes in Computer Science
ISBN 978-3-031-42913-2 ISBN 978-3-031-42914-9 (eBook)
https://doi.org/10.1007/978-3-031-42914-9

This Springer imprint is published by the registered company Springer Nature Switzerland AG
The registered company address is: Gewerbestrasse 11, 6330 Cham, Switzerland

Paper in this product is recyclable.

Preface

This volume contains a selection of the papers presented at the 27th European Conference on Advances in Databases and Information Systems (ADBIS 2023), held during September 4–7, 2023, in Barcelona, Spain.

The first ADBIS Conference was held in Saint Petersburg, Russia (1997). Since then, ADBIS has been continuously organized as an annual event. Its previous editions were held in: Poznan, Poland (1998); Maribor, Slovenia (1999); Prague, Czech Republic (2000); Vilnius, Lithuania (2001); Bratislava, Slovakia (2002); Dresden, Germany (2003); Budapest, Hungary (2004); Tallinn, Estonia (2005); Thessaloniki, Greece (2006); Varna, Bulgaria (2007); Pori, Finland (2008); Riga, Latvia (2009); Novi Sad, Serbia (2010); Vienna, Austria (2011); Poznan, Poland (2012); Genoa, Italy (2013); Ohrid, North Macedonia (2014); Poitiers, France (2015); Prague, Czech Republic (2016); Nicosia, Cyprus (2017); Budapest, Hungary (2018); Bled, Slovenia (2019); Lyon, France (2020); Tartu, Estonia (2021); Torino, Italy (2022). The official ADBIS portal - http://adbis.eu provides up to date information on all ADBIS Conferences, persons in charge, publications, and issues related to the ADBIS community.

The program of ADBIS 2023 was varied and included keynotes, research papers, tutorials, workshops, a doctoral consortium, as well as a diversity & inclusion panel and keynote. The conference attracted 77 paper submissions. All the papers went through a process of rigorous single-blind reviewing by at least three reviewers. When the review process was over, lengthy discussions on papers presenting conflicts were conducted. Moreover, under the newly introduced scheme of shepherding, five papers were conditionally accepted, under shepherding as short, and one as long. Eventually, the Program Committee selected 14 submissions as full contributions, appearing in this LNCS proceedings volume, and 25 papers as short ones, in Springer's Communications in Computer and Information Science (CCIS) series. The 14 papers that were accepted as full contributions amount to an acceptance rate of 18%.

The selected papers span a large spectrum of topics in the broader field of data management. We have organized the papers in groups, including (1) Index Management & Data Reconstruction, (2) Query Processing, (3) Advanced Querying Techniques, (4) Fairness and Explainability, (5) Data Science, (6) Temporal Graph Management, (7) Consistent Data Management, (8) Data Integration, (9) Data Quality, (10) Metadata Management.

Following a successful tradition, selected best papers of ADBIS 2023 will be invited for a special issue of the following journals: (a) Information Systems Frontiers (Springer) and Information Systems (Elsevier). Therefore, the PC chairs would like to express their sincere gratitude to the Information Systems Frontiers Editors-in- Chief: Ram Ramesh and H. Raghav Rao as well as the Information Systems Editors-in-Chief - Dennis Shasha, Gottfried Vossen, and Matthias Weidlich, for their approval of these special issues.

For this edition of ADBIS 2023, we had keynote talks by experts in different fields of the broader area of data management. The first keynote, by *Rosa M. Badia* (Barcelona

Supercomputing Center, Spain) had the topic of *Convergence of HPC, data analytics and AI through workflow methodologies.* The second keynote, by *Katja Hose* (TU Wien, Austria) dealt with the topic *No Intelligence without Knowledge.* A third keynote, by *Matthias Boehm* (Technische Universität Berlin, Germany) dealt with the topic of *System Infrastructure for Data-centric ML Pipelines.*

The following tutorials were included in the program: *Vincent T'kindt* (University of Tours, France) delivered a tutorial entitled *When operations research meets databases. Leopoldo Bertossi* (SKEMA Business School. Canada) delivered a tutorial entitled *Causality, Attribution-Scores and Repairs in Databases.* Finally, *Ekaterina Gorshkova* (Independent Software Consultant, Czech Republic) delivered a tutorial entitled *Data Mesh from an Industrial Perspective.*

The ADBIS 2023 conference hosted five affiliated workshops: 1st Workshop on Advanced AI Techniques for Data Management and Analytics (AIDMA), 4th Workshop on Intelligent Data - From Data to Knowledge (DOING), 2nd International Workshop on Knowledge Graphs Analysis on a Large Scale (K-GALS), 5th Workshop on Modern Approaches in Data Engineering and Information System Design (MADEISD), and 2nd Workshop on Personalization and Recommender Systems (PeRS). All five workshops addressed cutting-edge topics that call for extensive research both from a foundational as well as an application perspective to progress the state of the art. In total, 63 papers were submitted to these workshops, out of which 29 were selected for presentation and publication, yielding an overall acceptance rate of 46%. The accepted workshop papers are published in the accompanying CCIS volume of the conference.

The ADBIS community believes that diversity and a culture of support encourage retention and attraction of talent, promote diversity of thought and perspective, and help make the scientific community more flexible and responsive in times of change. As such, ADBIS 2023 encouraged authors and participants to consider D&I-aware communication guidelines when composing their papers and presentations. A pre-conference survey, collecting anonymous data, was used to understand our community and better promote D&I. A registration grant program was designed to support the participation at the conference of student and faculty authors starting their careers. As a consequence, 4 D&I awards were granted.

The conference program included three D&I plenary sessions. A data analytics activity with the Doctoral Consortium Chairs showed PhD students how to set equity-aware data-driven experiments. The D&I program proposed a keynote by Mercè Crosas (Barcelona Supercomputing Centre, Spain) on open data concerning computational social sciences and reproducibility, and the panel titled "Practices and methodologies for designing data science research projects with feminist and decolonial perspectives" with distinguished panellists from Europe and North America leading the design of D&I policies in their environment.

We would like to express our gratitude to every person who contributed to ADBIS 2023. The list is long, and we resort to grouping our thanks to the following groups of persons:

- Authors – without their hard work the conference would not be possible.
- Keynote speakers, Tutorial presenters, and panellists for sharing their wisdom with the audience of the conference.

- Members of the Program Committee and External Reviewers, for all the hard work in diligently reviewing, commenting, and discussing the submitted papers, as well as for their useful suggestions to the authors.
- Springer staff, for making it possible to publish the proceedings of ADBIS 2023 in the LNCS and CCIS series.
- Springer project coordinators, editorial, and book production staff for their help in the process of the proceedings preparation.
- The CMT staff for hosting the process of paper submission, reviewing, and collection of the final material.
- Members of the ADBIS Steering Committee - for their trust and support, and especially its chair Yannis Manolopoulos.
- Members of the Organizing Committee for all the work of taking care of their different tasks – with particular thanks to the people in the local organization, without whom the conference would have not been possible.

ADBIS 2023 was sponsored by Facultat d'Informàtica de Barcelona (FIB), Universitat Politècnica de Catalunya (UPC), and Springer.

July 2023

Alberto Abelló
Panos Vassiliadis
Oscar Romero
Robert Wrembel

Organization

General Chairs

Oscar Romero	Universitat Politècnica de Catalunya, Spain
Robert Wrembel	Poznan University of Technology, Poland

Program Committee Chairs

Alberto Abelló	Universitat Politècnica de Catalunya, Spain
Panos Vassiliadis	University of Ioannina, Greece

Workshop Chairs

Francesca Bugiotti	CentraleSupélec, France
Johann Gamper	Free University of Bozen-Bolzano, Italy

Doctoral Consortium Chairs

Genoveva Vargas-Solar	French Council of Scientific Research, France
Ester Zumpano	University of Calabria, Spain

Tutorials Chairs

Patrick Marcel	Université de Tours, France
Boris Novikov	National Research University – Higher School of Economics, Russia

Publicity Chair

Mirjana Ivanović	University of Novi Sad, Serbia

Proceedings and Website Chair

Sergi Nadal	Universitat Politècnica de Catalunya, Spain

Special Issue Chair

Ladjel Bellatreche	University of Poitiers, France

Diversity and Inclusion Chairs

Barbara Catania	Università degli Studi di Genova, Italy
Genoveva Vargas-Solar	French Council of Scientific Research, France

Organizing Committee

Besim Bilalli	Universitat Politècnica de Catalunya, Spain
Petar Jovanovic	Universitat Politècnica de Catalunya, Spain
Anna Queralt	Universitat Politècnica de Catalunya, Spain

Steering Committee

Chair

Yannis Manolopoulos	Open University of Cyprus, Cyprus

Members

Andreas Behrend	TH Köln, Germany
Ladjel Bellatreche	ENSMA Poitiers, France
Maria Bielikova	Kempelen Institute of Intelligent Technologies, Slovakia
Barbara Catania	University of Genoa, Italy
Tania Cerquitelli	Politecnico di Torino, Italy
Silvia Chiusano	Politecnico di Torino, Italy
Jérôme Darmont	University of Lyon 2, France
Johann Eder	Alpen-Adria-Universität Klagenfurt, Austria

Johann Gamper, Free University of Bozen-Bolzano, Italy
Tomáš Horváth, Eötvös Loránd University, Hungary
Mirjana Ivanović, University of Novi Sad, Serbia
Marite Kirikova, Riga Technical University, Latvia
Manuk Manukyan, Yerevan State University, Armenia
Raimundas Matulevicius, University of Tartu, Estonia
Tadeusz Morzy, Poznan University of Technology, Poland
Kjetil Nørvåg, Norwegian University of Science & Technology, Norway
Boris Novikov, National Research University – Higher School of Economics, Russia
George Papadopoulos, University of Cyprus, Cyprus
Jaroslav Pokorny, Charles University in Prague, Czech Republic
Oscar Romero, Universitat Politècnica de Catalunya, Spain
Sergey Stupnikov, Russian Academy of Sciences, Russia
Bernhard Thalheim, Christian Albrechts University Kiel, Germany
Goce Trajcevski, Iowa State University, USA
Valentino Vranić, Slovak University of Technology in Bratislava, Slovakia
Tatjana Welzer, University of Maribor, Slovenia
Robert Wrembel, Poznan University of Technology, Poland
Ester Zumpano, University of Calabria, Spain

Program Committee

Cristina D. Aguiar, Universidade de São Paulo, Brazil
Syed M. Fawad Ali, Accenture DACH, Germany
Bernd Amann, LIP6, Sorbonne Université, CNRS, France
Witold Andrzejewski, Poznan University of Technology, Poland
Sylvio Barbon, State University of Londrina, Brazil
Andreas Behrend, University of Bonn, Germany
Khalid Belhajjame, PSL, Université Paris-Dauphine, LAMSADE, France
Ladjel Bellatreche, ENSMA Poitiers, France
Josep L. Berral, Universitat Politècnica de Catalunya, Spain
Maria Bielikova, Kempelen Institute of Intelligent Technologies, Slovakia
Sandro Bimonte, INRAE, France
Paweł Boiński, Poznań University of Technology, Poland
Lars Ailo Bongo, UiT The Arctic University of Norway, Norway

Alejandro Maté	University of Alicante, Spain
Jose-Norberto Mazon	Universidad de Alicante, Spain
Sara Migliorini	University of Verona, Italy
Alex Mircoli	Università Politecnica delle Marche, Italy
Angelo Montanari	University of Udine, Italy
Lia Morra	Politecnico di Torino, Italy
Tadeusz Morzy	Poznan University of Technology, Poland
Kjetil Nørvåg	Norwegian University of Science and Technology, Norway
Boris Novikov	National Research University - Higher School of Economics, Russia
Omar Boussaid	Université Lumière Lyon 2, France
George Papastefanatos	Athena Research Center, Greece
Veronika Peralta	University of Tours, France
Giuseppe Polese	University of Salerno, Italy
Elisa Quintarelli	Università degli Studi di Verona, Italy
Milos Radovanovic	University of Novi Sad, Serbia
Franck Ravat	Institut de Recherche en Informatique de Toulouse, France
Stefano Rizzi	University of Bologna, Italy
María del Mar Roldán	University of Málaga, Spain
Gunter Saake	University of Magdeburg, Germany
Kai-Uwe Sattler	TU Ilmenau, Germany
Milos Savic	University of Novi Sad, Serbia
Marcos Sfair Sunye	Universidade Federal do Paraná, Brazil
Spiros Skiadopoulos	University of the Peloponnese, Greece
Claudia Steinberger	Universität Klagenfurt, Austria
Sergey A. Stupnikov	FRC CSC RAS, Russia
Bernhard Thalheim	CAU Kiel, Germany
Goce Trajcevski	Iowa State University, USA
Genoveva Vargas-Solar	French Council of Scientific Research, France
Goran Velinov	Ss. Cyril and Methodius University in Skopje, North Macedonia
José R. R. Viqueira	Universidade de Santiago de Compostela, Spain
Szymon Wilk	Poznan University of Technology, Poland
Ester Zumpano	University of Calabria, Spain

Additional Reviewers

Giorgos Alexiou
Paul Blockhaus
Andrea Brunello
Loredana Caruccio
Thanasis Chantzios
Marco Franceschetti
Luca Geatti
Balasubramanian Gurumurthy
Maude Manouvrier
Barbara Oliboni
Nicolas Ringuet
Vassilis Stamatopoulos

Keynotes and Tutorials

Convergence of HPC, Data Analytics and AI Through Workflow Methodologies

Rosa M. Badia

Barcelona Supercomputing Center, Spain

Abstract. With Exaflop systems already here, High-Performance Computing (HPC) involves ever larger and more complex supercomputers. At the same time, the user community is aware of the underlying performance and eager to leverage it by providing more complex application workflows to leverage them. What is more, current application trends aim to use data analytics and artificial intelligence combined with HPC modeling and simulation. However, the programming models and tools are different in these fields, and there is a need for methodologies that enable the development of workflows that combine HPC software, data analytics, and artificial intelligence. The eFlows4HPC project aims at providing a workflow software stack that fulfills this need. The project is also developing the HPC Workflows as a Service (HPCWaaS) methodology that aims at providing tools to simplify the development, deployment, execution, and reuse of workflows. The project showcases its advances with three application Pillars with industrial and social relevance: manufacturing, climate, and urgent computing for natural hazards. The keynote will present the actual progress and findings of the project.

No Intelligence Without Knowledge

Katja Hose

TU Wien, Austria

Abstract. Knowledge graphs and graph data in general are becoming more and more essential components of intelligent systems. This does not only include native graph data, such as social networks or Linked Data on the Web. The flexibility of the graph model and its ability to store data relationships explicitly enables the integration and exploitation of data from very diverse sources. However, to truly exploit their potential, it becomes crucial to provide intelligent systems with verifiable knowledge, reliable facts, patterns, and a deeper understanding of the underlying domains. This talk will therefore chart a number of challenges for exploiting graphs to manage and bring meaning to large amounts of heterogeneous data and discuss opportunities with, without, and for artificial intelligence emerging from research situated at the confluence of data management, knowledge engineering, and machine learning.

System Infrastructure for Data-Centric ML Pipelines

Matthias Boehm

Technische Universität Berlin, Germany

Abstract. The trend towards data-centric AI leads to increasingly complex, composite machine learning (ML) pipelines with outer loops for data integration and cleaning, data programming and augmentation, model and feature selection, hyper-parameter tuning and cross validation, as well as data validation and ML model debugging. Interestingly, state-of-the-art techniques for data integration, cleaning, and augmentation as well as model debugging are often based on machine learning themselves, which motivates their integration into ML systems. In this talk, we make a case for optimizing compiler infrastructure in Apache SystemDS and DAPHNE as two sibling open-source ML systems. We discuss recent feature highlights and how they all fit together. The covered topics range from linear-algebra-based data cleaning pipeline enumeration and slice finding; over lineage-based reuse and workload-aware redundancy exploitation; to federated learning, vectorized execution on heterogeneous HW devices, and extensibility.

When Operations Research Meets Databases

Vincent T'kindt

University of Tours, France

Abstract. In this tutorial we focus on the interplay between operations research and database theory, i.e., on how the optimization tools from operations research can contribute to the solution of optimization problems occurring in databases. Operations research is a field dealing with the solution of optimization and decision problems mathematically well formulated. The objective of this tutorial is to provide researchers in database theory with the basics of operations research, in order to enable them next to improve their own research.

Causality, Attribution-Scores and Repairs in Databases

Leopoldo Bertossi

SKEMA Business School, Canada

Abstract. The presenter, together with collaborators, introduced the notion of database repair some time ago. This area, including consistent query answering, has been extensively investigated. More recently, the presenter together with Babak Salimi unveiled a connection between database repairs and actual causality in databases, with actual causality as introduced by Helpern and Pearl. This connection allowed us to establish several new results on the computation and complexity of causes for query answers and their causal responsibility scores that reflect their causal weight. More recently, similar ideas have been applied by the presenter and coauthors to explanations for outcomes from classification with machine learning models. We will give an overview of all these connections and results.

Data Mesh from an Industrial Perspective

Ekaterina Gorshkova

Independent Software Consultant, Czech Republic

Abstract. The purpose of this tutorial is to introduce the reader to the concept of data mesh, explain how it can help solve data integration problems and show how this paradigm can be implemented using Apache Kafka technology. The audience for the tutorial could be anyone who is interested in understanding the benefits and challenges of implementing a data mesh approach in their organization.

Contents

Keynote Talk and Tutorials

Knowledge Engineering in the Era of Artificial Intelligence

Katja Hose(✉)

TU Wien, Vienna, Austria
katja.hose@tuwien.ac.at

Abstract. Knowledge engineering with respect to knowledge graphs and graph data in general is becoming a more and more essential component of intelligent systems. Such systems benefit from the wealth of structured knowledge, which does not only include native graph data, such as social networks or Linked Data on the Web, but also general knowledge describing particular topics of interest. Furthermore, the flexibility of the graph model and its ability to store data relationships explicitly enables the integration and exploitation of data from very diverse sources. Hence, to truly exploit their potential, it becomes crucial to provide intelligent systems with verifiable knowledge, reliable facts, patterns, and a deeper understanding of the underlying domains. This paper will therefore chart a number of current challenges in knowledge engineering and discuss opportunities.

1 Introduction

Knowledge in the form of knowledge graphs is everywhere. Their ability to model nearly arbitrary types of data have made knowledge graphs a universal tool for a broad range of applications. Hence, knowledge graph technologies have long been embraced as methods for storing and managing knowledge in a structured and easily accessible way. In fact, the term "knowledge graph" was coined by Google when they launched their Knowledge Graph in 2012. In the meantime, industry-scale knowledge graphs have grown to billions of entities and assertions about them [53]. While Google's Knowledge Graph fuels the information displayed in infoboxes that are shown along with the links to websites matching the user's keywords, Microsoft for instance does not only have a knowledge graph to power its search engine Bing but also manages the LinkedIn graph and the Academic graph – the latter covering information about people, publications, conferences, etc. Enterprise Knowledge Graphs [53] are another prominent example providing knowledge within the scope of a restricted organization or community. Other knowledge graphs, such as YAGO [29], DBpedia [42], or Wikidata [65], are freely accessible to the general public and mainly resulting from academic research and community efforts.

On the other hand, knowledge graphs also have roots in knowledge representation, description logics, and the Semantic Web. And indeed there is a broad body of related work in this context. The vision of the Semantic Web, for instance, as outlined by Sir Tim Berners-Lee et al. [12] sketched an environment where intelligent agents could gather information from the Web (Linked

A. Abelló et al. (Eds.): ADBIS 2023, LNCS 13985, pp. 3–15, 2023.
https://doi.org/10.1007/978-3-031-42914-9_1

Data) that comes along with semantics and links to other sources of information so that the agents would be able to understand the information that is being found and find more if needed. Such agents are also somewhat intelligent in the sense that they can use the information and apply logical reasoning to achieve a given task, and even independently "negotiate" with other agents to for example make appointments with a doctor. While this sounds similar to personal assistants (Siri, Echo, etc.), which are operating as black boxes on machine learning models and undisclosed data owned by a company, agents on the Semantic Web were supposed to work based on mostly Open Data accessible to everyone and in that way providing some kind of verifiability since, in principle, everybody could access the data that was used to solve a task.

Most recently, large language models, and in particular ChatGPT[1], have gained a lot of attention. Obviously, it is very appealing to simply formulate questions in natural language and receive elaborate and detailed replies that explain an extremely broad range of complex topics. While this system seems to be intelligent, it suffers from a similar problem as other large language models and machine learning approaches in general: the answer it returns is the most probable answer, it cannot be certain about its correctness. In the context of ChatGPT the latter is commonly referred to as hallucinations [11], i.e., the answer does not necessarily reflect reality but can be "made up". Furthermore, because of the way these systems are built, they operate as black boxes that cannot really explain how they arrived at a certain answer. Hence, verifying claims can so far only be done by accessing external information and methods. Moreover, reasoning capabilities of such systems are still limited. However, with advances in areas such as neurosymbolic AI, aiming to combine the strengths of logical reasoning and machine learning, we are slowly getting closer to arriving at intelligent systems.

In any case, approaches based on machine learning can only be as accurate, correct, and precise as the data they are trained upon; commonly referred to as the "garbage in, garbage out" principle. Hence, knowledge engineering plays an important role to improve or maintain the quality of the data delivered as input to such a system for training. Again, graphs and knowledge graphs have become very popular in this context; either providing raw input data or helping to access and extract additional information to enhance other datasets used for training. Finally, of course machine learning approaches help construct knowledge graphs, expand, and improve them; NLP techniques, for example, are used to extract facts from natural language text, machine learning plays an inherent role in entity resolution and matching, etc.

In this paper, we therefore claim that more advances in knowledge engineering are needed to provide systems with accurate knowledge so that they can become more intelligent. Hence, we review current challenges in modeling and storing knowledge (Sect. 2), querying knowledge (Sect. 3), knowledge quality and metadata (Sect. 4), and conclude the paper in Sect. 5.

[1] https://openai.com/chatgpt

2 Modeling and Storing Knowledge

While knowledge graphs are designed to capture information by representing entities (persons, places, concepts, etc.) as nodes, and relationships (bornIn, locatedIn, etc.) as directed edges between them, they can be modeled using alternative data models; typically these are either RDF[2,3] or property graphs. Depending on the chosen data model and the users' information needs, knowledge graphs are typically managed and queried by triple stores (in case of RDF) or graph stores (in case of property graphs). Given path and reachability queries [67], e.g., determining the shortest path between a given set of nodes, property graphs are typically preferable whereas triple stores have advantages for graph pattern matching involving the labels of multiple nodes and edges.

Whereas commercial systems are available for both data models, the research body supporting RDF is larger. This might be due to the fact that RDF and the corresponding query language SPARQL[4] have long been W3C standards and well defined whereas property graphs are mostly driven by commercial systems and lack common standards – there are some recent advances though to define such standards [8,15].

Still, even if there is a well-defined common data model, as it is the case for RDF, there are still several design options and alternative ways for representing and organizing the data. In general, the design space for RDF data representations can be organized in a three-dimensional space [59] defined by (i) subdivision (fragmentation and partitioning of the data), compression (how compressed is the data, how many bits are needed), and redundancy (does the system store multiple copies). In the end, it is the use case along with its query load, usage patterns, data characteristics, and other constraints that determine which design solution is the best. While some recommendations can be made based on an the expected workload [59], existing benchmarks often under-represent particular use cases, such as multi-hop traversals and existence checks, and therefore cannot help all use cases. Hence, developing more comprehensive benchmarks is one of the open challenges. Moreover, developing adaptive solutions that either manually or semi-automatically choose and adapt the design of an graph store to its needs – maybe in the sense of self-designing data structures [35] or self-organizing database systems [64], which both often involve machine learning approaches to guide optimizations – remains an interesting challenge for future work.

Apart from questions on how to store RDF in a single-server environment, there are many more use cases and storage paradigms for distributed environments. Such systems involve multiple servers in different configurations of autonomy [54] and range from cluster setups via federated scenarios to P2P systems. One of the well-understood challenges [60] in this context is that existing

---

[2] https://www.w3.org/RDF/

[3] In RDF edges are represented as triples (subject, predicate, object), where subject and object represent a pair of nodes and the predicate describes the relationship (edge label) between them.

[4] https://www.w3.org/TR/rdf-sparql-query/

ecosystems are designed for tabular data or relational databases, while we are still lacking an ecosystem for big graph processing covering all components of a pipeline incl. extracting and converting data into a native graph format, integrating different datasets into graphs, enabling OLTP as well as OLAP operations on graphs, and finally also providing input for graph-based applications, for instance specialized in machine learning, business intelligence, scientific computing, or augmented reality. There are still many challenges to build such an ecosystem. Naturally, scalability is one of them and the question how new hardware can be used for this purpose. But there are also interoperability issues between the different graph models, query languages, and standards that hamper efficient use of graph data. There is also still very little research on efficiently supporting dynamic and streaming graphs.

Some use cases require to integrate large amounts of heterogeneous data. This can include converting knowledge from one graph model to the other [2,9,40]. On the other hand, this also includes converting data originally provided in other formats into graphs, e.g., using R2ML [22] or RML [16], followed by some integration and homogenization to ensure quality. However, instead of transforming the data directly, knowledge graphs can also be used as a (virtual) integration layer – similar to traditional virtual data integration scenarios [54]. This scenario, often applied in the context of data fabrics and data lakes [50], uses knowledge graphs to provide an integrated interface where information can be accessed under a unified view. When the data then needs to be accessed, it can be searched, retrieved, and if needed converted on demand from the sources. Yet, finding and matching entities across different datasets is a challenging task.

When converting all data into an integrated knowledge graph directly, it can be queried in a single system – not only with standard queries. There are some works [23,51] on setting up semantic data warehouses incl. spatio-temporal extensions. The setup is similar to traditional data warehouses on relational data; there needs to be an ETL process as well as a way to define data cubes along with several analytical dimensions. The data can then be analyzed, cleansed, further data can be included, we can chose to keep track of metadata and provenance, links to external data sources can be created, and finally the data can be published, for instance on the Semantic Web. This is related in spirit to the lack of graph processing ecosystems mentioned above and therefore meets similar challenges.

Alternatively, knowledge can also be organized in federations of data sources accessible over the Web, i.e., each provider hosts a public endpoint (typically SPARQL) and makes it accessible and queryable to everyone. While SPARQL queries can be formulated over a combination of data from multiple remote sources, it is the task of the query optimizer (federation engine) to identify which publicly available sources should be queried with which part of the query to compute the final answer. An interesting observation here is that publishing data and making it available in this way is very easy as the publishers do not need to conform to a common integrated schema. However, this comes at the expense of query formulation and optimization, which then is considerably more complex. To formulate a query, users themselves have to know how the information in the different sources are connected – whereas this would typically be done when defining a common schema or table in a traditional relational database scenario. Likewise, the query optimizer has to decide for each query individually (on individual node-to-node connections) which data sources to select – whereas in a traditional relational setup such information could be captured by mappings valid for all tuples in a table, which therefore allows for preselecting sources.

Finally, solutions building upon P2P systems can be used to share data. Depending on the network structure and the replication rate, such systems can keep data available despite node failures. Some systems assume a fixed or previously known set of peers with some central agreement on where to store which kind of data [38] while others employ the unstructured P2P paradigm, where peers have the largest degree of autonomy [4,5], i.e., there is no global knowledge or control and peers are free to join with their data.

As discussed in this section, there are many ways to model, integrate, store, and manage knowledge. Each of these options has different advantages, disadvantages, and goals, e.g., if access efficiency is more important, then a single-server or cluster-system might be preferable – if the main goal is to keep the data available, then P2P systems are a good choice. Nevertheless, each of these systems faces a range of open challenges in each of the steps mentioned above.

3 Querying Knowledge

The way in which knowledge is queried very much depends on the chosen data model and the way the data is physically stored. Most commonly either Cypher or SPARQL queries are used as query languages. Then, similar to traditional relational database systems, it is the task of the query optimizer to find an efficient query execution plan to answer the query.

However, many users are not familiar with the details, content, and schema of a knowledge graph and therefore have difficulties formulating structured queries. To help such users, the literature has proposed exploratory query techniques and the query-by-example paradigm [44,45]. In this case, users do not formulate structured queries directly but provide the system with examples of potential answers – the system then tries to reverse engineer a query from the desired output, executes it, and presents the results to the user who can then iteratively refine the query until the information need is met. This is even possible for complex setups incl. analytical queries over statistical knowledge graphs [43]. Exploratory techniques for knowledge graphs cover a broad range of methods that include data profiling [1] as well as skyline queries [39].

Assuming that the user was able to formulate a structured query that expresses the information need, there is a broad range of query optimization techniques depending on the architecture of the system. In this respect, we can distinguish between several basic architectures: (i) centralized systems, where all data is stored on a single server, (ii) client/server architectures, where we have multiple clients querying the data on a single server, (iii) parallel systems, where we have a cluster of servers on which the knowledge graph is partitioned and queried exploiting parallel computation, (iv) federated systems, where we have multiple independent data sources with knowledge graphs and queries spanning multiple of these knowledge graphs, and (v) P2P systems, where we have a number of autonomous peers that might join and leave the network at any time.

Each of these architectures is tailored for providing access to knowledge in a different use case. While in all setups, the query optimizer aims at answering queries efficiently, optimization strategies in particular systems might have slightly different objective functions. In case of single-server centralized systems, the goal is typically to minimize query execution time by exploiting local indexing structures, access paths, caching, precomputation, etc. [18,18,32,34,37,48,66]. This is similar in spirit to parallel systems, where the goal is to additionally exploit parallel computations in a cluster of machines [17,31], e.g., by partitioning graphs to execute parts of a query independently and in parallel. Client/server architectures, on the other hand, also aim at answering queries efficiently. However, since such systems might suffer from times where more clients issue queries than the server can handle, the optimization goal is often not to focus on optimizing individual queries but instead the overall throughput, i.e., the number of queries that are successfully completed in a given period of time. This might entail that some expensive operations are "outsourced" from the server to the client, which might slow down the execution of an individual query but by freeing the resources on the server more queries can successfully complete within the same period of time [3,10].

Another optimization goal can be witnessed in systems building upon unstructured P2P architectures [4,7]. Since peers offering access to unique knowledge might leave the network at some point in time, the system stores copies of the data on other peers. Of course, the system tries to execute queries as efficiently as possible but more important than that is to keep the data available despite node failures. Due to the lack of central control, such systems then either resort to flooding the network with a particular query to find relevant data, i.e., sending messages to all known peers, or employing some kind of indexes that capture summaries of knowledge available at remote peers and query them directly.

Finally, federated setups [30,33,36,49,61,62] are commonly used on the Semantic Web and Linked Open Data [25][5], where hundreds of publicly accessibly servers provide query access to billions of facts covering a diverse range of topics. Such data sources typically offer access via SPARQL endpoints, i.e., they offer interfaces that given a SPARQL query provide corresponding answers executed over the local knowledge graph. Then, given a SPARQL query spanning multiple knowledge graphs, hosted on multiple remote SPARQL endpoints, the task of a federation engine is to analyze the given query, identify which endpoints provide relevant data for the different parts of the query, decide which parts of the query should be executed on which endpoint and in which order, have the subqueries executed on the chosen endpoints, receive the partial results, and combine them into the final answer to the query.

[5] http://cas.lod-cloud.net/

There are many factors that influence which architecture is the most appropriate for a particular use case, e.g., how many computing resources do we have, who owns the data, which kind of access is allowed, etc. On the other hand, the types of queries and the information need of users plays a decisive role. Connectivity and path queries or logic reasoning, for instance, benefit from different optimization strategies, data models, and indexes than regular graph pattern matching queries. Finally, if knowledge graphs are to be integrated into machine learning pipelines, then a centralized or cluster-based architecture and ecosystem is often preferable.

While there is plenty of related work in each of these setups and scenarios, there are still many open challenges; some of them with respect to traditional issues such as cardinality estimation, cost functions, etc. Other challenges originate from heterogeneity and interoperability of data models, interfaces, use cases, etc. The OneGraph vision [41], for instance, sketches a scenario where the data model no longer determines the query languages and would allow formulating Cypher queries over an RDF store. On the other hand, heterogeneous federations [27,47,48] have been explored with the goal of bridging the gap between different query interfaces and architectures. Still, in many aspects research is just about to begin looking into efficient solutions to the wide variety of challenges in this context.

4 Knowledge Quality and Metadata

Intuitively, any system – using machine learning or not – will not be able to produce high-quality results ("garbage in, garbage out") if it is not provided with high-quality input data. So far research on aspects such as data cleaning [14] and data profiling [1] has mainly focused on relational data and is not straightforward to apply to knowledge graphs. This is amplified by the issue that – in contrast to relational data – knowledge graphs are valid without the need for a strict schema or compliance to an ontology. Hence, a first step to ensure quality is to extract schemas from knowledge graphs and check conformance. Whereas there exist standards, such as OWL[6] and RDFS[7], to define ontologies and schema constraints for knowledge graphs modeled in RDF, similar work on property graphs has so far been hampered by the lack of a well-defined standard, which however might come soon [8].

Nevertheless, while OWL and RDFS have been developed for capturing the meaning of data by defining proper classes, hierarchies, and constraints, SHACL[8] has been proposed more recently as a standard to define constraints on the structure of knowledge graphs – without the need to define a proper full-fledged ontology and capture the meaning of the data. SHACL allows to define graph patterns, referred to as shapes, along with constraints that subgraphs matching the patterns/shapes should fulfill. While SHACL is becoming more and more

[6] http://www.w3.org/TR/owl2-overview/
[7] http://www.w3.org/TR/rdf-schema/
[8] https://www.w3.org/TR/shacl/

adopted by the community, it still remains a challenge to avoid having to define shapes manually [56] but instead being offered semi-automatic solutions for creating them given a knowledge graph as input. While mining shapes from large knowledge graphs meets scalability issues, it is also important to mine meaningful shapes [57] and avoid spurious ones, i.e., those that do not occur frequently or are fulfilled by only a small proportion of matching subgraphs. Once determined, such shapes can not only be used to create validation reports but they can also be used in a more interactive fashion in a similar way as mined association rules [20], e.g., to help experts find outliers and erroneous information so that the data can be corrected and the quality can be improved [58].

Another way of improving quality and trust in knowledge is to provide metadata. While metadata in property graphs can be expressed by adding attributes to nodes and edges, this is not straightforward for knowledge graphs. The latter require special constructs, such as reification, singleton properties [52], named graphs [13], or RDF-star[9]. While reification leads to a large increase in the number of triples (because subject, predicate, and object of the original triple are separated into their own triples), singleton properties (instantiating a unique subproperty for each triple with metadata) and named graph solutions (in the worst case creating a separate named graphs for each single triple) typically also suffer from scalability issues and require verbose query constructs since existing engines are not designed to efficiently support such use cases. On the other hand, RDF-star is proposing to nest triples, i.e., to use a complete triple on subject or object position of another triple. While this is very elegant from a modeling perspective, it poses several challenges on data organization and querying since nesting has not yet been a typical requirement. Still, many triple stores do already support RDF-star so that it can already be used in practice.

Provenance, in the sense of explaining the origin of data, is an important kind of metadata. In this sense it is often desired to capture information about who created the data, how and when it was obtained, how it was processed, etc. In RDF, such workflow provenance [19,24] can for instance be encoded using the PROV-O[10] ontology, which offers several classes with well defined meaning for this purpose. Another type of provenance, how-provenance [21,28], describes how an answer to a particular query was derived from a given input dataset. This approach allows to directly trace down the input tuples/triples/edges that were combined to derive a particular answer to a query – in addition, how-provenance also returns a polynomial describing how these tuples/triples/edges have been combined for a given query answer. In general, all flavors of provenance help explain answers to structured queries and in doing so increase the trust users can have in a system. To the best of our knowledge, however, there is currently no system for knowledge graphs combining workflow provenance with how-provenance.

Finally, it is also important to highlight that, although knowledge graphs are mostly considered to be static and not changing, in reality knowledge changes

[9] https://w3c.github.io/rdf-star/

[10] https://www.w3.org/TR/prov-o/

over time so that we can expect multiple versions of a knowledge graph. This is often referred to as knowledge evolution or dynamic knowledge graphs. However, in current practice, older versions of a knowledge graph typically disappear and only the latest version remains available. Instead of losing older versions, some systems attempt to capture previous versions and keep them retrievable [6,55, 63]. However, existing engines have not been designed for this kind of usage and the current state of the art still has deficiencies in efficiency and coverage of use cases.

While quality, provenance, and knowledge evolution are all very important topics discussed in the literature from different perspectives, most of today's engines do not yet come with efficient native support. Recent developments suggest that RDF-star might soon become the standard way of encoding and supporting metadata. Still, existing methods building upon it do not (yet) support the full spectrum and even though many triple stores already support RDF-star, they have limitations regarding full compliance and efficiency. Hence, there are many open challenges with supporting quality on all levels incl. optimizing the data layout, encoding and computing metadata, quality assurance and validation, query optimization, etc.

5 Conclusion

Knowledge engineering in all its different aspects discussed in this paper is the foundation for making machine learning based systems more intelligent and enabling Data Science [26,46]; it provides the structured and interconnected knowledge that enables intelligent systems to access, process, semantically integrate, interpret, discover, reason about knowledge and making informed decisions. In this sense, knowledge also builds the foundation for explainable systems that will help us get away from today's black box systems that demand blind trust. In this vibrant area, research is currently moving very fast in knowledge engineering as well as machine learning. While today we are still witnessing big challenges and open issues – as sketched in this paper – it will be interesting to witness how research will progress in the next couple of years.

References

1. Abedjan, Z., Golab, L., Naumann, F., Papenbrock, T.: Data Profiling. Morgan & Claypool Publishers, Synthesis Lectures on Data Management (2018)
2. Abuoda, G., Dell'Aglio, D., Keen, A., Hose, K.: Transforming RDF-star to property graphs: a preliminary analysis of transformation approaches. In: QuWeDa@ISWC. CEUR Workshop Proceedings, vol. 3279, pp. 17–32. CEUR-WS.org (2022)
3. Aebeloe, C., Keles, I., Montoya, G., Hose, K.: Star Pattern Fragments: Accessing Knowledge Graphs through Star Patterns. CoRR abs/ arXiv: 2002.09172 (2020)
4. Aebeloe, C., Montoya, G., Hose, K.: A decentralized architecture for sharing and querying semantic data. In: Hitzler, P., et al. (eds.) ESWC 2019. LNCS, vol. 11503, pp. 3–18. Springer, Cham (2019). https://doi.org/10.1007/978-3-030-21348-0_1

5. Aebeloe, C., Montoya, G., Hose, K.: Decentralized indexing over a network of RDF peers. In: Ghidini, C., et al. (eds.) ISWC 2019. LNCS, vol. 11778, pp. 3–20. Springer, Cham (2019). https://doi.org/10.1007/978-3-030-30793-6_1
6. Aebeloe, C., Montoya, G., Hose, K.: ColChain: collaborative linked data networks. In: WWW, pp. 1385–1396. ACM / IW3C2 (2021)
7. Aebeloe, C., Montoya, G., Hose, K.: The Lothbrok approach for SPARQL Query Optimization over Decentralized Knowledge Graphs. Semantic Web J. (2023)
8. Angles, R., et al.: PG-schema: schemas for property graphs. Proc. ACM Manag. Data **1**(2), 198:1–198:25 (2023)
9. Angles, R., Thakkar, H., Tomaszuk, D.: Mapping RDF databases to property graph databases. IEEE Access **8**, 86091–86110 (2020)
10. Azzam, A., Aebeloe, C., Montoya, G., Keles, I., Polleres, A., Hose, K.: WiseKG: balanced access to web knowledge graphs. In: WWW, pp. 1422–1434 (2021)
11. Bang, Y., et al.: A Multitask, Multilingual, Multimodal Evaluation of ChatGPT on Reasoning, Hallucination, and Interactivity. CoRR abs/ arXiv: 2302.04023 (2023)
12. Berners-Lee, T., Hendler, J., Lassila, O.: The semantic web. Sci. Am. **284**(5), 28–37 (2001)
13. Carroll, J.J., Bizer, C., Hayes, P.J., Stickler, P.: Named graphs. J. Web Semant. **3**(4), 247–267 (2005)
14. Chu, X., Ilyas, I.F., Krishnan, S., Wang, J.: Data cleaning: overview and emerging challenges. In: SIGMOD Conference, pp. 2201–2206. ACM (2016)
15. Deutsch, A., et al.: Graph pattern matching in GQL and SQL/PGQ. In: SIGMOD Conference, pp. 2246–2258. ACM (2022)
16. Dimou, A., Sande, M.V., Colpaert, P., Verborgh, R., Mannens, E., de Walle, R.V.: RML: a generic language for integrated rdf mappings of heterogeneous data. In: Proceedings of the Workshop on Linked Data on the Web co-located with the 23rd International World Wide Web Conference (WWW 2014), Seoul, Korea, April 8, 2014. CEUR Workshop Proceedings, vol. 1184. CEUR-WS.org (2014). https://ceur-ws.org/Vol-1184/ldow2014_paper_01.pdf
17. Galárraga, L., Hose, K., Schenkel, R.: Partout: a distributed engine for efficient RDF processing. In: WWW (Companion Volume), pp. 267–268. ACM (2014)
18. Galárraga, L., Ahlstrøm, K., Hose, K., Pedersen, T.B.: Answering provenance-aware queries on RDF data cubes under memory budgets. In: Vrandečić, D., et al. (eds.) ISWC 2018. LNCS, vol. 11136, pp. 547–565. Springer, Cham (2018). https://doi.org/10.1007/978-3-030-00671-6_32
19. Galárraga, L., Mathiassen, K.A.M., Hose, K.: QBOAirbase: the european air quality database as an RDF cube. In: ISWC (Posters, Demos & Industry Tracks). CEUR Workshop Proceedings, vol. 1963. CEUR-WS.org (2017)
20. Galárraga, L., Teflioudi, C., Hose, K., Suchanek, F.M.: Fast rule mining in ontological knowledge bases with AMIE+. VLDB J. **24**(6), 707–730 (2015)
21. Geerts, F., Unger, T., Karvounarakis, G., Fundulaki, I., Christophides, V.: Algebraic structures for capturing the provenance of SPARQL queries. J. ACM **63**(1), 7:1–7:63 (2016)
22. Group RW: R2RML: RDB to RDF Mapping Language. http://www.w3.org/TR/r2rml/, http://www.w3.org/2001/sw/rdb2rdf/ (2014)
23. Gür, N., Pedersen, T.B., Zimányi, E., Hose, K.: A foundation for spatial data warehouses on the semantic web. Semantic Web **9**(5), 557–587 (2018)
24. Hansen, E.R., Lissandrini, M., Ghose, A., Løkke, S., Thomsen, C., Hose, K.: Transparent integration and sharing of life cycle sustainability data with provenance. In: ISWC, pp. 378–394 (2020)

25. Harth, A., Hose, K., Schenkel, R. (eds.): Linked Data Management. Chapman and Hall/CRC (2014)
26. Helali, M., Vashisth, S., Carrier, P., Hose, K., Mansour, E.: Linked Data Science Powered by Knowledge Graphs. CoRR abs/ arXiv: 2303.02204 (2023)
27. Heling, L., Acosta, M.: Federated SPARQL query processing over heterogeneous linked data fragments. In: WWW, pp. 1047–1057. ACM (2022)
28. Hernández, D., Galárraga, L., Hose, K.: Computing how-provenance for SPARQL queries via query rewriting. Proc. VLDB Endow. **14**(13), 3389–3401 (2021)
29. Hoffart, J., Suchanek, F.M., Berberich, K., Lewis-Kelham, E., de Melo, G., Weikum, G.: YAGO2: exploring and querying world knowledge in time, space, context, and many languages. In: WWW (Companion Volume), pp. 229–232. ACM (2011)
30. Hose, K., Schenkel, R.: Towards benefit-based RDF source selection for SPARQL queries. In: SWIM, p. 2. ACM (2012)
31. Hose, K., Schenkel, R.: WARP: workload-aware replication and partitioning for RDF. In: ICDE Workshops, pp. 1–6. IEEE Computer Society (2013)
32. Ibragimov, D., Hose, K., Pedersen, T.B., Zimányi, E.: Towards exploratory OLAP over linked open data – a case study. In: Castellanos, M., Dayal, U., Pedersen, T.B., Tatbul, N. (eds.) BIRTE 2013-2014. LNBIP, vol. 206, pp. 114–132. Springer, Heidelberg (2015). https://doi.org/10.1007/978-3-662-46839-5_8
33. Ibragimov, D., Hose, K., Pedersen, T.B., Zimányi, E.: processing aggregate queries in a federation of SPARQL endpoints. In: Gandon, F., Sabou, M., Sack, H., d'Amato, C., Cudré-Mauroux, P., Zimmermann, A. (eds.) ESWC 2015. LNCS, vol. 9088, pp. 269–285. Springer, Cham (2015). https://doi.org/10.1007/978-3-319-18818-8_17
34. Ibragimov, D., Hose, K., Pedersen, T.B., Zimányi, E.: Optimizing aggregate SPARQL queries using materialized RDF views. In: Groth, P., et al. (eds.) ISWC 2016. LNCS, vol. 9981, pp. 341–359. Springer, Cham (2016). https://doi.org/10.1007/978-3-319-46523-4_21
35. Idreos, S., et al.: Design continuums and the path toward self-designing key-value stores that know and learn. In: CIDR. www.cidrdb.org (2019)
36. Jakobsen, A.L., Montoya, G., Hose, K.: How diverse are federated query execution plans really? In: Hitzler, P., et al. (eds.) ESWC 2019. LNCS, vol. 11762, pp. 105–110. Springer, Cham (2019). https://doi.org/10.1007/978-3-030-32327-1_21
37. Jakobsen, K.A., Andersen, A.B., Hose, K., Pedersen, T.B.: Optimizing RDF data cubes for efficient processing of analytical queries. In: COLD. CEUR Workshop Proceedings, vol. 1426. CEUR-WS.org (2015)
38. Kaoudi, Z., et al.: Atlas: Storing, updating and querying RDF(S) data on top of DHTs. J. Web Semant. **8**(4), 271–277 (2010)
39. Keles, I., Hose, K.: Skyline queries over knowledge graphs. In: Ghidini, C., et al. (eds.) ISWC 2019. LNCS, vol. 11778, pp. 293–310. Springer, Cham (2019). https://doi.org/10.1007/978-3-030-30793-6_17
40. Khayatbashi, S., Ferrada, S., Hartig, O.: Converting property graphs to RDF: a preliminary study of the practical impact of different mappings. In: GRADES-NDA@SIGMOD, pp. 10:1–10:9. ACM (2022)
41. Lassila, O., et al.: The OneGraph vision: challenges of breaking the graph model lock-in. Semantic Web **14**(1), 125–134 (2023)
42. Lehmann, J., et al.: Dbpedia - a large-scale, multilingual knowledge base extracted from wikipedia. Semantic Web **6**(2), 167–195 (2015)
43. Lissandrini, M., Hose, K., Pedersen, T.B.: Example-driven exploratory analytics over knowledge graphs. In: EDBT, pp. 105–117. OpenProceedings.org (2023)

44. Lissandrini, M., Mottin, D., Hose, K., Pedersen, T.B.: Knowledge graph exploration systems: are we lost? In: CIDR (2022). www.cidrdb.org
45. Lissandrini, M., Mottin, D., Palpanas, T., Velegrakis, Y.: Data Exploration Using Example-Based Methods. Morgan & Claypool Publishers, Synthesis Lectures on Data Management (2018)
46. Mansour, E., Srinivas, K., Hose, K.: Federated data science to break down silos. SIGMOD Rec. **50**(4), 16–22 (2021)
47. Montoya, G., Aebeloe, C., Hose, K.: Towards efficient query processing over heterogeneous RDF interfaces. In: ISWC (Best Workshop Papers). Studies on the Semantic Web, vol. 36, pp. 39–53. IOS Press (2018)
48. Montoya, G., Keles, I., Hose, K.: Analysis of the effect of query shapes on performance over LDF interfaces. In: QuWeDa@ISWC. CEUR Workshop Proceedings, vol. 2496, pp. 51–66. CEUR-WS.org (2019)
49. Montoya, G., Skaf-Molli, H., Hose, K.: The *Odyssey* approach for optimizing federated SPARQL queries. In: d'Amato, C., et al. (eds.) ISWC 2017. LNCS, vol. 10587, pp. 471–489. Springer, Cham (2017). https://doi.org/10.1007/978-3-319-68288-4_28
50. Nargesian, F., Pu, K.Q., Bashardoost, B.G., Zhu, E., Miller, R.J.: Data lake organization. IEEE Trans. Knowl. Data Eng. **35**(1), 237–250 (2023)
51. Nath, R.P.D., Hose, K., Pedersen, T.B., Romero, O.: SETL: a programmable semantic extract-transform-load framework for semantic data warehouses. Inf. Syst. **68**, 17–43 (2017)
52. Nguyen, V., Bodenreider, O., Sheth, A.P.: Don't like RDF reification?: making statements about statements using singleton property. In: WWW (2014)
53. Noy, N.F., Gao, Y., Jain, A., Narayanan, A., Patterson, A., Taylor, J.: Industry-scale knowledge graphs: lessons and challenges. Commun. ACM **62**(8), 36–43 (2019)
54. Özsu, M.T., Valduriez, P.: Principles of Distributed Database Systems, 4th edn. Springer (2020). https://doi.org/10.1007/978-3-030-26253-2
55. Pelgrin, O., Galárraga, L., Hose, K.: Towards fully-fledged archiving for RDF datasets. Semantic Web **12**(6), 903–925 (2021)
56. Rabbani, K., Lissandrini, M., Hose, K.: SHACL and ShEx in the wild: a community survey on validating shapes generation and adoption. In: WWW (Companion Volume), pp. 260–263. ACM (2022)
57. Rabbani, K., Lissandrini, M., Hose, K.: Extraction of validating shapes from very large knowledge graphs. Proc. VLDB Endow. **16**(5), 1023–1032 (2023)
58. Rabbani, K., Lissandrini, M., Hose, K.: SHACTOR: improving the quality of large-scale knowledge graphs with validating shapes. In: SIGMOD Conference Companion, pp. 151–154. ACM (2023)
59. Sagi, T., Lissandrini, M., Pedersen, T.B., Hose, K.: A design space for RDF data representations. VLDB J. **31**(2), 347–373 (2022)
60. Sakr, S., et al.: The future is big graphs: a community view on graph processing systems. Commun. ACM **64**(9), 62–71 (2021)
61. Schwarte, A., Haase, P., Hose, K., Schenkel, R., Schmidt, M.: FedX: a federation layer for distributed query processing on linked open data. In: Antoniou, G., et al. (eds.) ESWC 2011. LNCS, vol. 6644, pp. 481–486. Springer, Heidelberg (2011). https://doi.org/10.1007/978-3-642-21064-8_39
62. Schwarte, A., Haase, P., Hose, K., Schenkel, R., Schmidt, M.: FedX: optimization techniques for federated query processing on linked data. In: Aroyo, L., et al. (eds.) ISWC 2011. LNCS, vol. 7031, pp. 601–616. Springer, Heidelberg (2011). https://doi.org/10.1007/978-3-642-25073-6_38

63. Taelman, R., Mahieu, T., Vanbrabant, M., Verborgh, R.: Optimizing storage of RDF archives using bidirectional delta chains. Semantic Web (2021)
64. Varadarajan, R., Bharathan, V., Cary, A., Dave, J., Bodagala, S.: DBDesigner: a customizable physical design tool for vertica analytic database. In: ICDE, pp. 1084–1095. IEEE Computer Society (2014)
65. Vrandecic, D., Krötzsch, M.: Wikidata: a free collaborative knowledgebase. Commun. ACM **57**(10), 78–85 (2014)
66. Zervakis, L., Setty, V., Tryfonopoulos, C., Hose, K.: Efficient continuous multi-query processing over graph streams. In: EDBT, pp. 13–24. OpenProceedings.org (2020)
67. Zhang, C., Bonifati, A., Özsu, M.T.: An overview of reachability indexes on graphs. In: SIGMOD Conference Companion, pp. 61–68. ACM (2023)

Attribution-Scores in Data Management and Explainable Machine Learning

Leopoldo Bertossi(✉)

SKEMA Business School, Montreal, Canada
leopoldo.bertossi@skema.edu

Abstract. We describe recent research on the use of actual causality in the definition of responsibility scores as explanations for query answers in databases, and for outcomes from classification models in machine learning. In the case of databases, useful connections with database repairs are illustrated and exploited. Repairs are also used to give a quantitative measure of the consistency of a database. For classification models, the responsibility score is properly extended and illustrated. The efficient computation of Shap-score is also analyzed and discussed. The emphasis is placed on work done by the author and collaborators.

1 Introduction

In data management and artificial intelligence, and machine learning in particular, one wants *explanations* for certain results. For example, for query answers and inconsistency in databases. In machine learning (ML), one wants explanations for automated classification results, and automated decision making. Explanations, that may come in different forms, have been the subject of philosophical enquires for a long time, but, closer to our discipline, they appear under different forms in model-based diagnosis, and in causality as developed in artificial intelligence.

In the last few years, explanations that are based on *numerical scores* assigned to elements of a model that may contribute to an outcome have become popular. These *attribution scores* attempt to capture the degree of contribution of those components to an outcome, e.g. answering questions like these: What is the contribution of this tuple to the answer to this query? Or, what is the contribution of this feature value for the label assigned to this input entity?

In this article, we survey some of the recent advances on the definition, use and computation of the above mentioned score-based explanations for query answering in databases, and for outcomes from classification models in ML. This is not intended to be an exhaustive survey of the area. Instead, it is heavily influenced by our latest research. To introduce the concepts and techniques we will use mostly examples, trying to convey the main intuitions and issues.

Different scores have been proposed in the literature, and some that have a relatively older history have been applied. Among the latter we find the general *responsibility score* as found in *actual causality* [18,23,24]. For a particular kind of application, one

Paper associated to tutorial at ADBIS 2023.
L. Bertossi—Member of the Millennium Inst. for Foundational Research on Data (IMFD, Chile).

A. Abelló et al. (Eds.): ADBIS 2023, LNCS 13985, pp. 16–33, 2023.
https://doi.org/10.1007/978-3-031-42914-9_2

has to define the right causality setting, and then apply the responsibility measure to the participating variables.

In data management, responsibility has been used to quantify the strength of a tuple as a cause for a query result [6,35]. The responsibility score, *Resp*, is based on the notions of *counterfactual intervention* as appearing in actual causality. More specifically, (potential) executions of *counterfactual interventions* on a *structural logico-probabilistic model* [23] are considered, with the purpose of answering hypothetical questions of the form: *What would happen if we change ...?*.

Database repairs are commonly used to define and obtain semantically correct query answers from a database (DB) that may fail to satisfy a given set of integrity constraints (ICs) [5]. A connection between repairs and actual causality in DBs has been used to obtain complexity results and algorithms for responsibility [6] (see Sect. 2). Actual causality and responsibility can also be defined at the attribute-level in DBs (rather than at the tuple-level) [8]. We briefly describe this alternative in Sect. 2.1. On the basis of database repairs, a measure (or global score) to quantify the degree of inconsistency of a DB was introduced in [7]. We give the main ideas in Sect. 3.

The *Resp* score has also been applied to define scores for binary attribute values in classification [10]. However, it has to be generalized when dealing with non-binary features [9]. We describe this generalization in Sect. 4.1.

The Shapley value of *coalition game theory* [39,43] can be (and has been) used to define attribution scores, in particular in DBs. The main idea is that *several tuples together*, much like players in a coalition game, are necessary to produce a query result. Some may contribute more than others to the *wealth distribution function* (or simply, game function), which in this case becomes the query result, namely 1 or 0 if the query is Boolean, or a number in the case of an aggregation query. This use of Shapley value was developed in [29,30]. See also [14] for a more recent and general discussion of the use of the Shapley value in DBs.

The Shapley value has also been used to define explanation scores to feature values in ML-based classification, taking the form of the *Shap* score [33]. Since its computation is bound to be intractable in general, there has been recent research on classes of models for which *Shap* becomes tractable [3,4,47]. See Sect. 4.2.

There hasn't been much research on the use or consideration of domain knowledge (or domain semantics) in the definition and computation of attribution scores. In Sect. 5, we describe some problems that emerge in this context.

This article has [15] as a companion, which concentrates mostly on data management. It delves into the *Resp* score under ICs, and on the use of the Shapley value for query answering in DBs. That paper also discusses the *Causal Effect* score applied in data management [41]. It is also based on causality, as it appears mainly in *observational studies* [25,38,40].

2 Causal Responsibility in Databases

Before going into the subject, we recall some notions and notation used in data management. A relational schema $\mathcal{R}$ contains a domain of constants, $\mathcal{C}$, and a set of predicates of finite arities, $\mathcal{P}$. $\mathcal{R}$ gives rise to a language $\mathfrak{L}(\mathcal{R})$ of first-order (FO) predicate logic with built-in equality, $=$. Variables are usually denoted with $x, y, z, ...$; and finite sequences thereof with $\bar{x}$, ...; and constants with a, b, c, ..., etc. An *atom* is of the form $P(t_1, \ldots, t_n)$, with n-ary $P \in \mathcal{P}$ and $t_1, \ldots, t_n$ *terms*, i.e. constants, or variables. An atom is *ground* (a.k.a. a tuple) if it contains no variables. A database (instance), D, for $\mathcal{R}$ is a finite set of ground atoms; and it serves as an interpretation structure for $\mathfrak{L}(\mathcal{R})$.

A *conjunctive query* (CQ) is a FO formula, $\mathcal{Q}(\bar{x})$, of the form $\exists \bar{y}\ (P_1(\bar{x}_1) \wedge \cdots \wedge P_m(\bar{x}_m))$, with $P_i \in \mathcal{P}$, and (distinct) free variables $\bar{x} := (\bigcup \bar{x}_i) \smallsetminus \bar{y}$. If $\mathcal{Q}$ has n (free) variables, $\bar{c} \in \mathcal{C}^n$ is an *answer* to $\mathcal{Q}$ from D if $D \models \mathcal{Q}[\bar{c}]$, i.e. $Q[\bar{c}]$ is true in D when the variables in $\bar{x}$ are componentwise replaced by the values in $\bar{c}$. $\mathcal{Q}(D)$ denotes the set of answers to $\mathcal{Q}$ from D. $\mathcal{Q}$ is a *Boolean conjunctive query* (BCQ) when $\bar{x}$ is empty; and when *true* in D, $\mathcal{Q}(D) := \{\mathit{true}\}$. Otherwise, it is *false*, and $\mathcal{Q}(D) := \emptyset$. We will consider only conjunctive queries, which are the most common the data management.

Integrity constraints (ICs) are sentences of $\mathfrak{L}(\mathcal{R})$ that a DB is expected to satisfy. Here, we consider mainly *denial constraints* (DCs), i.e. of the form $\kappa\colon \neg\exists \bar{x}(P_1(\bar{x}_1) \wedge \cdots \wedge P_m(\bar{x}_m))$, where $P_i \in \mathcal{P}$, and $\bar{x} = \bigcup \bar{x}_i$. If an instance D does not satisfy the set Σ of ICs associated to the schema, we say that D is *inconsistent*, denoted with $D \not\models \Sigma$.

Now we move into the proper subject of this section. In data management we need to understand and compute *why* certain results are obtained or not, e.g. query answers, violations of semantic conditions, etc.; and we expect a database system to provide *explanations*. Here, we will consider explanations that are based on *actual causality* [23,24]. They were introduced in [35,36], and will be illustrated by means of an example.

Example 1. Consider the database D, and the Boolean conjunctive query (BCQ)

R	A	B
	a	b
	c	d
	b	b

S	C
	a
	c
	b

$\mathcal{Q}\colon\ \exists x \exists y (S(x) \wedge R(x,y) \wedge S(y))$, for which $D \models \mathcal{Q}$ holds, i.e. the query is true in D. We ask about the causes for $\mathcal{Q}$ to be true. A tuple $\tau \in D$ is *counterfactual cause* for $\mathcal{Q}$ (being true in D) if $D \models \mathcal{Q}$ and $D \smallsetminus \{\tau\} \not\models \mathcal{Q}$. In this example, $S(b)$ is a counterfactual cause for $\mathcal{Q}$: If $S(b)$ is removed from D, $\mathcal{Q}$ is no longer true.

Removing a single tuple may not be enough to invalidate the query. Accordingly, a tuple $\tau \in D$ is an *actual cause* for $\mathcal{Q}$ if there is a *contingency set* $\Gamma \subseteq (D \smallsetminus \{\tau\})$, such that τ is a counterfactual cause for $\mathcal{Q}$ in $D \smallsetminus \Gamma$. That is, $D \models \mathcal{Q}$, $D \smallsetminus \Gamma \models \mathcal{Q}$, but $D \smallsetminus (\Gamma \cup \{\tau\}) \not\models \mathcal{Q}$. In this example, $R(a,b)$ is not a counterfactual cause for $\mathcal{Q}$, but it is an actual cause with contingency set $\{R(b,b)\}$: If $R(b,b)$ is removed from D, $\mathcal{Q}$ is still true, but further removing $R(a,b)$ makes $\mathcal{Q}$ false. □

Notice that every counterfactual cause is also an actual cause, with empty contingent set. Actual causes that are not counterfactual causes need company to invalidate a query result. Now, we ask how strong are tuples as actual causes. To answer this question, we appeal to the *responsibility* of an actual cause τ for $\mathcal{Q}$ [35], defined by:

$$\mathit{Resp}_{D}^{\mathcal{Q}}(\tau) \ :=\ \frac{1}{|\Gamma| \,+\, 1},$$

where $|\Gamma|$ is the size of a smallest contingency set, Γ, for τ, and 0, otherwise.

Example 2. (Example 1 cont.) The responsibility of $R(a,b)$ is $\frac{1}{2} = \frac{1}{1+1}$ (its several smallest contingency sets have all size 1). $R(b,b)$ and $S(a)$ are also actual causes with responsibility $\frac{1}{2}$; and $S(b)$ is actual (counterfactual) cause with responsibility $1 = \frac{1}{1+0}$. □

We can see that causes are, in this database context, tuples that come with their responsibilities as "scores". It turns out that there are precise connections between *database repairs* and tuples as actual causes for queries answers in databases. These connections where exploited to obtain complexity results for responsibility [6] (among other uses, e.g. to obtain answer-set programs for the specification and computation of actual causes and responsibility [8]).

The notion of *repair* of a relational database was introduced in order to formalize the notion of *consistent query answering* (CQA) [2,5]: If a database D is inconsistent in the sense that is does not satisfy a given set of integrity constraints, *ICs*, and a query $\mathcal{Q}$ is posed to D, what are the meaningful, or consistent, answers to $\mathcal{Q}$ from D? They are sanctioned as those that hold (are returned as answers) from *all* the *repairs* of D. The repairs of D are consistent instances D' (over the same schema of D), i.e. $D' \models ICs$, and *minimally depart* from D.

Example 3. Let us consider the following set of *denial constraints* (DCs) and a database D, whose relations (tables) are shown right here below. D is inconsistent, because it violates the DCs: it satisfies the joins that are prohibited by the DCs.

$$\neg\exists x \exists y (P(x) \land Q(x,y))$$
$$\neg\exists x \exists y (P(x) \land R(x,y))$$

P	A
	a
	e

Q	A	B
	a	b

R	A	C
	a	c

We want to repair the original instance by *deleting tuples* from relations. Notice that, for DCs, insertions of new tuple will not restore consistency. We could change (update) attribute values though, a possibility that has been investigated in [8].

Here we have two *subset-repairs*, a.k.a. *S-repairs*. They are subset-maximal consistent subinstances of D: $D_1 = \{P(e), Q(a,b), R(a,c)\}$ and $D_2 = \{P(e), P(a)\}$. They are consistent, subinstances of D, and any proper superset of them (still contained in D) is inconsistent. (In general, we will represent database relations as set of tuples.)

We also have *cardinality repairs*, a.k.a. *C-repairs*. They are consistent subinstances of D that minimize the *number* of tuples by which they differ from D. That is, they are maximum-cardinality consistent subinstances. In this case, only D_1 is a C-repair. Every C-repair is an S-repair, but not necessarily the other way around. □

Let us now consider a BCQ

$$\mathcal{Q}\colon \exists \bar{x}(P_1(\bar{x}_1) \land \cdots \land P_m(\bar{x}_m)), \quad (1)$$

which we assume is true in a database D. It turns out that we can obtain the causes for $\mathcal{Q}$ to be true D, and their contingency sets from database repairs. In order to do this, notice that $\neg\mathcal{Q}$ becomes a DC

$$\kappa(\mathcal{Q})\colon \neg\exists \bar{x}(P_1(\bar{x}_1) \land \cdots \land P_m(\bar{x}_m)). \quad (2)$$

$\mathcal{Q}$ holds in D iff D is inconsistent w.r.t. $\kappa(\mathcal{Q})$.

S-repairs are associated to causes with minimal contingency sets, while C-repairs are associated to causes for $\mathcal{Q}$ with minimum contingency sets, and maximum responsibilities [6]. In fact, for a database tuple $\tau \in D$:

(a) τ is actual cause for $\mathcal{Q}$ with subset-minimal contingency set Γ iff $D \smallsetminus (\Gamma \cup \{\tau\})$ is an S-repair (w.r.t. $\kappa(\mathcal{Q})$), in which case, its responsibility is $\frac{1}{1+|\Gamma|}$.
(b) τ is actual cause with minimum-cardinality contingency set Γ iff $D \smallsetminus (\Gamma \cup \{\tau\})$ is C-repair, in which case, τ is a maximum-responsibility actual cause.

Conversely, repairs can be obtained from causes and their contingency sets [6]. These results can be extended to unions of BCQs (UBCQs), or equivalently, to sets of denial constraints.

One can exploit the connection between causes and repairs to understand the computational complexity of the former by leveraging existing results for the latter. Beyond the fact that computing or deciding actual causes can be done in polynomial time in data for CQs and unions of CQs (UCQs) [6,35], one can show that most computational problems related to responsibility are intractable, because they are also intractable for repairs, in particular, for C-repairs (all this in data complexity) [32]. In particular, one can prove [6]:

(a) The *responsibility problem*, about deciding if a tuple has responsibility above a certain threshold, is NP-complete for UCQs. However, on the positive side, this problem is *fixed-parameter tractable* (equivalently, belongs to the *FPT* class) [21], with the parameter being the inverse of the threshold.
(b) Computing $\mathit{Resp}_{D}^{\mathcal{Q}}(\tau)$ is $\mathit{FP}^{NP(log(n))}$-complete for BCQs. This the *functional*, non-decision, version of the responsibility problem. The complexity class involved is that of computational problems that use polynomial time with a logarithmic number of calls to an oracle in *NP*.
(c) Deciding if a tuple τ is a most responsible cause is $P^{NP(log(n))}$-complete for BCQs. The complexity class is as the previous one, but for decision problems [1].

For further illustration, property (b) right above tells us that there is a database schema and a Boolean conjunctive query $\mathcal{Q}$, such that computing the responsibility of tuple in an instance D as an actual cause for $\mathcal{Q}$ being true in D is $\mathit{FP}^{NP(log(n))}$-hard in $|D|$. This is due to the fact that, through the C-repair connection, determining the responsibility of a tuple becomes the problem on hyper-graphs of determining the size of a minimum vertex cover that contains the tuple as a vertex (among all vertex covers that contain the vertex) [6, sec. 6]. This latter problem was first investigated in the context of C-repairs w.r.t. denial constraints. Those repairs can be characterized in terms of hyper-graphs [32] (see Sect. 3 for a simple example). This hyper-graph connection also allows to obtain the FPT result in (a) above, because we are dealing with hyper-edges that have a fixed upper bound given by the number of atoms in the denial constraint associated to que conjunctive query $\mathcal{Q}$ (see [6, sec. 6.1] for more details).

Notice that a class of database repairs determines a *possible-world semantics* for database consistency restoration. We could use in principle any reasonable notion of distance between database instances, with each choice defining a particular *repair semantics*. S-repairs and C-repairs are examples of repair semantics. By choosing alternative repair semantics and an associated notion of counterfactual intervention, we can *define* also alternative notions of actual causality in databases and associated responsibility measures [8]. We will show an example in Sect. 2.1.

2.1 Attribute-Level Causal Responsibility in Databases

Causality and responsibility in databases can be extended to the attribute-value level; and can also be connected with appropriate forms of database repairs [6,8]. We will develop this idea in the light of an example, with a particular repair-semantics, and we will apply it to define attribute-level causes for query answering, i.e. we are interested in attribute values in tuples rather than in whole tuples. The repair semantics we use here is very natural, but others could be used instead.

Example 4. Consider the database D, with tids, and query $\mathcal{Q}\colon \exists x \exists y(S(x) \wedge R(x,y) \wedge S(y))$, of Example 1 and the associated denial constraint $\kappa(\mathcal{Q}) : \neg\exists x \exists y(S(x) \wedge R(x,y) \wedge S(y))$.

R	A	B
t_1	a	b
t_2	c	d
t_3	b	b

S	C
t_4	a
t_5	c
t_6	b

Since $D \not\models \kappa(\mathcal{Q})$, we need to consider repairs of D w.r.t. $\kappa(\mathcal{Q})$. Repairs will be obtained by "minimally" changing attribute values by NULL, as in SQL databases, which cannot be used to satisfy a join. In this case, minimality means that *the set* of values changed by NULL is minimal under set inclusion. These are two different minimal-repairs:

R	A	B
t_1	a	b
t_2	c	d
t_3	b	b

S	C
t_4	a
t_5	c
t_6	NULL

R	A	B
t_1	a	NULL
t_2	c	d
t_3	b	NULL

S	C
t_4	a
t_5	c
t_6	b

It is easy to check that they do not satisfy $\kappa(\mathcal{Q})$. If we denote the changed values by the tid with the position where the changed occurred, the first repair is characterized by the set $\{t_6[1]\}$, whereas the second, by the set $\{t_1[2], t_3[2]\}$. Both are minimal since none of them is contained in the other.

Now, we could also introduce a notion of *cardinality-repair*, keeping those where the number of changes is a minimum. In this case, the first repair qualifies, but not the second. These repairs identify (actually, define) the value in $t_6[1]$ as a maximum-responsibility cause for $\mathcal{Q}$ being true (with responsibility 1). Similarly, $t_1[2]$ and $t_3[2]$ become actual causes, that do need contingent companion values, which makes them take a responsibility of $\frac{1}{2}$ each. □

We should emphasize that, under the semantics illustrated with the example, we are considering attribute values participating in joins as interesting causes. A detailed treatment of the underlying repairs semantics with its application to attribute-level causality can be found in [8]. In a related vein, one could also consider as causes other attribute values in a tuple that participate in a query (being true), e.g. that in $t_3[1]$, but making them *non-prioritized* causes. One could also think of adjusting the responsibility measure in order to give to these causes a lower score.

More generally, in order to define actual causality and responsibility at the attribute level, we could -alternatively- consider all the attributes in tuples (in joins or not), and all the possible updates drawn from an attribute domain for a given value in a given tuple in the database, so as for potential contingency sets for it. If the attribute domains contain NULL, this update semantics would generalize the one illustrated with Example 4. As earlier in this section, the purpose of all this would be to counterfactually (and minimally) invalidate the query answer. On the repair counterpart, we could consider *update-repairs* [48] to restore the satisfaction of the associated denial constraint.

In this more general situation, where we may depart from the "binary" case so far (i.e. the tuple is or not in the DB, or a value is replaced by a null or not), we have to be more careful: Some value updates on a tuple may invalidate the query and some other may not. So, what if only one such update invalidates the query, but all the others do not? This would not be reflected in the responsibility score as defined so far. We could generalize the responsibility score by bringing into the picture the average (or expected) value of the query (of 1s or 0s for true or false) in the so-intervened database. This idea has been developed in the context of explanation scores for ML-based classification [9]. See Sect. 4.1 for details.

3 Measuring Database Inconsistency

A database D is expected to satisfy a given set of integrity constraints (ICs), Σ, that come with the database schema. However, databases may be inconsistent in that those ICs are not satisfied. A natural question is: *To what extent, or how much inconsistent is D w.r.t. Σ, in quantitative terms?*. This problem is about defining a *global numerical score* for the database, to capture its "degree of inconsistency". This number can be interesting *per se*, as a measure of data quality (or a certain aspect of it), and could also be used to compare two databases (for the same schema) w.r.t. (in)consistency.

In [7], a particular and natural *inconsistency measure* (IM) was introduced and investigated. More specifically, the measure was inspired by the g_3-measure used for functional dependencies (FDs) in [28], and reformulated and generalized *in terms of a class of database repairs*. Once a particular measure is defined, an interesting research program can be developed on its basis. In particular, around its mathematical properties, and the algorithmic and complexity issues related to its computation. In [7], in addition to algorithms, complexity results, approximations for hard cases of the IM computation, and the dynamics of the IM under updates, *answer-set programs* were proposed for the computation of this measure.

In the rest of this section, we will use the notions and notation introduced in Example 3. For a database D and a set of *denial constraints* Σ (this is not essential, but to fix ideas), we have the classes of subset-repairs (or S-repairs), and cardinality-repairs (or C-repairs), denoted $\mathit{Srep}(D,\Sigma)$ and $\mathit{Crep}(D,\Sigma)$, resp. The consider the following inconsistency measure:

$$\mathit{inc\text{-}deg}^{C}(D,\Sigma) := \frac{|D| - \mathit{max}\{|D'| \ : \ D' \in \mathit{Crep}(D,\Sigma)\}}{|D|}.$$

It is easy to verify that replacing C-repairs by S-repairs gives the same measure. It is clear that $0 \leq \mathit{inc\text{-}deg}^{C}(D,\Sigma) \leq 1$, with value 0 when D consistent. Notice that one could use other repair semantics instead of C-repairs [7].

Example 5. (Example 3 cont.) Here, $Srep(D, \Sigma) = \{D_1, D_2\}$ and $Crep(D, \Sigma) = \{D_1\}$. It holds: $\textit{inc-deg}^C(D, \Sigma) = \frac{4-|D_1|}{4} = \frac{1}{4}$. □

It turns out that complexity and efficient computation results, when they exist, can be obtained via C-repairs, for which we end up confronting graph-theoretic problems. Actually, C-repairs are in one-to-one correspondence with maximum-size independent sets in hyper-graphs, whose vertices are the database tuples, and hyper-edges are formed by tuples that jointly violate an IC [32].

Example 6. Consider the database $D = \{A(a), B(a), C(a), D(a), E(a)\}$, which is inconsistent w.r.t. the set of DS:

$$\Sigma = \{\neg\exists x(B(x) \wedge E(x)),\ \neg\exists x(B(x) \wedge C(x) \wedge D(x)),\ \neg\exists x(A(x) \wedge C(x))\}.$$

We obtain the following *conflict hyper-graph* (CHG), where tuples become the nodes, and a hyper-edge connects tuples that together violate a DC:

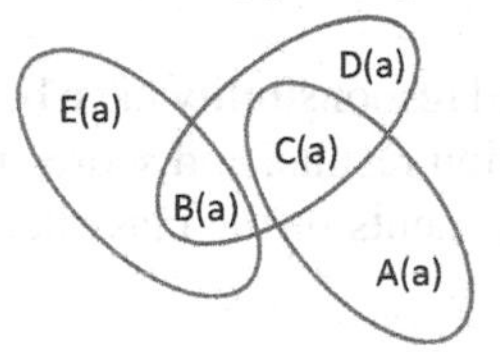

S-repairs are maximal independent sets: $D_1 = \{B(a), C(a)\}$, $D_2 = \{C(a), D(a), E(a)\}$, $D_3 = \{A(a), B(a), D(a)\}$; and the C-repairs are D_2, D_3. □

There is a connection between C-repairs and *hitting-sets* (HS) of the hyper-edges of the CHG [32]: The removal from D of the vertices in a minimum-size *hitting set* produces a C-repair. The connections between hitting-sets in hyper-graphs and C-repairs can be exploited for algorithmic purposes, and to obtain complexity and approximation results [7]. In particular, the complexity of computing $\textit{inc-deg}^C(D, \Sigma)$ for DCs belongs to $FP^{NP(log(n))}$, in data complexity. Furthermore, there is a relational schema and a set of DCs Σ for which computing $\textit{inc-deg}^C(D, \Sigma)$ is $FP^{NP(log(n))}$-complete.

Inconsistency measures have been introduced and investigated in knowledge representation, but mainly for propositional theories [22,26]. In databases, it is more natural to consider the different nature of the combination of a database, as a structure, and ICs, as a set of first-order formulas. It is also important to consider the asymmetry: databases are inconsistent or not, not the combination. Furthermore, the relevant issues that are usually related to data management have to do with algorithms and computational complexity; actually, in terms of the size of the database (that tends to be by far the largest parameter).

Scores for *individual tuples* about their contribution to inconsistency can be obtained through responsibility scores for query answering, because every IC gives rise to a "violation view" (a query). Also Shapley values have been applied to measure the contribution of tuples to inconsistency of a database w.r.t. FDs [31].

4 Attribution Scores in Machine Learning

Scores as explanations for outcomes from machine learning (ML) models have also been proposed. In this section we discuss some aspects of two of them, the *Resp* and the *Shap* scores.

Example 7. Consider a client of bank who is applying for loan. The bank will process his/her application by means of a ML system that will decide if the client should be given the loan or not. In order for the application to be computationally processed, the client is represented as an *entity*, say $\mathbf{e} = \langle \mathsf{john}, 18, \mathsf{plumber}, 70\mathsf{K}, \mathsf{harlem}, 10\mathsf{K}, \mathsf{basic}\rangle$, that is, a finite record of *feature values*. The set of features is $\mathcal{F} = \{\mathsf{Name}, \mathsf{Age}, \mathsf{Activity}, \mathsf{Income}, \mathsf{Debt}, \mathsf{EdLevel}\}$.

The bank uses a *classifier*, $\mathcal{C}$, which, after receiving input $\mathbf{e}$, returns a *label*: *Yes* or *No* (or 0 or 1). In this case, it returns *No*, indicating that the loan request is rejected. The client (or the bank executive) asks *"Why?"*, and would like to have an explanation. What kind of explanation? How could it be provided? From what? □

There are different ways of building such a classifier, and depending on the resulting model, the classifier may be less or more "interpretable". For example, complex neural networks are considered to be difficult to interpret, and become "black-boxes", whereas more explicit models, such as decision trees, are considered to be much more interpretable, and "open-boxes" for that matter.

In situations such as that in Example 7, actual causality and responsibility have been applied to provide *counterfactual explanations* for classification results, and scores for them. In order to do this, having access to the internal components of the classifier is not needed, but only its input/output relation.

Example 8. (Example 7 cont.) The entity $\mathbf{e} = \langle \mathsf{john}, 18, \mathsf{plumber}, 70\mathsf{K}, \mathsf{harlem}, 10\mathsf{K}, \mathsf{basic}\rangle$ received the label 1 from the classifier, indicating that the loan is not granted.[1] In order to identify counterfactual explanations, we intervene the feature values replacing them by alternative ones from the feature domains, e.g. $\mathbf{e}_1 = \langle \mathsf{john}, \underline{25}, \mathsf{plumber}, 70\mathsf{K}, \mathsf{harlem}, 10\mathsf{K}, \mathsf{basic}\rangle$, which receives the label 0. The counterfactual version $\mathbf{e}_2 = \langle \mathsf{john}, 18, \mathsf{plumber}, \underline{80\mathsf{K}}, \underline{\mathsf{brooklyn}}, 10\mathsf{K}, \mathsf{basic}\rangle$ also get label 0. Assuming, in the latter case, that none of the single changes alone switch the label, we could say that $\mathsf{Age} = 25$, so as $\mathsf{Income} = 70\mathsf{K}$ with contingency $\mathsf{Location} = \mathsf{harlem}$ (and the other way around) in $\mathbf{e}$ are (minimal) counterfactual explanations, by being actual causes for the label.

We could go one step beyond, and define responsibility scores: $\mathit{Resp}(\mathbf{e}, \mathsf{Age}) := 1$, and $\mathit{Resp}(\mathbf{e}, \mathsf{Income}) := \frac{1}{2}$ (due to the additional, required, contingent change). This choice does reflect the causal strengths of attribute values in $\mathbf{e}$. However, it could be the case that only by changing the value of Age to 25 we manage to switch the label, whereas for all the other possible values for Age (while nothing else changes), the label is always *No*. It seems more reasonable to redefine responsibility by considering an average involving all the possible labels obtained in this way. □

An application of the responsibility score similar to that in [6,35] works fine for explanation scores when features are *binary*, i.e. taking two values, e.g. 0 or 1 [10]. Even in this case, responsibility computation can be intractable [10]. However, as mentioned in the previous example, when features have more than two values, it makes sense to extend the definition of the responsibility score.

[1] We are assuming that classifiers are *binary*, i.e. they return labels 0 or 1. For simplicity and uniformity, but without loss of generality, we will assume that label 1 is the one we want to explain.

4.1 The Generalized *Resp* Score

In [9], a generalized *Resp* score was introduced and investigated. We describe it next in intuitive terms, and appealing to Example 8.

1. For an entity $\mathbf{e}$ classified with label $L(\mathbf{e}) = 1$, and a feature $F^\star$, whose value $F^\star(\mathbf{e})$ appears in $\mathbf{e}$, we want a numerical responsibility score $Resp(\mathbf{e}, F^\star)$, characterizing the causal strength of $F^\star(\mathbf{e})$ for outcome $L(\mathbf{e})$. In the example, $F^\star = \mathsf{Salary}$, $F^\star(\mathbf{e}) = \mathsf{70K}$, and $L(\mathbf{e}) = 1$.
2. While we keep the original value for Salary fixed, we start by defining a "local" score for a fixed contingent assignment $\Gamma := \bar{w}$, with $F^\star \notin \Gamma \subseteq \mathcal{F}$. We define $\mathbf{e}^{\Gamma,\bar{w}} := \mathbf{e}[\Gamma := \bar{w}]$, the entity obtained from $\mathbf{e}$ by changing (or redefining) its values for features in Γ, according to $\bar{w}$.

   In the example, it could be $\Gamma = \{\mathsf{Location}\}$, and $\bar{w} := \langle \mathsf{brooklin} \rangle$, a contingent (new) value for Location. Then, $\mathbf{e}^{\{\mathsf{Location}\},\langle \mathsf{brooklin}\rangle} = \mathbf{e}[\mathsf{Location} := \mathsf{brooklin}] = \langle \mathsf{john}, \mathsf{25}, \mathsf{plumber}, \mathsf{70K}, \underline{\mathsf{brooklin}}, \mathsf{10K}, \mathsf{basic}\rangle$.

   We make sure (or assume in the following) that $L(\mathbf{e}^{\Gamma,\bar{w}}) = L(\mathbf{e}) = 1$ holds. This is because, being these changes only contingent, we do not expect them to switch the label by themselves, but only and until the "main" counterfactual change on $F^\star$ is made.

   In the example, we assume $L(\mathbf{e}[\mathsf{Location} := \mathsf{brooklin}]) = 1$. Another case could be $\mathbf{e}^{\Gamma',\bar{w}'}$, with $\Gamma' = \{\mathsf{Activity}, \mathsf{Education}\}$, and $\bar{w}' = \langle \mathsf{accountant}, \mathsf{medium}\rangle$, with $L(\mathbf{e}^{\Gamma',\bar{w}'}) = 1$.
3. Now, for each of those $\mathbf{e}^{\Gamma,\bar{w}}$ as in the previous item, we consider all the different possible values v for $F^\star$, with the values for all the other features fixed as in $\mathbf{e}^{\Gamma,\bar{w}}$.

   For example, starting from $\mathbf{e}[\mathsf{Location} := \mathsf{brooklin}]$, we can consider $\mathbf{e}_1' := \mathbf{e}[\mathsf{Location} := \mathsf{brooklin}; \underline{\mathsf{Salary} := \mathsf{60K}}]$ (which is the same as $\mathbf{e}^{\mathsf{Location},\langle \mathsf{brooklin}\rangle}[\mathsf{Salary} := \mathsf{60K}]$), obtaining, e.g. $L(\mathbf{e}_1') = 1$. However, for $\mathbf{e}_2' := \mathbf{e}[\mathsf{Location} := \mathsf{brooklin}; \underline{\mathsf{Salary} := \mathsf{80}}]$, we now obtain, e.g. $L(\mathbf{e}_2') = 0$, etc.

   For a fixed (potentially) contingent change $(\Gamma, \bar{w})$ on $\mathbf{e}$, we consider the difference between the original label 1 and the expected label obtained by further modifying the value of $F^\star$ (in all possible ways). This gives us a *local* responsibility score, local to $(\Gamma, \bar{w})$:

$$Resp(\mathbf{e}, F^\star, \underline{\Gamma, \bar{w}}) := \frac{L(\mathbf{e}) - \mathbb{E}(\, L(\mathbf{e}') \mid F(\mathbf{e}') = F(\mathbf{e}^{\Gamma,\bar{w}}),\ \forall F \in (\mathcal{F} \smallsetminus \{F^\star\}\,)}{1 + |\Gamma|}$$

$$= \frac{1 - \mathbb{E}(\, L(\mathbf{e}^{\Gamma,\bar{w}}[F^\star := v]) \mid v \in Dom(F^\star)\,)}{1 + |\Gamma|}. \quad (3)$$

   This local score takes into account the size of the contingency set Γ.

   We are assuming here that there is a probability distribution over the entity population $\mathcal{E}$. It could be known from the start, or it could be an empirical distribution obtained from a sample. As discussed in [9], the choice (or whichever distribution that is available) is relevant for the computation of the general *Resp* score, which involves the local ones (coming right here below).
4. Now, generalizing the notions introduced in Sect. 2, we can say that the value $F^\star(\mathbf{e})$ is an *actual cause* for label 1 when, for some $(\Gamma, \bar{w})$, (3) is positive: at least one change of value for $F^\star$ in $\mathbf{e}$ changes the label (with the company of $(\Gamma, \bar{w})$).

When $\Gamma = \emptyset$ (and then, $\bar{w}$ is an empty assignment), and (3) is positive, it means that at least one change of value for $F^\star$ in $\mathbf{e}$ switches the label by itself. As before, we can say that $F^\star(\mathbf{e})$ is a *counterfactual cause*. However, as desired and expected, it is not necessarily the case anymore that counterfactual causes (as original values in $\mathbf{e}$) have all the same causal strength: $F_i(\mathbf{e}), F_j(\mathbf{e})$ could be both counterfactual causes, but with different values for (3), for example if changes on the first switch the label "fewer times" than those on the second.

5. Now, we can define the global score, by considering the "best" contingencies $(\Gamma, \bar{w})$, which involves requesting from Γ to be of minimum size:

$$Resp(\mathbf{e}, F^\star) \ := \max_{\Gamma, \bar{w}\colon\, |\Gamma| \text{ is min. \& (3)} > 0} Resp(\mathbf{e}, F^\star, \Gamma, \bar{w}). \tag{4}$$

This means that we first find the minimum-size contingency sets Γ's for which, for an associated set of value updates $\bar{w}$, (3) becomes greater that 0. After that, we find the maximum value for (3) over all such pairs $(\Gamma, \bar{w})$. This can be done by starting with $\Gamma = \emptyset$, and iteratively increasing the cardinality of Γ by one, until a $(\Gamma, \bar{w})$ is found that makes (3) non-zero. We stop increasing the cardinality, and we just check if there is another $(\Gamma', \bar{w}')$ that gives a greater value for (3), with $|\Gamma'| = |\Gamma|$. By taking the maximum of the local scores, we have an existential quantification in mind: there must be a good contingency $(\Gamma, \bar{w})$, as long as Γ has a minimum cardinality.

With the generalized score, the difference between counterfactual and actual causes is not as relevant as before. In the end, and as discussed under Item 4. above, what matters is the size of the score. Accordingly, we can talk only about "counterfactual explanations with responsibility score r". In Example 8, we could say "$\mathbf{e}_2$ is a (minimal) counterfactual for $\mathbf{e}$ (implicitly saying that it switches the label), and the value 60K for Salary is a counterfactual explanation with responsibility $Resp(\mathbf{e}, \mathsf{Salary})$". Here, $\mathbf{e}_2$ is possibly only one of those counterfactual entities that contribute to making the value for Salary a counterfactual explanation, and to its (generalized) *Resp* score.

The generalized *Resp* score was applied for different financial data [9], and experimentally compared with the *Shap* score [33,34], which can also be applied with a black-box classifier, using only the input/output relation. Both were also experimentally compared, with the same data, with a the FICO-score [17] that is defined for and applied to an open-box model, and computes scores by taking into account components of the model, in this case coefficients of nested logistic regressions.

The computation cost of the *Resp* score is bound to be high in general since, in essence, it explicitly involves in (3) all possible subsets of the set of features; and in (4), also the minimality condition which compares different subsets. Actually, for binary classifiers and in its simple, binary formulation, *Resp* is already intractable [10]. In [9], in addition to experimental results, there is a technical discussion on the importance of the underlying distribution on the population, and on the need to perform optimized computations and approximations.

4.2 The Shap Score and Its Tractable Computation

The *Shap* score was introduced in explainable ML in [33], as an application of the general *Shapley value* of *coalition game theory* [43], which we briefly describe next.

Consider a set of players $\mathcal{S}$, and a *wealth-distribution function* (or *game function*), $\mathcal{G}: \mathcal{P}(\mathcal{S}) \rightarrow \mathbb{R}$, that maps subsets of $\mathcal{S}$ to real numbers. The Shapley value of player $p \in \mathcal{S}$ quantifies the contribution of p to the game, for which all different coalitions are considered; each time, with p and without p:

$$Shapley(\mathcal{S}, \mathcal{G}, p) := \sum_{S \subseteq \mathcal{S} \setminus \{p\}} \frac{|S|!(|\mathcal{S}| - |S| - 1)!}{|\mathcal{S}|!} (\mathcal{G}(S \cup \{p\}) - \mathcal{G}(S)). \quad (5)$$

Here, $|S|!(|D| - |S| - 1)!$ is the number of permutations of $\mathcal{S}$ with all players in S coming first, then p, and then all the others. In other words, this is the *expected contribution* of p under all possible additions of p to a partial random sequence of players, followed by random sequences of the rest of the players.

The Shapley value emerges as the only quantitative measure that has some specified properties in relation to coalition games [39]. It has been applied in many disciplines. For each particular application, one has to define a particular and appropriate game function $\mathcal{G}$. In particular, it has been applied to assign scores to logical formulas to quantify their contribution to the inconsistency of a knowledge base [26], to quantify the contribution of database tuples to making a query true [29, 30], and to quantify contributions to the inconsistency of a database [31].

In different application and with different game functions, the Shapley value turns out to be computationally intractable, more precisely, its time complexity is $\#P$*-hard* in the size of the input, c.f., for example, [30]. Intuitively, this means that it is at least as hard as any of the problems in the class $\#P$ of problems about counting the solutions to decisions problems (in *NP*) that ask about the existence of a certain solution [37, 46]. For example, SAT is the decision problem asking, for a propositional formula, if there exists a truth assignment (a solution) that makes the formula true. Then, $\#SAT$ is the computational problem of counting the number of satisfying assignments of a propositional formula. Clearly, $\#SAT$ is at least as hard as SAT (it is good enough to count the number of solutions to know if the formula is satisfiable), and SAT is the prototypical *NP*-complete problem, and furthermore, $\#SAT$ is $\#P$-hard, actually, $\#P$-complete since it belongs to $\#P$. As a consequence, computing the Shapley value can be at least as hard as computing the number of solutions for SAT; a clear indication of its high computational complexity.

As already mentioned, the *Shap* score is a particular case of the Shapley value in (5). In this case, the players are the features F in $\mathcal{F}$, or, more precisely, the values $F(\mathbf{e})$ they take for a particular entity $\mathbf{e}$, for which we have a binary classification label, $L(\mathbf{e})$, we want to explain. The explanation takes the form of a numerical score for $F(\mathbf{e})$, reflecting its relevance for the observed label. Since all the feature values contribute to the resulting label, features values can be seen as players in a coalition game.

The game function, for a subset S of features, is the *expected (value of the) label* over all possible entities whose values coincide with those of $\mathbf{e}$ for the features in S:

$$\mathcal{G}_{\mathbf{e}}(S) := \mathbb{E}(L(\mathbf{e}') \mid \mathbf{e}' \in \mathcal{E} \text{ and } \mathbf{e}'_S = \mathbf{e}_S), \quad (6)$$

where $\mathbf{e}'_S, \mathbf{e}_S$ denote the projections of $\mathbf{e}'$ and $\mathbf{e}$ on S, resulting in two subrecords of feature values. We can see that the game function depends on the entity at hand $\mathbf{e}$.

With the game function in (5), the *Shap* score for a feature value $F^\star(\mathbf{e})$ in $\mathbf{e}$ is:

$$Shap(\mathcal{F}, \mathcal{G}_{\mathbf{e}}, F^\star) := \sum_{S \subseteq \mathcal{F} \setminus \{F^\star\}} \frac{|S|!(|\mathcal{F}| - |S| - 1)!}{|\mathcal{F}|!} [\mathbb{E}(L(\mathbf{e}')|\mathbf{e}'_{S \cup \{F^\star\}} = \mathbf{e}_{S \cup \{F^\star\}}) - \mathbb{E}(L(\mathbf{e}')|\mathbf{e}'_S = \mathbf{e}_S)]. \quad (7)$$

Example 9. (example 8 cont.) For the same fixed entity $\mathbf{e} = \langle \mathsf{john}, 18, \mathsf{plumber}, 70\mathsf{K}, \mathsf{harlem}, 10\mathsf{K}, \mathsf{basic}\rangle$ and feature $F^\star = \mathsf{Salary}$, one of the terms in (7) is obtained by considering $S = \{\mathsf{Location}\} \subseteq \mathcal{F}$:

$$\frac{|1|!(7-1-1)!}{7!} \times (\mathcal{G}_{\mathbf{e}}(\{\mathsf{Location}\} \cup \{\mathsf{Salary}\}) - \mathcal{G}_{\mathbf{e}}(\{\mathsf{Location}\})) = \frac{1}{42} \times (\mathcal{G}_{\mathbf{e}}(\{\mathsf{Location}, \mathsf{Salary}\}) - \mathcal{G}_{\mathbf{e}}(\{\mathsf{Location}\})),$$

with, e.g., $\mathcal{G}_{\mathbf{e}}(\{\mathsf{Location}, \mathsf{Salary}\}) = \mathbb{E}(L(\mathbf{e}') \mid \mathbf{e}' \in \mathcal{E},\ \mathsf{Location}(\mathbf{e}') = \mathsf{harlem}$, and $\mathsf{Salary}(\mathbf{e}') = 70\mathsf{K})$, that is, the expected label over all entities that have the same values as $\mathbf{e}$ for features Salary and Location. Then, $\mathcal{G}_{\mathbf{e}}(\{\mathsf{Location}, \mathsf{Salary}\}) - \mathcal{G}_{\mathbf{e}}(\{\mathsf{Location}\})$ is the expected difference in the label between the case where the values for Location and Salary are fixed as for $\mathbf{e}$, and the case where only the value for Location is fixed as in $\mathbf{e}$, measuring a local contribution of $\mathbf{e}$'s value for Salary. After that, all these local differences are averaged over all subsets S of $\mathcal{F}$, and the permutations in which they participate. □

We can see that, so as the *Resp* score, *Shap* is a *local* explanation score, for a particular entity at hand $\mathbf{e}$. Since the introduction of *Shap* in this form, some variations have been proposed. So as for *Resp*, *Shap* depends, via the game function, on an underlying probability distribution on the entity population $\mathcal{E}$. The distribution may impact not only the *Shap* scores, but also their computation [9].

Boolean classifiers, i.e. propositional formulas with binary input features and binary labels, are particularly relevant, *per se* and because they can represent other classifiers by means of appropriate encodings. For example, the circuit in Fig. 1 can be seen as a binary classifier that can be represented by means of a propositional formula that, depending on the binary values for x_1, x_2, x_3, x_4, also returns a binary value.

Boolean classifiers, as logical formulas, have been extensively investigated. In particular, much is known about the satisfiability problem of propositional formulas, SAT, and also about the *model counting* problem, i.e. that of counting the number of satisfying assignments, denoted $\#SAT$. In the area of *knowledge compilation*, the complexity of $\#SAT$ and other problems in relation to the syntactic form of the Boolean formulas have been investigated [19,20,42]. Boolean classifiers turn out to be quite relevant to understand and investigate the complexity of *Shap* computation.

The computation of *Shap* is bound to be expensive, for similar reasons as for *Resp*. For the computation of both, all we need is the input/output relation of the classifier, to compute labels for different alternative entities (counterfactuals). However, in principle, far too many combinations have to go through the classifier. Actually, under the *product probability distribution* on $\mathcal{E}$ (which assigns independent probabilities to the feature values), even with an explicit, open classifier for binary entities, the computation of *Shap* can be intractable.

In fact, as shown in [9], for Boolean classifiers in the class $Monotone2CNF$, of negation-free propositional formulas in conjunctive normal form with at most two atoms per clause, *Shap* can be #*P*-hard. This is obtained via a polynomial reduction from $\#Monotone2CNF$, the problem of counting the number of satisfying assignments for a formula in the class, which is known to be #*P*-complete [46]. For example, if the classifier is $(x_1 \vee x_2) \wedge (x_2 \vee x_3)$, which belongs to $\#Monotone2CNF$, the entity $\mathbf{e}_1 = \langle 1, 0, 1 \rangle$ (with values for x_1, x_2, x_3, in this order) gets label 1, whereas the entity $\mathbf{e}_2 = \langle 1, 0, 0 \rangle$ gets label 0. The number of satisfying truth assignments, equivalently, the number of entities that get label 1, is 5, corresponding to $\langle 1, 1, 1 \rangle$, $\langle 1, 0, 1 \rangle$, $\langle 0, 1, 1 \rangle$, $\langle 0, 1, 0 \rangle$, and $\langle 1, 1, 0 \rangle$.

Given that *Shap* can be $\#P$-hard, a natural question is whether for some classes of open-box classifiers one can compute *Shap* in polynomial time in the size of the model and input. The idea is to try to take advantage of the internal stricture and components of the classifier -as opposed to only the input/output relation of the classifier- in order to compute *Shap* efficiently. We recall from results mentioned earlier in this section that having an open-box model does not guarantee tractability of *Shap*. Natural classifiers that have been considered in relation to a possible tractable computation of *Shap* are decision trees and random forests [34].

The problem of tractability of *Shap* was investigated in detail in [3], and with other methods in [47]. They briefly describe the former approach in the rest of this section. Tractable and intractable cases were identified, with algorithms for the tractable cases. (Approximations for the intractable cases were further investigated in [4].) In particular, the tractability for decision trees and random forests was established, which required first identifying the right abstraction that allows for a general proof, leaves aside contingent details, and is also broad enough to include interesting classes of classifiers.

In [3], it was proved that, for a Boolean classifier L (identified with its label, the output gate or variable), the uniform distribution on $\mathcal{E}$, and $\mathcal{F} = \{F_1, \ldots, F_n\}$:

$$\#SAT(L) \; = \; 2^{|\mathcal{F}|} \times (L(\mathbf{e}) - \sum_{i=1}^{n} Shap(\mathcal{F}, G_{\mathbf{e}}, F_i)). \tag{8}$$

This result makes, under the usual complexity-theoretic assumptions, impossible for *Shap* to be tractable for any circuit L for which $\#SAT$ is intractable. (If we could compute *Shap* fast, we could also compute $\#SAT$ fast, assuming we have an efficient classifier.) This excludes, as seen earlier in this section, classifiers that are in the class $Monotone2CNF$. Accordingly, only classifiers in a more amenable class became candidates, with the restriction that the class should be able to accommodate interesting classifiers. That is how the class of *deterministic and decomposable Boolean circuits* (dDBCs) became the target of investigation.

Each $\vee$-gate of a dDBC can have only one of the disjuncts true (determinism), and for each $\wedge$-gate, the conjuncts do not share variables (decomposition). Nodes are labeled with $\vee$, $\wedge$, or $\neg$ gates, and input gates with features (propositional variables) or binary constants. An example of such a classifier, borrowed from [3], is shown in Figure 1. It has the $\wedge$ gate at the top that returns the output label.

Model counting is known to be tractable for dDBCs. However, this does not imply (via (8) or any other obvious way) that *Shap* is tractable. It is also the case that relaxing any of the determinism or decomposability conditions makes model counting $\#P$-hard [4], preventing *Shap* from being tractable.

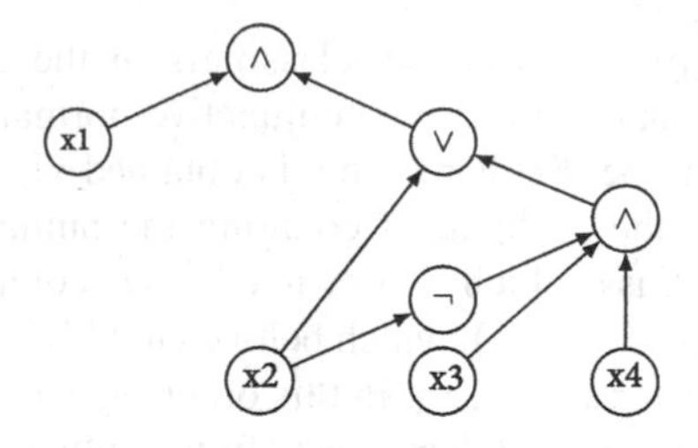

Fig. 1. A decomposable and deterministic Boolean classifier

It turns out that *Shap* computation is tractable for dDBCs (under the uniform and the product distribution), from which we also get the tractability of *Shap* for free for a vast collection of classifiers that can be efficiently compiled into (or represented as) dDBCs; among them we find: Decision Trees (even with non-binary features), Random Forests, Ordered Binary Decision Diagrams (OBDDs) [16], Deterministic-Decomposable Negation Normal-Forms (dDNNFs), Binary Neural Networks via OBDDs or Sentential Decision Diagrams (SDDs) (but with an extra, exponential, but FPT compilation cost) [12, 13, 44], etc.

For the gist, consider the binary decision tree (DT) on the LHS of Fig. 2. It can be inductively and efficiently compiled into a dDBC [4, appendix A]. The leaves of the DT become labeled with propositional constants, 0 or 1. Each node, n, is compiled into a circuit $c(n)$, and the final dDBC corresponds to the compilation, $c(r)$, of the root node r, in this case, $c(n7)$ for node $c7$. Fig. 2 shows on the RHS, the compilation $c(n5)$ of node $n5$ of the DT. If the decision tree is not binary, it is first binarized, and then compiled [4, sec. 7].

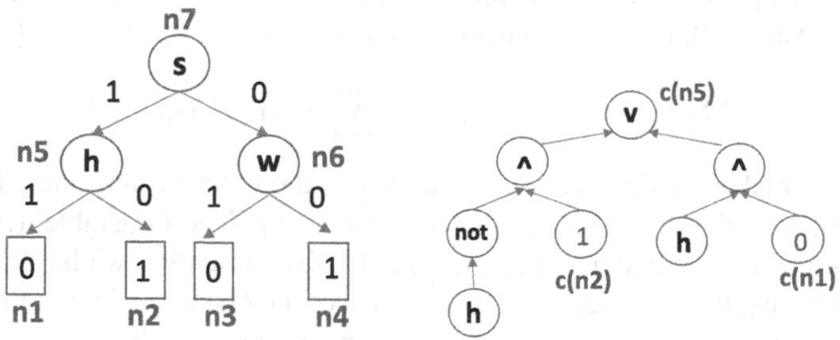

Fig. 2. A decision tree and part of its compilation into an dDBC

5 Looking Ahead: Domain Knowledge

There are different approaches and methodologies to provide explanations in data management and artificial intelligence, with causality, counterfactuals and scores being prominent approaches that have a relevant role to play.

Much research is still needed on the use of *contextual, semantic and domain knowledge* in explainable data management and explainable machine learning, in particular, when it comes to *define and compute* attribution scores. Some approaches may be more appropriate in this direction, and we have argued that declarative, logic-based specifications can be successfully exploited [10]. For a particular application, maybe only some counterfactuals make sense, are reasonable or are useful; the latter becoming *actionable* or *resources* in that they may indicate feasible and achievable conditions (or courses of action) under which we could obtain a different result [27,45].

Domain knowledge may come in different forms, among them:

1. In data management, via semantic constraints, in particular ICs, and ontologies that are fed by data sources or describe them. For example, satisfied ICs might make unnecessary (or prohibited) to consider certain counterfactual interventions on data. In this context, we might want to make sure that sub-instances associated to teams of tuples for Shapley computation satisfy a given set of ICs. Or, at the attribute level, that we do not violate constraints preventing certain attributes from taking null values.
2. In the context of machine learning, we could define an *entity schema*, basically a wide-predicate $Ent(Feat_1, \ldots, Feat_n)$, on which certain constraints could be imposed. In our running example, we would have the entity schema $Ent(\mathsf{Name}, \mathsf{Age}, \mathsf{Activity}, \mathsf{Income}, \mathsf{Debt}, \mathsf{EdLevel})$; and depending on the application domain, *local constraints*, i.e. at the single entity level, such as: (a) $\neg(\mathsf{Age} < 6 \wedge \mathsf{EdLevel} = \mathsf{phd})$, a denial constraint prohibiting that someone who is less than 6 years old has a PhD; (b) Or an implication $\mathsf{Activity} = \mathsf{none} \longrightarrow \mathsf{Income} = 0$; etc. Counterfactual interventions should be restricted by these constraints [10, sec. 6], or, if satisfied, they could be used to speed up a score computation. In [10, sec. 7], we showed how the underlying probability distribution, needed for *Resp* or *Shap*, can be conditioned by logical constraints of this kind.
   We could also have *global constraints*, e.g. requiring that any two entities whose values coincide for certain features must have their values for other features coinciding as well. This would be like a functional dependency in databases. This may have an impact of the subteams that are considered for *Shap*, for example.
3. Domain knowledge could also come in the form of *probabilistic or statistical constraints*, e.g. about the stochastic independence of certain features, or an explicit stochastic dependency of others via conditional distributions. In this direction, we could have a whole Bayesian network representing (in)dependencies among features. We could also have constraints indicating that certain probabilities are bounded above by a given number; etc. This kind of knowledge would have impact on attribution scores that are defined and computed on the basis of a given probability distribution.

The challenge becomes that of bringing these different forms of domain knowledge (and others) into the definitions or the computations of attribution scores.

Acknowledgments. Part of this work was funded by ANID - Millennium Science Initiative Program - Code ICN17002; and NSERC-DG 2023-04650.

References

1. Arora, S., Barak, B.: Computational Complexity. Cambridge University Press, Cambridge (2009)

2. Arenas, M., Bertossi, L., Chomicki, J.: Consistent query answers in inconsistent databases. In: Proceedings of ACM PODS 1999, pp. 68–79 (1999)
3. Arenas, M., Barcelo, P., Bertossi, L., Monet, M.: The tractability of SHAP-scores over deterministic and decomposable boolean circuits. In: Proceedings of AAAI (2021)
4. Arenas, M., Barcelo, P., Bertossi, L., Monet, M.: On the complexity of SHAP-score-based explanations: tractability via knowledge compilation and non-approximability results. J. Mach. Learn. Res. **24**(63), 1–58 (2023)
5. Bertossi, L.: Database repairing and consistent query answering. In: Synthesis Lectures in Data Management. Morgan & Claypool (2011)
6. Bertossi, L., Salimi, B.: From causes for database queries to repairs and model-based diagnosis and back. Theory Comput. Syst. **61**(1), 191–232 (2017)
7. Bertossi, L.: Repair-based degrees of database inconsistency. In: Proceedings of LPNMR 2019, LNCS, vol. 11481, pp. 195–209. Springer, Heidelberg (2019). https://doi.org/10.1007/978-3-030-20528-7_15
8. Bertossi, L.: Specifying and computing causes for query answers in databases via database repairs and repair programs. Knowl. Inf. Syst. **63**(1), 199–231 (2021)
9. Bertossi, L., Li, J., Schleich, M., Suciu, D., Vagena, Z.: Causality-based explanation of classification outcomes. In: Proceedings of the Fourth Workshop on Data Management for End-To-End Machine Learning, DEEM@SIGMOD 2020, pp. 6:1–6:10 (2020)
10. Bertossi, L.: Declarative approaches to counterfactual explanations for classification. Theory Pract. Logic Program. **23**(3), 559–593 (2023). https://arxiv.org/abs/2011.07423
11. Bertossi, L.: Attribution-scores and causal counterfactuals as explanations in artificial intelligence. In: Reasoning Web: Causality, Explanations and Declarative Knowledge, LNCS, vol. 13759, pp. 1–23. Springer, Heidelberg (2023). https://doi.org/10.1007/978-3-031-31414-8_1
12. Bertossi, L., Leon, J.E.: Compiling neural network classifiers into Boolean circuits for efficient shap-score computation. In: Proceedings of AMW 2023, CEUR-WS Proceedings, vol. 3409 (2023)
13. Bertossi, L., Leon, J.E.: Efficient Computation of Shap Explanation Scores for Neural Network Classifiers via Knowledge Compilation (2023). https://arxiv.org/abs/2303.06516
14. Bertossi, L., Kimelfeld, B., Livshits, E., Monet, M.: The shapley value in database management. ACM Sigmod Rec. **52**(2), 6–17 (2023)
15. Bertossi, L.: From database repairs to causality in databases and beyond. In: To Appear in Special Issue of Springer TLDKS dedicated to 'Bases des Donnees Avances' (BDA 20022) (2023). https://arxiv.org/abs/2306.09374
16. Bryant, R.E.: Graph-based algorithms for Boolean function manipulation. IEEE Tran. Comput. C **35**, 677–691 (1986)
17. Chen, C., Lin, K., Rudin, C., Shaposhnik, Y., Wang, S., Wang, T.: An interpretable model with globally consistent explanations for credit risk (2018). https://arxiv.org/abs/1811.12615
18. Chockler, H., Halpern, J.: Responsibility and blame: a structural-model approach. J. Artif. Intell. Res. **22**, 93–115 (2004)
19. Darwiche, A., Marquis, P.: A knowledge compilation map. J. Artif. Intell. Res. **17**, 229–264 (2002)
20. Darwiche, A.: On the tractable counting of theory models and its application to truth maintenance and belief revision. J. Appl. Non-Class. Logics **11**(1–2), 11–34 (2011)
21. Flum, J., Grohe, M.: Parameterized Complexity Theory. TTCSAES, Springer, Heidelberg (2006). https://doi.org/10.1007/3-540-29953-X
22. Grant, J., Martinez, M.V. (eds.): Measuring Inconsistency in Information. Studies in Logic, vol. 73. College Publications, Suwanee (2018)
23. Halpern, J., Pearl, J.: Causes and explanations: a structural-model approach. Part I: causes. Brit. J. Phil. Sci. **56**(4), 843–887 (2005)

24. Halpern, J.Y.: A modification of the Halpern-Pearl definition of causality. In: Proceedings of IJCAI 2015, pp. 3022–3033 (2015)
25. Holland, P.W.: Statistics and causal inference. J. Am. Stat. Assoc. **81**(396), 945–960 (1986)
26. Hunter, A., Konieczny, S.: On the measure of conflicts: shapley inconsistency values. Artif. Intell. **174**(14), 1007–1026 (2010)
27. Karimi, A-H., Barthe, G., Schölkopf, B., Valera, I.: A survey of algorithmic recourse: contrastive explanations and consequential recommendations. ACM Comput. Surv. **55**(5), 95:1–95:29 (2023)
28. Kivinen, J., Mannila, H.: Approximate inference of functional dependencies from relations. Theor. Comput. Sci. **149**, 129-149 (1995)
29. Livshits, E., Bertossi, L., Kimelfeld, B., Sebag, M.: The shapley value of tuples in query answering. Logical Methods Comput. Sci. **17**(3), 22.1-22.33 (2021)
30. Livshits, E., Bertossi, L., Kimelfeld, B., Sebag, M.: Query games in databases. ACM Sigmod Rec. **50**(1), 78–85 (2021)
31. Livshits, E., Kimelfeld, B.: The shapley value of inconsistency measures for functional dependencies. In: Proceedings of ICDT 2021, pp. 15:1–15:19 (2019)
32. Lopatenko, A., Bertossi, L.: Complexity of consistent query answering in databases under cardinality-based and incremental repair semantics. In: Schwentick, T., Suciu, D. (eds.) ICDT 2007. LNCS, vol. 4353, pp. 179–193. Springer, Heidelberg (2006). https://doi.org/10.1007/11965893_13
33. Lundberg, S., Lee, S.: A unified approach to interpreting model predictions. In: Proceedings of Advances in Neural Information Processing Systems, pp. 4765–4774 (2017)
34. Lundberg, S., et al.: From local explanations to global understanding with explainable AI for trees. Nat. Mach. Intell. **2**(1), 2522–5839 (2020)
35. Meliou, A., Gatterbauer, W., Moore, K.F., Suciu, D.: The complexity of causality and responsibility for query answers and non-answers. In: Proceedings of VLDB 2010, pp. 34–41 (2010)
36. Meliou, A., Gatterbauer, W., Halpern, J.Y., Koch, C., Moore, K.F., Suciu, D.: Causality in databases. IEEE Data Eng. Bull. **33**(3), 59–67 (2010)
37. Papadimitriou, Ch.: Computational Complexity. Addison-Wesley, Boston (1994)
38. Pearl, J.: Causality: Models, Reasoning and Inference, 2nd edn. Cambridge University Press, Cambridge (2009)
39. Roth, A.E. (ed.): The Shapley Value: Essays in Honor of Lloyd S. Shapley. Cambridge University Press, Cambridge (1988)
40. Rubin, D.B.: Estimating causal effects of treatments in randomized and nonrandomized studies. J. Educ. Psychol. **66**, 688–701 (1974)
41. Salimi, B., Bertossi, L., Suciu, D., Van den Broeck, G.: Quantifying causal effects on query answering in databases. In: Proceedings of 8th USENIX Workshop on the Theory and Practice of Provenance (TaPP) (2016)
42. Gomes, C.P., Sabharwal, A., Selman, B.: Model counting. In: Handbook of Satisfiability, pp. 993–1014. IOS Press (2009)
43. Shapley, L.S.: A Value for n-Person Games. Contrib. Theory Games **2**(28), 307–317 (1953)
44. Shi, W., Shih, A., Darwiche, A., Choi, A.: On tractable representations of binary neural networks. In: Proceedings of KR, pp. 882–892 (2020)
45. Ustun, B., Spangher, A., Liu, Y.: Actionable recourse in linear classification. In: Proceedings of FAT 2019, pp. 10–19 (2019)
46. Valiant, L.G.: The complexity of enumeration and reliability problems. Siam J. Comput. **8**(3), 410–421 (1979)
47. Van den Broeck, G., Lykov, A., Schleich, M., Suciu, D.: On the tractability of SHAP explanations. J. Artif. Intell. Res. **74**, 851–886 (2022)
48. Wijsen, J.: Database repairing using updates. ACM Trans. Database Syst. **30**(3), 722–768 (2005)

When Operations Research Meets Databases

Vincent T'kindt(✉)

Laboratoire d'Informatique Fondamentale et Appliquée (EA 6300), Université de Tours, 64 Avenue Jean Portalis, 37200 Tours, France
tkindt@univ-tours.fr

Abstract. Overview and relevance: In this tutorial we focus on the interplay between operations research and database theory, i.e. on how the optimization tools from operations research can contribute to the solution of optimization problems occurring in database. Operations research is a field dealing with the solution of optimization and decision problems mathematically well formulated. The objective of this tutorial is to provide researchers in database theory with the basics of operations research, in order to enable them next to improve their own researches. This tutorial is targeted for researchers having never heard about operations research. Attendees should have a background in databases and information systems. Knowledge of complexity and decidability is a plus. The tutorial is divided into three main parts:

1. Introduction to operations research: elements of complexity, exact methods, heuristics.
2. Application to the horizontal partitioning in data warehouses.
3. Application to exploratory data analysis.

To the best of our knowledge, this is the first time that operations research for database theory is proposed for a tutorial at ADBIS. Neither did this appear in the tutorials of top tier data management conferences (VLDB, SIGMOD/PODS, ICDE, EDBT/ICDT) in the last 3 years. Recent PODS tutorials cover very specific topics (e.g., Modern Lower Bound Techniques in Database Theory and Constraint Satisfaction by Dániel Marx or Fine-Grained Complexity Analysis of Queries: From Decision to Counting and Enumeration by Arnaud Durand at PODS 2020).

Keywords: Operations Research · Optimization · Databases

1 Introduction

This tutorial focuses on some intersections between two research fields: database theory and operations research (OR). More precisely, we consider how OR can be used to solve optimization problems occurring in databases. The reader may wonder what is OR? It can be roughly defined as dealing with the development of advanced analytical methods to solve decision or optimization problems. Thus, OR is at the crossroad of mathematics and computer science and relates to the

A. Abelló et al. (Eds.): ADBIS 2023, LNCS 13985, pp. 34–41, 2023.
https://doi.org/10.1007/978-3-031-42914-9_3

design of relevant optimization/decision models and of efficient solution algorithms. Officially, OR dates back to the beginning of the XXth century but has really grown since World War II. It has many applications like in business, computer science, engineering, finance or logistics.

In this tutorial, we assume that the reader is familiar with database theory but not with OR. Therefore, the aim is to introduce the reader to OR and illustrate on problem examples taken from database theory how OR can help. Section 2 presents the basics of OR in three major steps: (i) elements of complexity, (ii) exact algorithms, (iii) heuristic algorithms. Section 3 introduces two applications on which OR can be applied: the problem of horizontal partitioning in data warehouse and the problem of exploratory data analysis.

2 Operations Research for Dummies

2.1 Elements of Complexity

Assume we are given an optimization problem (P) generically defined as follows:

$$\begin{gathered}\text{Minimize } f(s)\\ \text{subject to}\\ s \in \mathcal{S} \subset \mathbb{R}^n\end{gathered}$$

with S the set of solutions and $f : \mathcal{S} \mapsto \mathbb{R}$ the objective function to minimize. We will focus on problems (P) for which set $\mathcal{S}$ is a discrete subset of $\mathbb{R}^n$. Notice that maximization problems are equivalent to minimization problems: maximizing a function $f(x)$ is equivalent to minimizing $-f(x)$. For the sake of clarity, we will consider minimization problems except in few illustrative cases.

Let us consider a first, and very classic, problem.

Traveling Salesperson Problem (TSP)

Data: A set $\mathcal{C}$ of n cities, and for any cities i and j their associated distance $d_{i,j}$

Objective: Find a tour s such that any city $j \in \mathcal{C}$ appears exactly once in s and $\sum_{i=1}^{n} d_{s[i],s[i+1]}$ is minimum, with $d_{s[n],s[n+1]} = d_{s[n],s[1]}$ and $s[k]$ refers to the city in position k in s

In the TSP, the set $\mathcal{S}$ of solutions is defined as the set of permutations over $\{1, ..., n\}$ and $f(s) = \sum_{i=1}^{n} d_{s[i],s[i+1]}$.

When a problem (P) has to be solved, the first task is to determine its complexity status. It is important as it conditions the way to solve it. Complexity theory offers a large panel of results to answer this question ([1]). Basically, and to simplify the explanations, an optimization problem (P) is either:

1. optimally solvable in polynomial time of the *input size* (we say it belongs to class $\mathcal{P}$), or
2. optimally solvable but in super-polynomial time of the *input size* (we say it belongs to class $\mathcal{NP}$-hard).

Notice that the *input size* is a measure of the number of elements in the data. For instance, in TSP it is the number of cities. Thus, for any problem $(P) \in \mathcal{P}$ there exists an *exact* algorithm which finds an optimal solution in $O(p(n))$ time with n the input size and p a polynomial. Consequently, solving (P) requires to find such an algorithm. The situation is tougher for $\mathcal{NP}$-hard problems as we know that it does not exist polynomial-time exact algorithms. There exists two kinds of $\mathcal{NP}$-hard problems: weakly $\mathcal{NP}$-hard problems and strongly $\mathcal{NP}$-hard problems. The former have the property that they can be solved by an exact algorithm running in *pseudo-polynomial time*, i.e. in $O(p(n, ||I||))$ with $||I||$ a measure of the magnitude of the data and p a polynomial. For instance, in TSP, we can set $||I|| = \sum_{i,j} d_{i,j}$. The hardest problems are those which are strongly $\mathcal{NP}$-hard problems, as any exact algorithm requires an *exponential* time to compute an optimal solution, i.e. requires $O(c^n), c \in \mathbb{N}$, or $O(n!)$ time. Solving an $\mathcal{NP}$-hard problem (P) leaves us two options:

1. Design an *exact* algorithm to compute an optimal solution. In this case, the super-polynomial running time may prevent the algorithm from solving real-life instances of (P) .
2. Design a *heuristic* algorithm that is a polynomial-time algorithm computing a solution *as close as possible to an optimal one.* This last notion relates to the art of heuristics and challenges the scientists: thus, the aim is to provide heuristics as effective as possible.

The TSP introduced above can be shown to be strongly $\mathcal{NP}$-hard ([1]) and, so, cannot be solved by a pseudo-polynomial exact time algorithm. However, nowadays, this problem has been extensively studied and very efficient exact algorithms exist. The best one is the Concorde solver ([2]) which is capable of solving instances with up to 86000 cities. Numerous heuristic algorithms have been also proposed in the literature: the most efficient is the LKH heuristic which runs in $O(n^{2.2})$ time and provides solutions at most at 0.162% of the optimal solutions on the TSPlib ([3]). Now, let us introduce a second very classic problem.

`Knapsack Problem (KP)`

Data: A set $\mathcal{O}$ of n objects and, for each object i a weight w_i and a profit p_i. A knapsack of total capacity W.

Objective: Find a selection s of objects such that $s \subseteq O$, $\sum_{i \in s} w_i \leq W$ and $\sum_{i \in s} p_i$ is maximum.

This problem can be shown to be weakly $\mathcal{NP}$-hard and can be optimally solved in $O(nW)$ time ([4]). It is computationally easier to solve than TSP as instances with up to hundreds of thousands of objects can be optimally solved.

In the next sections, we focus on the solution of $\mathcal{NP}$-hard problems and present brief overviews of exact and heuristic algorithms to solve them.

2.2 Exact Approaches for Hard Problems

Polynomial problems, i.e. those belonging to class $\mathcal{P}$, are not considered in this section as they are mainly solved by dedicated algorithms not falling into the approaches introduced hereafter. We rather focus on some general techniques that are relevant for $\mathcal{NP}$-hard problems.

Mathematical programming ([5]) is certainly the first one to consider. It consists in defining a mathematical model of (P) that is next solved by a black-box solver. To illustrate, let us consider KP, and let $x_j, j = 1...n$, be boolean variables with $x_j = 1$ if object j is selected in the knapsack and 0 otherwise. Thus, KP can be formulated as follows: This model is linear and involves n variables

$$\begin{aligned} &\text{Maximize } \textstyle\sum_{j=1}^{n} p_j x_j \\ &\text{subject to} \\ &\textstyle\sum_{j=1}^{n} w_j x_j \leq W \\ &x_j \in \{0; 1\}, \forall j = 1..n \end{aligned}$$

and a single constraint. Experimentally, it can solve instances with up to several hundreds of objects. Notice that, as long as it is possible, linear models are worth being setup as solvers are much more efficient than when solving non-linear ones. Mathematical programming is very interesting as a first approach to build an exact algorithm at the cost of, only, creating a model. It is worth mentioning *Constraint Programming* ([6]) which also offers a good alternative to derive exact algorithms.

Dynamic Programming is a general technique which relies on recursively decomposing a problem into sub-problems and solving them independently ([7]). Let us illustrate this on TSP: we denote by $\texttt{OPT}[\ell, \mathcal{C}']$ the optimal solution value when a set of cities $\mathcal{C}'$ has to be visited after city ℓ. As all cities belong to an optimal tour, we can assume that we start, e.g., with city number 1. We have:

$$\texttt{OPT}[\ell, \mathcal{C}'] = \min_{j \in \mathcal{C}'} \left(\texttt{OPT}[j, \mathcal{C}' \backslash \{j\}] + d_{\ell,j} \right)$$
$$\text{and}$$
$$\texttt{OPT}[\ell, \emptyset] = d_{\ell,1}.$$

Solving TSP for a given set of cities $\mathcal{C} = \{1, .., n\}$ requires to compute $\texttt{OPT}[1, \mathcal{C} \backslash \{1\}]$. From a computational point of view, computing this quantity implies computing all possible $\texttt{OPT}[\ell, \mathcal{C}'], \forall \ell = 2, ...n$ and $\forall \mathcal{C}' \subseteq \mathcal{C} \backslash \{1\}$, thus leading to a dynamic programming algorithm running in $O(n 2^n)$ time and space. Obviously, for this problem, this exact algorithm has only a theoretical interest as it is far from being as fast as the Concorde solver. However, notice that on KP the best algorithm is based on dynamic programming.

To complete this big picture of exact approaches, we can quote *branching algorithms* ([8], [9]) like *branch-and-bound*, *branch-and-cut* or *branch-and-price* algorithms that all use a search tree to explore the solution space. They proceed by decomposing problem (P) into sub-problems which are solved independently from each other. For instance, on TSP, the root node s^0 of the search tree corresponds to the initial problem: a collection $\mathcal{C}^0 = \mathcal{C}$ of cities have to be visited and the sequence of already visited cities $\pi^0 = \emptyset$. Next, from s^0, n children nodes

s^j are created by taking each $j \in \mathcal{C}^0$ and appending it at the end of π^0. Thus, node s^j has $\mathcal{C}^j = \mathcal{C}^0 - \{j\}$ and $\pi^j = (j)$. This process is iteratively repeated until reaching leave nodes s^k with $\mathcal{C}^k = \emptyset$: sequences π^k are solutions and the ones with the minimum total distance are optimal solutions. For sure, the number of created nodes being exponential, techniques need to be embedded to prune those not leading to an optimal solution. A classic one is called the *bounding procedure* and exploits the knowledge of a global *upper bound* computed before starting the branching process, and a *lower bound* computed at each node.

Usually, branching algorithms are very effective in solving hard optimization problems but require a real expertise on the problem.

2.3 Heuristic Approaches

Heuristic algorithms apply to $\mathcal{NP}$-hard optimization problems and they enable to provide good approximate solutions in a reasonable, polynomial, running time. Very numerous categories of algorithms can be found in the literature and the aim of this section is not to present them all. We rather focus on some very basic ones which may be relevant in the context of database theory.

The simplest heuristics that can be designed are called *greedy heuristics* and compute a solution by iteratively taking a decision according to a given priority rule. To illustrate this principle, let us consider a well-known heuristic for KP which, to any object j, associate a priority $\alpha_j = \frac{p_j}{w_j}$. Next, this heuristic sorts the objects by decreasing value of their priority and selects the k first ones such that:

$$\sum_{j=1}^{k} w_{[j]} \leq W < \sum_{j=1}^{k+1} w_{[j]},$$

with $w_{[j]}$ the weight of the j-th object inthe sorting.

This heuristic runs in $O(n\log(n))$ time. As an example, consider $n = 5$ objects with a profit vector $p = [4; 3; 2; 6; 1]$ and a weight vector $w = [1; 1; 1; 4; 4]$. The total capacity of the knapsack is $W = 6$. Thus, the vector of priorities is $\alpha = [4; 3; 2; \frac{3}{2}; \frac{1}{4}]$ and we already have $\alpha_j \leq \alpha_{j+1}, \forall j = 1..4$. The heuristic selects the 3 first objects for a total weight of $3 \leq W = 6$ and a total profit of 9. Notice that it is also possible to conceive greedy heuristics with dynamic priorities, i.e. priorities which are updated each time a decision is taken. Greedy heuristics are usually basic ones which can be easily improved by more elaborated heuristics but at the cost of an increase in the computational complexity. Their main advantage is to be able to scale up to very large size instances.

Local search heuristics ([10]) are also well-known heuristics often considered as first heuristics. They take as input an existing solution of problem (P) and improve it by means of the iterative use of *neighborhood operators*. Let us denote by s^t the solution at iteration t and by $\mathcal{N}(s^t)$ the neighborhood of s^t. To define $\mathcal{N}(s^t)$ we need to use a neighborhood operator which generates solutions "close" to s^t. Then, the local search heuristic selects $s^{t+1} \in \mathcal{N}(s^t)$ such that $f(s^{t+1})$

improves over $f(s^t)$. Ideally, we take s^{t+1} as being the best solution in $\mathcal{N}(s^t)$. This process iterates until no improving solution is found.
To illustrate this principle, let us consider KP and the 5-object instance introduced above. We denote by $s^0 = \{1, 2, 3\}$ the solution computed by the greedy heuristic. For KP we can use the `Swap` neighborhood operator which builds a solution s from s^0 by exchanging an object in s^0 with one in $\mathcal{O} \backslash s^0$: all possible uses of `Swap` on s^0 defines $\mathcal{N}(s^0)$. We obtain:

$$\mathcal{N}(s^0) = \{\{1, 2, 4\}; \{1, 2, 5\}; \{1, 3, 4\}; \{1, 3, 5\}; \{2, 3, 4\}; \{2, 3, 5\}\},$$

which are all feasible with respect to the knapsack constraint. The best solution in $\mathcal{N}(s^0)$ is $s^1 = \{1, 2, 4\}$ with a total profit of 13. Iterating this process, shows that in $\mathcal{N}(s^1)$ there are no improving solutions, so the algorithm stops.
It is easy to remark that such local search heuristics are more time consuming as the exploration of neighborhoods may involve generating a lot of solutions. In the above example, the time complexity of the heuristic is in $O(Tn^2)$, with T the number of iterations. Consequently, some local searches may have sometimes problems to scale up to very large size instances.
Notice that a lot of different local search heuristics exist like tabu search or nature-inspired heuristics. Even if they will be evocated during this tutorial, they are out-of-the-scope of an introduction to operations research.

More complex heuristics exist like *matheuristics*. They rely on mathematical programming and make use of the ever improving efficiency of black-box solvers. Various forms of matheuristics can be found in operations research: there are those which build a solution like greedy heuristics, and there are those improving an initial solution like local search heuristics. As a general definition we can state ([11]) that matheuristics are the hybridization of mathematical programming with metaheuristics. It is not a rigid paradigm but rather a concept framework for the design of mathematically sound heuristics. For instance, solving the mathematical model of KP given in Sect. 2.2 within a fixed time limit leads to a matheuristic: the black-box solver returns the best solution it found without any guarantee of optimality. Hopefully, more complex ways of taking advantage of a mathematical model have been studied in the literature ([12]).

To complete this panorama of heuristic approaches, we mention *branching heuristics* which rely on branching algorithms as introduced in Sect. 2.2. There are numerous variants like beam search algorithms, limited discrepancy search or branch-and-greed algorithms to quote a few. All share the common feature of pruning heuristically the search tree representing the solution space in such a way that the number of explored nodes is polynomial in the input size. For instance, beam search algorithms select at each level of the search tree the w most promising nodes, with w a parameter of the algorithm. On the TSP example, a beam search would first create the root node s^0 and the n children nodes s^j: we are at level 1 of the search tree as only one city has been scheduled. Thus, the w most promising nodes (according to bounds) are selected. Next, at level 2, all possible children nodes of these w nodes are created and only w are, in their turn, selected for further branching. In the case of TSP, there are n cities to schedule, implying that the number of created nodes is in $O(n^2 w)$.

Branching heuristics are known to be a little more time consuming than local searches and much less than matheuristics, but almost as good as the latter.

3 Two Applications Coming from Database Theory

We first focus on the problem of horizontal partitioning in data warehouses and we elaborate on the work of [13].

Assume we are given a simple data warehouse modeled as a star relational scheme with a single dimension table and m fact tables. We are also given a set $\mathcal{Q}$ of selection queries Q_j which are frequently ran on this data warehouse. The aim of horizontal partitioning is to split the fact tables and next the dimension table so that the processing of queries in $\mathcal{Q}$ is as fast as possible.

Each query Q_j is defined by a set of n_j predicates $i, p_{j,i,k} = A_{i,k}\theta V$ with $A_{i,k}$ an attribute of dimension table k, $\theta \in \{=, <, >, \leq, \geq\}$ and $V \in Domain(A_{i,k})$. We also assume that to each query Q_j is associated a weight w_j reflecting its frequency: more often the query is ran, higher is the weight. Finally, to limit the fragmentation of the data warehouse, we assume that a bound W on the number of partitions to generate from the tables, is given. This problem is $\mathcal{NP}$-hard in the strong sense ([13]) thus ruling out the existence of an exact polynomial-time algorithm to solve it.

Next, we focus on a second application, namely Exploratory Data Analysis (EDA) when considering comparison queries. Assume we are given a database and the set $\mathcal{Q}$ of all possible comparison queries. The EDA problem consists in identifying a series of comparison queries to run in order to maximize the interest of the provided insights. We elaborate on the work of [14] and [15].

To each query $Q_j \in \mathcal{Q}$ is associated a processing time t_j and an interestingness score v_j: the latter is a measure of the interest of the query for the user. To each couple (Q_i, Q_j) is also associated a distance $d_{i,j}$ which reflects the cognitive distance of browsing from Q_i to Q_j. Thus, the problem is to determine a permutation π of $n' \leq n = |\mathcal{Q}|$ queries that maximizes the total interestingness $\sum_{Q_j \in \pi} v_j$ while minimizing the total processing time $\sum_{Q_j \in \pi} t_j$ and the total cognitive distance $\sum_{j=1}^{n'-1} d_{Q_{\pi[j]}, Q_{\pi[j+1]}}$, with $Q_{\pi[j]}$ the j-th query in π. This is a multiobjective problem which is $\mathcal{NP}$-hard in the strong sense.

For both problems, we will discuss during the tutorial on how a mathematical programming formulation can be setup and how a black-box solver performs on it. Heuristic algorithms will be also dealt with.

Biography

Vincent T'kindt is full professor at the University of Tours. His research relates to the field of operations research and more specifically to scheduling theory to which he heavily contributed over the years. He pioneered the development of multicriteria scheduling, on which subject he published a reference book, and the development of moderate exponential-time algorithms for scheduling problems. He also contributes to researches on the interplay of operations research with other fields like machine learning, pattern recognition and database theory. He is the coauthor of four books, more than 50 publications in leading international journals and more than 200 communications in conferences. He served as editor of international journals in operations research and is member of the advisory board of the European working group on project management and scheduling.

References

1. Garey, M.R., Johnson, D.S.: Computers and Intractability: A Guide to the Theory of NP-Completeness. Freeman, San Francisco, USA (1979)
2. Appelgate, D.L., Bixby, R.E., Chvatal, V., Cook, W.J.: The Travelling Salesman Problem: A Computational Study. Princeton Press (2007)
3. Tinós, R., Helsgaun, K., Whitley, D.: Efficient recombination in the Lin-Kernighan-Helsgaun traveling salesman heuristic. In: Auger, A., Fonseca, C.M., Lourenço, N., Machado, P., Paquete, L., Whitley, D. (eds.) PPSN 2018. LNCS, vol. 11101, pp. 95–107. Springer, Cham (2018). https://doi.org/10.1007/978-3-319-99253-2_8
4. Kellerer, H., Pferschy, U., Pisinger, D.: Knapsack problems. Springer (2004). https://doi.org/10.1007/978-3-540-24777-7
5. Nemhauser, G.L., Wolsey, L.A.: Integer and Combinatorial Optimization. Wiley (1988)
6. Rossi, F., van Beek, P., Walsh, T.: Handbook of Constraint Programming. Elsevier Science (2006)
7. Bellman, R.: Dynamic programming. Dover publication (2003)
8. Morrison, D.R., Jacobson, S.H., Saupe, J.J., Sewell, E.C.: Branch-and-Bound algorithms: a survey of recent advances in searching, branching and pruning. Discret. Optim. **19**, 79–102 (2016)
9. Feillet, D.: A tutorial on column generation and branch-and-price for vehicle routing problems. 4OR **8**, 407–424 (2010)
10. Aarts, E., Lenstra, J.-K.: Local Search in Combinatorial Optimization. Princeton University Press (2003)
11. Maniezzo, V., Stutzle, T., Voss, S.: Matheuristics: Hybridizing Metaheuristics and Mathematical Programming. Springer (2009). https://doi.org/10.1007/978-1-4419-1306-7
12. T'kindt, V.: The marriage of matheuristics and scheduling. Scheduling seminar (2023). www.schedulingseminar.com
13. Bellatreche, Ladjel, Boukhalfa, Kamel, Richard, Pascal: Data partitioning in data warehouses: hardness study, heuristics and oracle validation. In: Song, Il-Yeol., Eder, Johann, Nguyen, Tho Manh (eds.) DaWaK 2008. LNCS, vol. 5182, pp. 87–96. Springer, Heidelberg (2008). https://doi.org/10.1007/978-3-540-85836-2_9
14. Chanson, A., Labroche, N., Marcel, P., Rizzi, S., T'kindt, V.: Automatic generation of comparison notebooks for interactive data exploration. In: Proceedings of the 25th International Conference on Extending Database Technology (EDBT), 29th March-1st April (2022), ISBN 978-3-89318-085-7 on OpenProceedings.org
15. Chanson, A., T'kindt, V., Labroche, N., Marcel: matheuristics for solving the traveling analyst problem. In: 23rd Conference of the French Society on Operations Research and Decision Aid (ROADEF 2022), Lyon (2022)

Query Processing and Data Exploration

Using a Key-Value Index-Store for Cross-Model Join Queries over Heterogeneous Data Sources

Gajendra Doniparthi(✉)

Heterogeneous Information Systems Group, RPTU Kaiserslautern-Landau, Kaiserslautern, Germany
gajendra.doniparthi@cs.rptu.de

Abstract. With the increasing diversity in the data management landscape, integrating data from heterogeneous sources emerged as a viable solution for disparate bio-science, medical, and healthcare data. Cross-model query processing is a common feature of systems integrating multiple data sources from different models. This paper proposes using an independent key-value index store to maintain system-level indexes and a complementing query processing mechanism to execute cross-model queries over multiple data sources. A separate index store is beneficial, mainly when using data sources with limited indexing and join query processing capabilities. We show that the average cross-model query execution time is faster when using indexes on large-sized data sets.

Keywords: data integration · cross-model queries · multi-omics · bitmap indexes

1 Introduction

Data management systems that use virtual integration methods to integrate heterogeneous data stores are designed around the individual data stores and their respective query processing capabilities [11]. The primary focus is on generating optimized query execution plans while leveraging the indexing and join capabilities of the underlying database engines. These systems must compensate for the limited join capabilities of the underlying data sources to execute complex cross-model queries, which might require query re-writing into individual sub-queries and joining the intermediate results at the middleware layer [5]. In particular, when using NoSQL data stores that offer limited join capabilities or data stores with primitive indexing support, novel indexing and joining techniques at the system level can help improve the overall cross-model query execution times.

Consider two data sources, DS1 and DS2, where DS1 contains the list of bio-science experiments, and DS2 has the collections of bio-entities (data output from individual experiments). There are several options to answer cross-model join queries like *list of all bio-entities having the property search_engine_score greater than 0.5 that are generated by experiments with buffer_condition X.*

A. Abelló et al. (Eds.): ADBIS 2023, LNCS 13985, pp. 45–58, 2023.
https://doi.org/10.1007/978-3-031-42914-9_4

1. using sub-query pushdown, where one of the sub-queries (say, *list of all experiments with a buffer_condition X*) is pushed to DS1. The intermediate query result from the first sub-query is used to re-write the other sub-query (*list of bio-entities with search_engine_score* > 0.5) such that the actual join happens in DS2, which is similar to performing a semi-join in DS2 by sending the join keys from DS1.
2. or pushing both the sub-queries to individual data stores and joining the intermediate results at the middleware layer.
3. or using table migration, where one of the smaller collections (say, experiments from DS1) is entirely migrated to DS2 before executing the join query directly in DS2.

Assume that the middleware has access to global index structures on experiment collections in DS1 for the attribute *buffer_condition* = *X* and bio-entity collections in DS2 for the attribute *search_engine_score*. In the options mentioned above, the middleware can directly get the list of experiments (IDs) from the global index for the join operation without actually executing the sub-query in DS1. Similarly, the second sub-query can be refined using the other global index on the attribute *search_engine_score*. In addition, we still take advantage of the native optimizers, indexes and query processing engines of the underlying data stores to process the sub-queries that can further improve the overall cross-model query execution.

We explore the usage of global indexes in an application setting that requires integrating the data from several sources. In our design, we mainly focus on the additional index store, the global indexes, and the query optimization and execution around the indexes. We use the term cross-model even for cross-engine queries across data stores of the same data model. Also, we exclude the optimization part and restrict our discussion in this paper only to index management and cross-model query execution. We make the following contributions:

- we introduce our multi-omics research data management application that performs data integration from heterogeneous data sources.
- we present the distinct features: a separate index store, system-level indexes, and the query processing mechanism that leverages the indexes.
- we evaluate the effectiveness of using global indexes for cross-model queries and compare with the state-of-the-art.

We briefly introduce our multi-omics system architecture in Sect. 3 to keep the paper self-contained. Section 4 presents the index data structures and how they are generated and maintained within a key-value index store. Later, we discuss the query execution mechanism that uses these indexes for cross-model queries. In the experiments section, we present our experimental results evaluated using cross-model bench-marking queries to measure the query execution time against the state-of-the-art and summarize our work.

2 Related Work

We present some relevant approaches that perform data integration involving multiple data sources and cross-model querying. Polystores and distributed query processing frameworks are well-known classes of systems that enable query processing across heterogeneous data stores and support multiple query interfaces. Among the most notable is the BigDAWG polystore system that provides an integrated view and query capabilities on multiple heterogeneous data sources [3]. BigDAWG evaluates cross-engine queries in two ways: it performs shuffle joins by migrating data to different data stores, or splits the input query into sub-queries per data store and integrates the intermediate results.

In the Cloud Multi-Datastore polystore system, the querying language CloudMDsSQL provides a high level of control and breaks down the cross-engine query into sub-queries targeting specific data stores [6]. MuSQLE [4] is another polystore system for SQL-based analytics over heterogeneous data sources. It supports a generic SQL engine API that undertakes the integration of underlying database engines. It uses Spark data frames to provide functions to push an SQL sub-query into respective execution engines. QUEPA [7] introduces augmented search to let users query the polystore without knowing the structure of data in the underlying data stores. The QUEPA system represents the data in key-value pairs and captures the relationships between the data objects from individual data stores in undirected graphs. Augmented search includes automatic expansion of the query result over one data store with the data relevant to the query but stored elsewhere in the polystore.

Open-source distributed query processing frameworks such as Apache Drill support querying data from multiple data sources [8]. It enables SQL-based querying over both the relational and NoSQL data stores. However, Drill can only leverage indexes created in the data sources for index-based query plans that use indexes versus full table scans to access data. PrestoDB [9] is another well-known analytical query execution engine over heterogeneous data sources. It provides engine-specific connectors similar to Apache Drill and provides a distributed execution model for querying data from external data sources.

In summary, the above approaches either do not use an additional indexing layer at all, or, in the case of Drill, has severe limitations in index usage. This gap is filled by our research presented in this paper.

3 Architecture

We briefly overview the system design before describing the architecture in detail. A research project from an individual-omics[1] study that follows FAIR data principles[2] can be packaged into research contexts which contain the measurement data, experimental metadata, annotations, tools, and scripts. The integrated analysis of data provides opportunities to understand the biological systems and their relationships. We define *data containers* as individual data stores that follow a specific data model and provide uniform access to data through APIs.

[1] *omics* - an informal reference to a field of studies ending in -omics, such as Genomics, Proteomics, or Metabolomics.

[2] *FAIR* - is the acronym for the research data that is Findable, Accessible, Interoperable, and Reusable.

A standard program at the source generates data containers from individual research contexts. Once attached to the multi-omics application, the data from individual containers are indexed and ready for multi-omics analysis. Figure 1 shows the detailed architecture. We model the raw data in the containers as collections of objects with attribute values. Each object within the collection gets identified with a unique contiguously increasing sequence number. The data is maintained in partitions of non-overlapping subsets such that the indexes are created for each partition. The global schema and the local schema from individual containers are maintained in the catalog, which is required for query parsing and optimization. The data references within and across the object collections are explicitly identified from the schema. The query parser receives the incoming user queries through the query API and then performs the syntax checks and validations against domain query constraints. The query optimizer is responsible for the planning and optimization, where the parsed query gets translated into a system-specific execution plan. To prepare the optimized query plans, it identifies the query type, the data containers involved, and the list of indexes required from the index store accordingly.

An independent key-value store maintains the index data structures to retrieve objectIDs of respective data objects from individual data containers. The index generator creates and maintains the global index data structures. It ensures that indexes are created only for partitions on a particular attribute value from the object collection. Also, the metadata of each index is maintained in the catalog. The execution node follows steps in the plan to execute any select-(filter)-join-project query and the materialization module help retrieve data from individual data containers and formats the final query result.

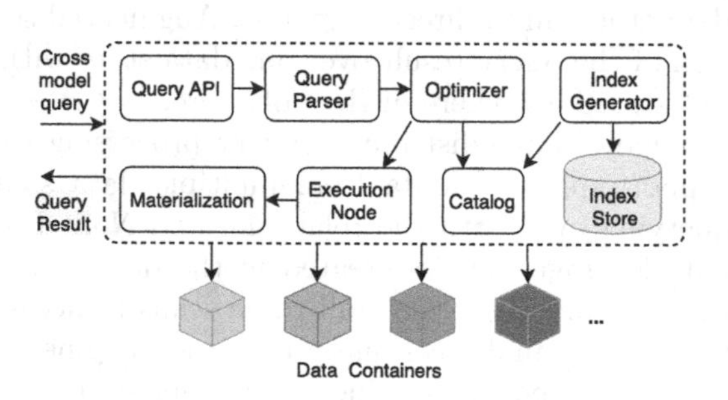

Fig. 1. Detailed system architecture.

4 Key-Value Index Store

The key-value model is the most simple form of data modeling where the string *keys* get mapped to the *values* of either native data types or data structures such as sets, lists, or bitmaps. When augmented with relevant objectIDs of data tuples or documents, the query optimizer and the application middleware can use these data structures during optimization and query processing to improve the overall query performance.

We classify the index data structures into two categories. The *base indexes* are generated on the attributes from individual data partitions of object collections from each data container. The value type of the base indexes can be lists, hash maps, or bitmaps that can directly return the corresponding objectIDs from a specific data partition. The *indexes on indexes* are the tree data structures built on top of the base indexes within a data partition and return a pointer, i.e., the *key* of the respective base index. The base indexes can be used for evaluating the containment or equality predicates, and the tree indexes can be effectively used for range predicates.

The *key* of an index data structure of a particular index type contains the hierarchical order of data container → collection → partition → attribute → attribute-value and also encoded with index specific information. For example, in the case of tree indexes, the attribute value will be the value range. In this paper, we only consider the atomic data types such as number, string, or boolean for the attribute values, not the compound data types of arrays and nested objects.

Base Indexes. We use the bitmap indexes example to demonstrate the base index creation process. Also, the process is similar for other indexes such as lists and hash maps. Consider the bitmap indexes in Fig. 2 where the attribute t has a total of three indexes per partition, one for each distinct value of the domain. By partitioning the data into non-overlapping subsets of size 5, we maintain one bitmap index for each distinct value of the domain for each partition. Each bit vector's size would be the same as the partition size, and the individual bit vector becomes the *value* of a specific base index *key* in the index store. Since the IDs are contiguously increasing sequence numbers, it is trivial to compute the actual ID, given the bit position in the bit vector and the partition number. For a fixed segment size s and the bit position k in a bitmap index of a particular data partition $p = 1, 2, \ldots$, the corresponding ID in the base data can be computed as $((p-1) * s) + k$.

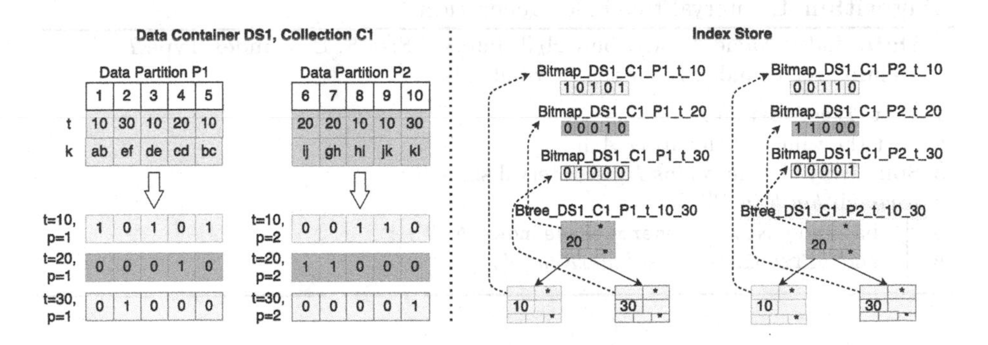

Fig. 2. Bitmap indexes per partition.

The hierarchical order in naming the individual index makes it easy to reconstruct the index key for a specific attribute value, given the data store and collection names. The bit vectors are handy when evaluating Boolean expressions, particularly the equality predicates. For the example in Fig. 2, to find the matching IDs satisfying the predicate expression $t = 10\ OR\ t = 30$ from the collection C1 and data container DS1, the application need to retrieve a total of four bitmap index-keys for the attribute t corresponding to the values 10 and 30. Since the partition sizes are the same, bitmap indexes allow us to perform a direct bit-wise Boolean *OR* operation between the respective bit vectors by partition to compute the resulting IDs 1, 2, 3, 5, 8, 9, and 10.

Tree Indexes. We build *indexes on indexes* as tree structures augmented with the pointers to the respective base indexes of a data partition. The payload of each node in the tree index structure would be the *key* of a particular base index in the index store. The tree indexes can effectively evaluate the range predicates for data types such as integers and strings. The tree index would return the respective *keys* of the individual base index structures that can directly give the corresponding objectIDs. As shown in Fig. 2, the tree nodes consist of the values of the attribute t, and each node-key pointing to the base index (in this case, a bitmap index) that can directly return the positions of IDs.

We introduce interval tree indexes to work with range queries on data types like float. The main difference between the standard tree index structure and the interval tree index is that the node keys contain bounded intervals instead of a single value. We consider only closed intervals and include both endpoints. When we extend a tree structure for non-overlapping interval ranges, each node n in the tree holds the element x, which is assumed to be part of the interval $[x_i, x_j]$ with x_i being the low-end point and x_j becoming the high-end point. We outline the interval tree creation process in Algorithm 1. We first sort the

---

Algorithm 1: Interval tree index generation

Data: Index Name N, Attribute $A_k[]$, Interval Size S, Base Index Type T
Result: I-Tree index N for the Attribute A_k

1 `initITreeIndex(`N`)` ;
2 Sort the values of the array $A_k[]$;
3 Split array $A_k[]$ into bins $B[]$ of interval sizes S ;
4 **foreach** *bin b in* $B[]$ **do**
5   baseIndexKey ← `generateBaseIndex(`b, T`)` ;
6   `addToITreeIndex(`*baseIndexKey*, N`)` ;

---

data by the attribute's value when generating the interval tree index structure for an attribute within a data partition. We then use binning to split the data into intervals of pre-defined sizes. Also, we ensure that the repeated values are enclosed in the same interval to avoid overlaps. Binning divides the attribute's values into an arbitrary number of bins, each containing a disjoint range of values. Each bin's distinct value effectively produces a primary base index. For each interval, we first generate the individual base indexes of the data partition size with the set of corresponding objectIDs, before augmenting with the tree nodes. The node keys consist of the intervals of the attribute's values, and the payload is the pointer to the base indexes for each interval range. The search algorithm finds all the intervals intersecting the given query interval q and returns a base index per partition.

The index updates can be performed in real-time or at periodic intervals while keeping track of the changes. As we generate and maintain the indexes per data partitions, updates and deletes are confined only to the specific partitions. Any changes to the values of the attributes of the data require updating the respective indexes depending on the type of indexes existing for the attribute. Since the tree indexes are mere pointers to the underlying base indexes of values or value ranges of an attribute, they would require no changes unless the data changes require a new base index.

Binding Attributes: We leverage the relationships in the base data while generating the indexes to speed up the cross-model query execution. The global schema must be known before-hand to identify the attributes being referenced. The binding attributes are the natural join partners among the data collections across the data stores. For data stores with limited join capabilities, the binding attributes between the individual collections of the same data store can be used to perform a regular join operation within the collections of the data source. We also maintain binding attribute cross-references in the index store where the *key* represents a distinct domain value of the binding attribute. The *value* is the list of data partitions referenced by the value of the binding attribute.

5 Cross-Model Query Execution

In order to hide the heterogeneity of data sources from the end-users, our multi-omics application accepts user queries in the form of application-specific standard JSON query format. However, the user queries can also be submitted in standard SQL, where an additional SQL parser decomposes it into the JSON query format before identifying the data stores, collections, binding attributes for joining, and indexes needed to execute the query. Typical analytical queries include cross-model joins between two heterogeneous data sources.

The optimizer then transforms the JSON query plan into an internal JSON execution plan with pre-defined logical sections in the sequence in which the execution node executes the plan as shown in the Algorithm 2. For user queries involving a single data source, the optimizer simply rewrites the query in its native format so that the execution node can push it directly to the source. However, when using the indexes, the optimizer follows a pipeline of tasks to generate the optimized operator tree before writing it in a JSON execution plan. Apart from the traditional optimization techniques such as join ordering and selection/projection push-down, the optimization pipeline also includes the steps of range condition creation to effectively use the interval trees and index selection to pick the most suitable indexes for each predicate. In some instances, pushing the sub-queries and joining the intermediate results can be faster than using the indexes, needing to evaluate the combined cost of executing the sub-queries within the data stores against the cost of using the indexes. We consider it as part of our future work.

Algorithm 2: Cross-model query optimization and execution

```
    Data: Input Query Q_i
    Result: Query result Q_R
 1  if not cross-model query then
 2  |   Q_s ← toNativeFormat (Q_i) ;
 3  |   execSubQueryPushdown (Q_s) ;
 4  else
 5  |   if exist(indexes) then
 6  |   |   Q_t ← generateOptimizedOpTree (Q_i) ;
 7  |   |   Q_e ← generateIndexExecPlan (Q_t) ;
 8  |   |   execIndexPlan (Q_e) ;
 9  |   else
10  |   |   Q_s ← generateSubQueryPlan (Q_i) ;
11  |   |   execSubQueryPushdown (Q_s) ;
12  def execIndexPlan(Q_e):
13  |   S ← initQueryResultSpace () ;
14  |   p ← readPrimarySource(Q_e) ;
15  |   evalSubQuery(p) ;
16  |   loadDataPartitions(S, p) ;
17  |   DS[] ← readOtherDataSources(Q_e) ;
18  |   foreach DataSource d in DS[] do
19  |   |   P[] ← fetchPredicates(d) ;
20  |   |   foreach Predicate p in P[] do
21  |   |   |   applyPredicate (S, p) ;
22  |   Q_R ← materialize(S) ;
```

The query optimizer ensures that the number of sub-queries needed to be pushed down to the individual data stores is minimized to take advantage of the indexes. A sample index execution plan is given in Listing 1.1. Consider the example of two data sources, DS1 and DS2, containing the list of bio-science experiments and bio-entities collections, respectively. To execute the cross-model join query and fetch list of all bio-entities having the property *q_value* between 0.0001 & 0.0003 and *abundance* between 400.5 & 480.8, that are generated by experiments with buffer_condition X, the system pushes the first sub-query and fetch the list of all objectIDs of the experiments with a buffer_condition X from

DS1 (PostgreSQL) to initialize the query result space, which is an in-memory data structure with the list of objectIDs and binding attribute values. It then uses the binding attribute cross-references to identify the referencing data partitions of bio-entities collections from DS2 (MongoDB). It progressively filters the objectIDs of DS2 by evaluating the predicate conditions on the attribute values, where the query result space gets reduced after evaluating each predicate condition. Lastly, the query result space with the remaining list of objectIDs is materialized into the final query result.

Using interval trees and data partitions in predicate evaluation might introduce false positives, thus affecting the precision of the intermediate query result [2]. The materialization module performs the condition re-check for false positives before materializing the final query result.

Listing 1.1. A snippet of index execution plan with two data sources

```
1      "queryResultSpace": {[{
2            "subQuery": "SELECT id AS sample_id FROM samples WHERE buffer = 'X'",
3            "dataSourceID": "postgresql",
4            "bindingAttributes": ["sample_id"],
5            ....
6        },{ "dataSourceID": "mongo",
7            "collectionName": "protein",
8            "bindingAttributes": ["sample_id"],
9        "predicates": [
10           { "attributeName": "q-value",
11              "indexType": "itree",
12              "interval": [0.0001, 0.0003]},
13           {  "attributeName": "abundance",
14              "indexType": "itree",
15              "interval": [400.5, 480.8]}],
16              ....
```

6 Experiments

Our experiments primarily measure the effectiveness of using system-level indexes for cross-model queries. We use our multi-omics application and evaluate the query performance against the state-of-the-art BigDAWG and Apache Drill. We consider a set of benchmark queries of varying degrees of query selectivity, data types, and predicate expressions on a synthetic data set for our evaluation. We compare the average query execution times against the state-of-the-art that performs joins for the cross-model queries. Our tests also include measuring the storage space requirements for indexes, and we present the correlation between the partition size, index sizes, and the query execution time.

We use two bio-science data sets for evaluation in our experiments: PRIDE and Synthetic data. In general, bio-science experiments follow specific description standards to capture the *metadata*, which is the minimum required information to describe an experiment to understand the biological and technical origins of the data. Depending on the bio-science domain and the technology, the instruments generate chunks of unstructured instrument-specific data which is then processed through quantification and identification software to produce *raw data* containing lists of bio-entities such as proteins, metabolites, etc.

The PRIDE data contains the raw data (≈1 million protein instances) extracted from selected Proteomics experiments from the PRoteomics IDEntifications Database [10]. It contains a set of measurement/contextual attributes of various data types that follow either a uniform or skewed distribution for each bio-entity in the raw data collections. Since the raw data from PRIDE requires considerable manual processing in transforming from domain specific formats to standardized document format, we use synthetic data, a more significant data set with 20 million protein instances, the same as the PRIDE data in standardized format. We use one bio-sample relation to represent the proteomics experiment metadata and collections of bio-entities (for example, proteins) in JSON format for raw data. There is no loss of generality since the central aspect of our evaluation is to execute cross-model joins between the participating data stores in a heterogeneous data store setting. We list the individual benchmark queries and their characteristics under Table 1. They execute cross-model joins between the bio-samples relation and the proteins collection in JSON format to retrieve the matching protein instances. We run the queries thrice and average the measured execution times. We vary the contextual/measurement attributes of different data types, predicate types, and the complexity of the predicate expression from the raw data to have different query selectivities.

Table 1. Benchmark queries

Query#	Data Type	Predicate Type	Operations	Query Selectivity
Q1	Integer	Point	AND	0.0335
Q2	Integer	Point	OR	0.3996
Q3	Integer, String	Point	AND, OR	0.1864
Q4	Float	Range	AND	0.00001
Q5	Float	Range	OR	0.0076
Q6	Integer, String, Float	Point, Range	AND, OR	0.0965
Q7	Integer, String, Float	Point, Range	AND, OR	0.0311
Q8	Integer, String, Float	Point, Range	AND, OR	0.2947
Q9	Integer, String, Float	Point, Range	AND, OR	0.6488
Q10	Integer, String, Float	Point, Range	OR	0.9475

We use the *de.NBI* cloud (German Network for Bio-Informatics Infrastructure) that runs on the OpenStack platform for our experiments. The virtual machine configuration is eight vCPUs, 32 GB RAM, and 200 GB attached storage disk volume and runs on Rocky Linux v8.5. We deploy the data stores of our multi-omics application in docker (v20.10.18) containers and use PostgreSQL for the structured metadata, MongoDB for the semi-structured bio-entities (proteins) in the raw data, and Redis for index store. As shown in Fig. 3, our multi-omics application and Apache Drill share the same data stores. We use the publicly available dockerized BigDAWG application which comes with its own PostgreSQL and Vertica containers. We copy the raw data in relational format to the Vertica container before executing the benchmark queries in BigDAWG.

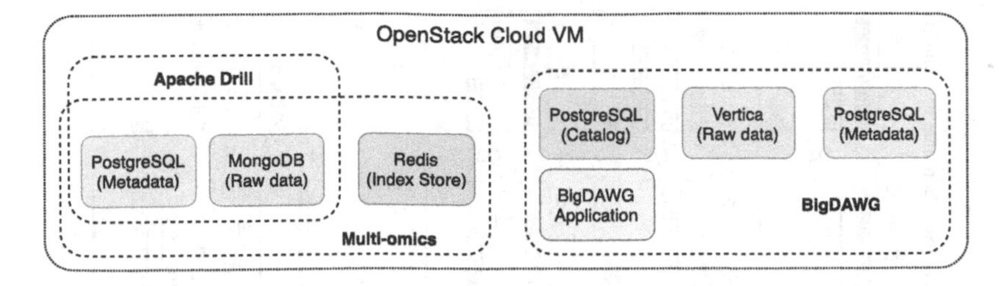

Fig. 3. Setup in *de.NBI* cloud VM

Our multi-omics application maintains the indexes created on measurement/contextual attributes in the Redis index store deployed with the custom modules. We use bitmaps for base indexes and the tree indexes over the base indexes. The index node runs in a basic cluster setup of three primaries and three secondary nodes. We do not use any native indexes on the measurement/contextual attributes in the data stores with the raw data. However, to compare the impact of using the native indexes, we separately run the benchmark queries in Apache Drill with the native indexes created on the individual attributes of the raw data in the MongoDB instance.

6.1 Results

We observe that Apache Drill executed the cross-model queries by joining the intermediate results from the sub-queries at the middleware layer, and BigDAWG used table migration since the bio-samples relation in the PostgreSQL is smaller than the protein relation in the Vertica instance. Figure 4 shows the average query execution times across the systems between the two data sets.

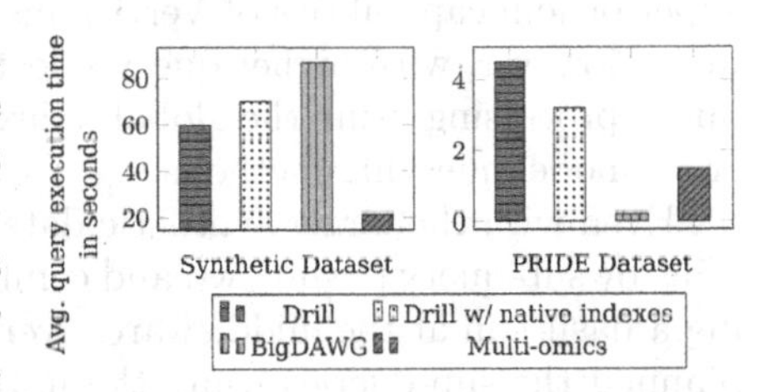

Fig. 4. Average query execution times.

We run a set of low selectivity (<2%) queries on the smaller PRIDE data set and measure the average query execution times when there is a limited amount of data movement between the data stores and the respective middleware layer. We observe that the table migration method of cross-model execution is effective. BigDAWG is faster overall when the Vertica instance performs the joins. In other words, Apache Drill is slower due to the overhead in sub-query push-down and joining the intermediate results. However, Apache Drill with native indexes performed better as the sub-queries pushed to MongoDB executed slightly faster using the native indexes. Our multi-omics application is slower than BigDAWG because of the higher average query initialization time. However, it is faster than both variants of Apache Drill.

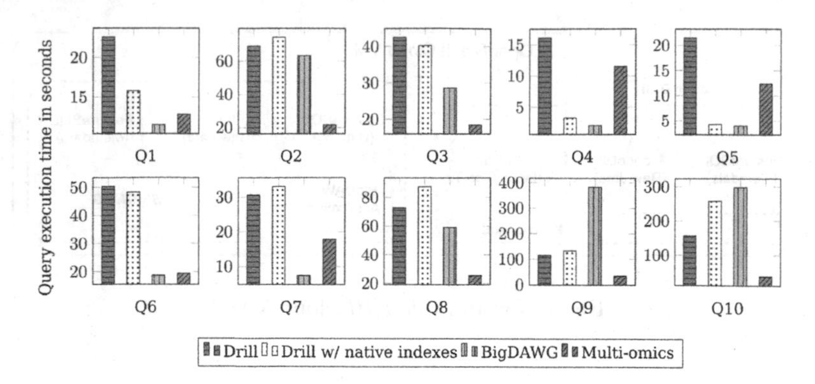

Fig. 5. Query execution times across polystores per each benchmark query.

In the case of the more extensive synthetic data set, our multi-omics application outperformed the state-of-the-art in average query execution time. In particular, we observe that the benchmark queries Q9 and Q10 influenced the average query execution times of both Apache Drill and BigDAWG. We give the break-up of query execution times for individual benchmark queries in Fig. 5. As expected, BigDAWG performed better in the lower query selectivity scenarios for the benchmark queries Q1, Q4, Q5, Q6, and Q7. The reason being the table migration method of cross-model join is powerful when combined with the superior join capabilities of Vertica instance with the raw data in relational format. However, with higher query selectivity, i.e., Q2, Q3, Q8, Q9, and Q10, our query processing using the global indexes is very effective compared to the other cross-model querying methods.

Even with the larger synthetic data set, Apache Drill performed cross-model joins by sub-query push-down and combined the intermediate results by performing a hash-join at the middleware layer. When available, the MongoDB instance planned the sub-queries using the native indexes. Our multi-omics application outperformed Apache Drill without the native indexes for all benchmark queries. With the native indexes, Apache Drill is faster only when executing the low selectivity queries Q4 and Q5.

The query execution times of Apache Drill are proportional to the selectivity, with the higher selectivity queries taking longer execution time. In comparison, BigDAWG query execution times followed a different pattern, where higher selectivity queries took longer execution times, and the low selectivity queries were relatively faster. Like Apache Drill, the pattern of query execution times of multi-omics application is also proportional to the selectivity, although the query execution times are relatively faster overall. Our approach of using the bitmap base indexes at the system level is quite powerful, particularly when executing the cross-model queries on large-scale data sets. For lower selectivity queries, the time taken to initialize the query result space is significant out of the total execution time. However, for benchmark queries with complex predicate expressions and larger query result set sizes, expectantly, predicate evaluation and query result materialization took more time in comparison.

To find the optimal data partition sizes of indexes that can yield the best average query execution times, we evaluated the benchmark queries on the synthetic data set by varying the partition sizes of 2^{12}, 2^{13}, 2^{14}, 2^{15}, and 2^{16}. We generate the base bitmap and tree indexes for each partition size on a set of measurement attributes of varying data types from the protein collection. We run the benchmark cross-model queries to measure the average query execution time. We use the roaring bitmap compression [1] for the bitmap indexes in the index store. Figure 6 shows that the average query execution time improves as we increase the partition size. However, the average time it took to build the compressed indexes increased with the partition size. Similarly, the overall storage space for the indexes in the index store decreased since the number of compressed indexes required to be maintained fewer with a larger partition size. The optimal partition size is close to 2^{16}.

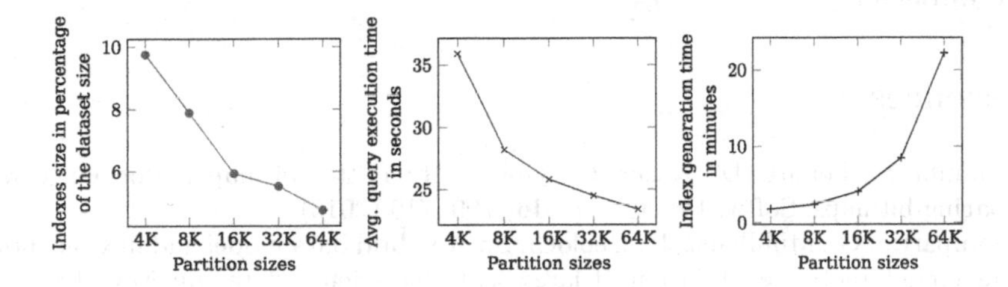

Fig. 6. Index sizes vs. average query execution time vs. index generation time. Measurements taken by varying the data partition sizes.

6.2 Limitations

Our query execution is similar to the sub-query push-down and is orthogonal to the existing polystores; however, adapting in its current form requires changes to the optimizers and execution engines. Also, our system currently supports only basic query types such as containment, point, and range queries. It requires additional processing at the middleware layer to accommodate complex analytical queries. When using global indexes, we must consider the additional storage space requirements for the index store, indexes compression, and the time it takes to generate them. The storage space required for the indexes highly depends on the cardinality of the data and the number of attributes indexed. Lastly, the indexes can be generated and tuned per the underlying data and the workloads. In particular, the partition size can be adapted rather than using a fixed size.

7 Conclusion and Future Work

This paper presents a strong case for system-level indexes for integrating data over heterogeneous data sources. We designed a multi-omics application architecture around a global index store. We took the query processing to the middleware layer, thus, reducing the dependencies on the underlying database engines during cross-model query execution. Our experiments show that multi-omics application using the index store is comparable to the state-of-the-art. We demonstrated that the index-store and system-level indexes make the cross-model queries faster when querying from more extensive bio-science data sets and evaluating complex Boolean predicate expressions. In particular, our multi-omics application performs better using more extensive data sets and queries with higher query selectivity.

As an avenue of future work, we fine-tune the architecture and accommodate complex analytical query processing. We can bring parallelism in index generation and query execution while making the indexes self-tuning to the data and query workloads.

References

1. Chambi, S., Lemire, D., Kaser, O., Godin, R.: Better bitmap performance with roaring bitmaps. Softw. Pract. Exp. **46**, 709–719 (2016)
2. Doniparthi, G., Mühlhaus, T., Deßloch, S.: A hybrid data model and flexible indexing for interactive exploration of large-scale bio-science data. In: New Trends in Database and Information Systems - ADBIS 2021 Short Papers, Estonia, 24–26 August 2021, Proceedings (2021)
3. Gadepally, V., et al.: The BigDAWG polystore system and architecture. CoRR (2016)
4. Giannakouris, V., Papailiou, N., Tsoumakos, D., Koziris, N.: MuSQLE: distributed SQL query execution over multiple engine environments. In: 2016 IEEE International Conference on Big Data Washington DC, USA, 5–8 December 2016
5. Kaoudi, Z., Quiané-Ruiz, J.: Cross-platform data processing: use cases and challenges. In: 34th IEEE International Conference on Data Engineering, ICDE 2018, Paris, France, 16–19 April 2018
6. Kolev, B., Pau, R., Levchenko, O., Valduriez, P., Jiménez-Peris, R., Pereira, J.: Benchmarking polystores: the CloudMdsQL experience. In: 2016 IEEE International Conference on Big Data Washington DC, USA, 5–8 December 2016
7. Maccioni, A., Torlone, R.: Augmented access for querying and exploring a polystore. In: 34th IEEE International Conference on Data Engineering, ICDE 2018, Paris, France, 16–19 April 2018
8. Online: Apache drill. http://drill.apache.org/, webpage. Accessed 26 Apr 2023
9. Online: PrestoDB. http://prestodb.io/, webpage. Accessed 26 Apr 2023
10. Online: Proteomics identifications database. http://www.ebi.ac.uk/pride/, webpage. Accessed 26 Apr 2023
11. Tan, R., Chirkova, R., Gadepally, V., Mattson, T.G.: Enabling query processing across heterogeneous data models: a survey. In: 2017 IEEE International Conference on Big Data Boston, MA, USA, 11–14 December 2017

Reconstructing Spatiotemporal Data with C-VAEs

Tiago F.R. Ribeiro[1], Fernando Silva[1], and Rogério Luís de C. Costa[2(✉)]

[1] CIIC, ESTG, Polytechnic of Leiria, Leiria, Portugal
{tiago.f.ribeiro,fernando.silva}@ipleiria.pt
[2] CIIC, Polytechnic of Leiria, Leiria, Portugal
rogerio.l.costa@ipleiria.pt

Abstract. The continuous representation of spatiotemporal data commonly relies on using abstract data types, such as *moving regions*, to represent entities whose shape and position continuously change over time. Creating this representation from discrete snapshots of real-world entities requires using interpolation methods to compute in-between data representations and estimate the position and shape of the object of interest at arbitrary temporal points. Existing region interpolation methods often fail to generate smooth and realistic representations of a region's evolution. However, recent advancements in deep learning techniques have revealed the potential of deep models trained on discrete observations to capture spatiotemporal dependencies through implicit feature learning.
In this work, we explore the capabilities of Conditional Variational Autoencoder (C-VAE) models to generate smooth and realistic representations of the spatiotemporal evolution of moving regions. We evaluate our proposed approach on a sparsely annotated dataset on the burnt area of a forest fire. We apply compression operations to sample from the dataset and use the C-VAE model and other commonly used interpolation algorithms to generate in-between region representations. To evaluate the performance of the methods, we compare their interpolation results with manually annotated data and regions generated by a U-Net model. We also assess the quality of generated data considering temporal consistency metrics.
The proposed C-VAE-based approach demonstrates competitive results in geometric similarity metrics. It also exhibits superior temporal consistency, suggesting that C-VAE models may be a viable alternative to modeling the spatiotemporal evolution of 2D moving regions.

Keywords: Region Interpolation Problem · Moving Regions · Conditional Variational Autoencoder · Continuous Representation

1 Introduction

Spatiotemporal data describe phenomena, storing the time of occurrence and data on the position, shape, or dimensions of entities of interest. It is commonly used to monitor and analyze changes in land cover, optimize transportation routes, or track iceberg movement, for example. Typically, such data is stored

A. Abelló et al. (Eds.): ADBIS 2023, LNCS 13985, pp. 59–73, 2023.
https://doi.org/10.1007/978-3-031-42914-9_5

using discrete snapshots, associating a timestamp to some representation of the entity's shape and position. Nevertheless, some applications benefit from using a continuous representation of the spatiotemporal evolution of modeled entities.

The continuous representation frequently relies on abstract data types, such as *moving regions* and *moving lines*, and associates discrete representations of the modeled entities to functions that represent their evolution [18,27]. It has some advantages over the discrete model, such as compression capabilities, but also has its own challenges, as creating the continuous representation of an entity requires the specification of a method to generate the entity's shape and position in-between representations. For instance, in Fig. 1, the first and last snapshots are actual images of a real-world phenomenon (forest fire burnt area), and intermediary polygons recreate the spatiotemporal evolution of the burnt area. Commonly, such a recreation employs region interpolation functions, but current methods often fail to create realistic representations of the evolution of 2D regions [6]. On the other hand, recent works in the deep learning field have proved the capability of such models to learn from implicit features.

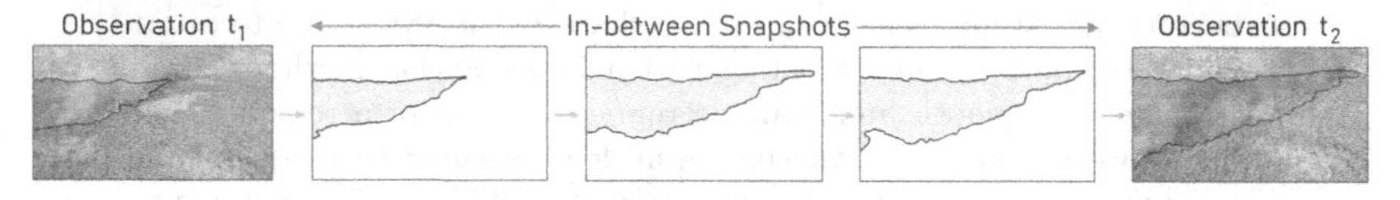

Fig. 1. Continuous representation model requires a method to recreate the spatiotemporal evolution of a region, such as the progression of the burned area.

In this work, we explore the potential of deep learning-based models to create representations of the spatiotemporal evolution of 2D regions by employing Conditional Variational Autoencoders (C-VAEs) [26]. Variational Autoencoders (VAEs) [12] are neural networks consisting of encoders and decoders that learn the probabilistic distribution of the latent space (*i.e.* the low-dimensional representation generated by the encoder). C-VAEs extend VAEs by having conditioning information on the latent space [26]. We assess our methods' capabilities of generating realistic representations by using them to track the dynamics of the propagation of forest fires, generating in-between observations from a discrete set of snapshots of a controlled fire captured by a stationary drone. We compare our model to two widely referenced interpolation methods from the literature ([17,25]), considering distinct scenarios and compression methods, and use geometric similarity metrics to compare generated polygons with ground truth data (manually annotated and automatically segmented). We also evaluate the quality of generated geometries by using temporal consistency metrics.

Hence, the contributions of this work include (**I**) employing C-VAE neural network to generate the spatiotemporal representation of real-world data; (**II**) introducing a specific Temporal Consistency metric to validate the generated in-between observations; (**III**) a systematic evaluation of the proposals considering distinct compression methods, geometric similarity and temporal consistency metrics and (**IV**) a comparison with interpolation methods from the literature.

In the next section, we review some background and related work. Section 3 outlines the proposed network, compression methods, and performance metrics. Then, Sect. 4 discusses experimental results, and Sect. 5 presents the conclusions.

2 Background and Related Work

Spatiotemporal data is often discretely encoded, with spatial attributes such as location and shape associated with time instants. A typical continuous representation of real-world spatiotemporal phenomena data is through *moving regions*. Formally, a *region* is a set of non-intersecting line segments connecting a distinct set of points and forming a closed loop that represents the external faces of a polygon. A region may contain holes, which are also delimited by line segments in a closed loop and, crucially, do not intersect the external faces [27]. *Moving regions* serve as abstract data types used to describe the spatiotemporal evolution of objects of interest, *i.e.*, how their shape and position change over time [9]. *Moving regions* are described as a series of sequentially stored regions (*interval regions*) such that an interval region represents the movement of an object over a time interval between two defined instants, called *slices* [27].

2.1 Moving Regions Evolution Representation

Tøssebro *et al.* [27] proposed a framework in which *moving regions* can be represented from observations stored in spatiotemporal databases. This work was, since then, expanded by different authors using the same principles [8,10,18]. The primary objective is to produce representations that continuously maintain topological validity while ensuring consistency with the underlying spatiotemporal database systems. The *Region Interpolation Problem* (RIP) is the challenge of creating a *moving region* from a set of observations. Specifically, considering two observations at instants t_1 and t_2, the objective is to identify some interpolating function f capable of generating a valid representation of the moving object, its position and shape at any time point between t_1 and t_2 [6].

Other interpolation approaches have also been suggested. For instance, if we regard a *moving region* as a polyhedron, where time takes the place of a third dimension (height) [10], techniques used to interpolate polygons representing sections of a volumetric object, such as human organs in tomographic imaging, may be adapted to generate in-between regions. The so-called *Shape-Based* interpolation is an example of one such algorithm. Contrary to the abovementioned methods, this algorithm operates with raster data [11,25]. Its process can be described in a sequence of steps, as follows. Let x_1 and x_2 be the 2D snapshots that contain the shape of the region at instants t_1 and t_2. For each selected snapshot, a binary image $y_k, k \in \{1, 2\}$ is generated by segmenting the region of interest. Next, a grey-level distance map $z_k, k \in \{1, 2\}$ is generated for each binary image y_k by mapping the Euclidean distance to the boundary of the region. The distance values inside the region are set to positive, and the outside ones are set to negative. The maps z_k are then reconstructed by using linear interpolation at the pixel level. The shape of the region at a given point in time $t_i, t_1 < t_i < t_2$ is found by identifying the zero-crossings of the interpolated distance maps. Finally, these contours generate the region of interest in t_i.

An alternative strategy to interpolating regions involves applying algorithms that can learn the phenomena representations. Deep learning models have shown promising results in various image interpolation applications. For instance, in [21], Oring *et al.* explore interpolation techniques for raster images depicting polyhedra at different angles and other geometric representations with deformable objects, by *smoothly* interpolating the latent space of Autoencoder models. Similarly, Cristovao *et al.* and Mi *et al.* propose equivalent methods for interpolating raster images with three-dimensional objects at various angles, as well as snapshots representing moving objects, using the latent space interpolation of various Generative Latent Variable models [5,19]. In this work, we apply C-VAEs to generate in-between representations of the evolution of 2D regions.

2.2 Conditional Variational Autoencoders

An **Autoencoder** (AE) [12] is a neural network that takes a high-dimensional input $\mathbf{x} \in \mathbb{R}^D$, such as an image, and maps it to a compact, low-dimensional representation $\mathbf{z} \in \mathbb{R}^d$, referred to as the *latent space* z, which is typically a vector. This compressed representation is then used by a decoder to reconstruct the original input. The AE architecture can be decomposed into three components: the encoder $f_\phi(\cdot)$, which maps the input to the latent space, a decoder $g_\theta(\cdot)$, which maps the latent representation back to input space, and a bottleneck $\mathbf{z}$ that stores the compressed codes. The encoder and decoder are often implemented as neural networks with learnable parameters ϕ and θ, respectively.

The **Variational Autoencoder** (VAE) [14,22], an AE development, consist of an encoder $q_\phi(z|x)$ and a decoder $p_\theta(x|z)$. Unlike regular AEs, VAEs learn a probabilistic distribution of the latent space instead of a deterministic mapping. VAEs are trained to minimize the evidence lower bound (ELBO) on $\log p(x)$, where $p(x)$ is the data generating distribution. The ELBO can be expressed as:

$$\mathcal{L}(\theta, \phi, x) = \mathbb{E}q\phi(z|x)[\log p_\theta(x|z)] - D_{KL}(q_\phi(z|x)||p(z)) \quad (1)$$

Here, $p(z)$ is a chosen prior distribution, such as a multivariate Gaussian distribution with mean zero and identity covariance matrix. The encoder predicts the mean $\mu_\phi(x)$ and standard deviation $\sigma_\phi(x)$ for a given input x, and a latent sample $\hat{z}$ is drawn from $q_\phi(z|x)$ using the reparameterization trick: $\hat{z} = \mu_\phi(x) + \sigma_\phi(x) * \epsilon$, where $\epsilon \sim \mathcal{N}(0, I)$. By choosing a multivariate Gaussian prior, the KL divergence term can be calculated analytically. The first term in the ELBO equation is typically approximated by calculating the reconstruction error between many samples of x and their corresponding reconstructions $\hat{x} = D_\theta(E_\phi(x))$. New samples, not present in the training data, can be synthesized by first drawing latent samples from the prior, $z \sim p(z)$, and then drawing data samples from $p_\theta(x|z)$, which is equivalent to passing the latent samples through the decoder, $D_\theta(z)$. The VAEs architecture enables better interpolation than traditional AEs because they learn a continuous latent space that can be easily sampled to generate new data, leading to smooth transitions in the generated outputs [1].

Conditional Variational Autoencoders (C-VAEs) [26] extend VAEs to learn a conditional distribution $p_\theta(x|y)$ where y represents some conditioning information, such as class labels. C-VAEs consist of an encoder $q_\phi(z|x,y)$ and decoder $p_\theta(x|z,y)$, both of which take in the conditioning information y. C-VAEs are also trained to minimize the ELBO on $\log p_\theta(x|y)$.

The ELBO for C-VAEs is similar to that of VAEs, but conditioned on y:

$$\mathcal{L}(\theta,\phi,x,y) = \mathbb{E}q\phi(z|x,y)[\log p_\theta(x|z,y)] - D_{KL}(q_\phi(z|x,y)||p(z|y)) \quad (2)$$

where $p(z)$ is a chosen prior distribution. The encoder predicts the mean $\mu_\phi(x,y)$ and standard deviation $\sigma_\phi(x,y)$ for a given input (x,y), and a latent sample $\hat{z}$ is drawn from $q_\phi(z|x,y)$ as follows: $\epsilon \sim \mathcal{N}(0,I)$ then $z = \mu_\phi(x,y) + \sigma_\phi(x,y) * \epsilon$. The first term in the loss function is typically approximated by calculating the reconstruction error, such as Mean Squared Error or Binary Cross-Entropy losses, between many samples of x and $\hat{x} = D_\theta(E_\phi(x,y))$.

3 Spatiotemporal Data Compression and Reconstruction

By capturing the evolution of real-world objects through images or videos, for example, we are generating a sample of the space-time evolution of an entity. This discrete sample leads to a compressed representation, but also to data loss. For the reconstruction of such data, a method capable of representing the evolution of the object of interest is required [16]. Several factors can impact the quality of the reconstruction, including the used compression technique [2].

3.1 C-VAE Based Interpolation

In addition to generating new samples conditioned on y, C-VAEs have the ability to perform conditional image editing in the latent space. Given two different conditioning inputs, x_1 and x_2, one can interpolate in the latent space between the corresponding latent codes z_1 and z_2 to generate novel images that smoothly blend the two conditioning factors. This capability makes C-VAEs suitable to interpolate different discrete or continuous codified representations.

Our approach consists in training a C-VAE model on the set of samples to be interpolated, conditioned by the timestamp of each sample. For applications that operate with discrete variables, *i.e.* limited number of classes, it is common to encode the classes with one-hot encoding and then concatenate that vector to the input as well as the latent space.

Since we are dealing with continuous phenomena, and the conditioning variable representing the interpolating instant (`label`) is continuous, we chose to simply use the frame number of the original video, normalize it to a range between 0 and 1, and then concatenate the resulting value y to the latent space z and the input x, as shown in Fig. 2.

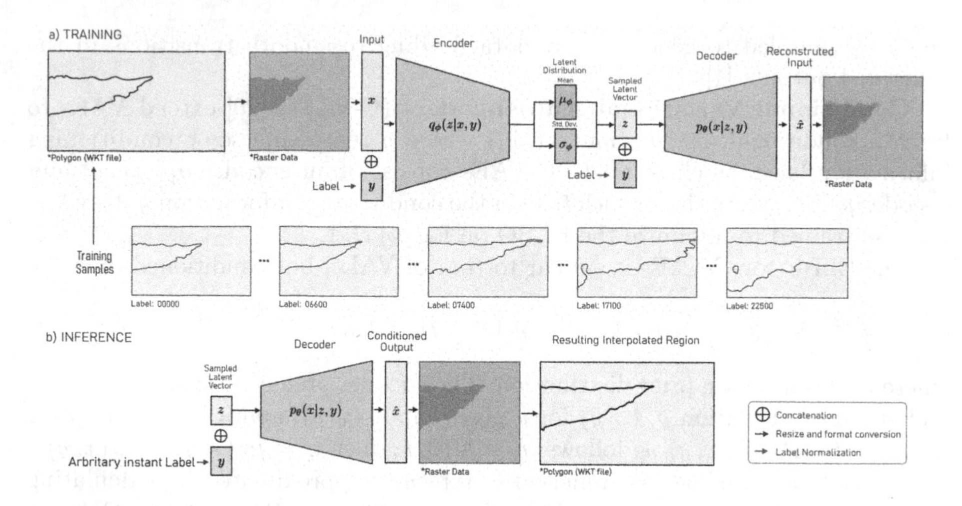

Fig. 2. Employed C-VAE Architecture. a) each region stored in WKT format is converted to raster image to be processed by the model b) a new image is generated conditioned by a label and converted to WKT format.

During the inference stage, we sample both the latent space variable z and the conditioning variable y to generate new samples from the learned conditional distribution $p_\theta(x|y)$. More specifically, we first sample a random vector ϵ from a standard normal distribution, then use it to compute a sample z from the learned approximate posterior distribution $q_\phi(z|x, y)$ using the reparameterization trick. Next, we define a specific conditioning variable y_i, representing an arbitrary instant within the length of the video, and concatenate it with the sampled z to the decoder network $p_\theta(x|z, y)$ to generate a new sample $\hat{x}_i$, an estimated raster image for the instant i.

3.2 Compression Methods

The sampling strategy reduces the amount of spatiotemporal data stored but also determines the representation used for interpolation, thus impacting the performance of reconstruction methods [2].

Periodic Sampling. As a first compression approach, we consider the regions corresponding to the burned area as a sequence of observations $x_t = \{x_1, x_2, ..., x_n\}$, where n is the length of the sequence, ordered in time with a given sampling frequency f_s, corresponding to the video frame rate. Then we sample the sequence periodically using some downsampling factor $d \in \mathbb{N}$. This results in a new sequence of observations $w_t = \{w_1, w_2, ..., w_m\}$, where $m = \lfloor n/d \rfloor$ is the length of the downsampled sequence. Each observation y_i corresponds to the original observation x_{id}, where i is the index of the downsampled sequence and d is the downsampling factor. This approach reduces the size of the sequence by a factor of d, however, this method may discard relevant samples.

Distance-Based Sampling. As a second strategy, we follow the method suggested in [2]. This method downsamples a sequence of observations by selecting representative points that are dissimilar from each other. It takes a set of observations and a distance function that calculates the dissimilarity between two observations, along with a threshold value α. The algorithm initializes the downsampled sequence with the first observation and iteratively adds subsequent observations to the sequence only if they are more dissimilar than the threshold value from the last selected observation. This process continues until all observations have been considered, or the sequence length reaches the desired length. The distance function can be any metric that calculates dissimilarity, such as Jaccard distance, Hausdorff distance, or a combination of several metrics.

3.3 Quality Evaluation Metrics

To evaluate the performance of the spatiotemporal representation, we use two similarity metrics between generated images and ground truth data. We also propose a temporal consistency metric that measures the consistency between generated representations considering phenomena-specific features.

Jaccard Index . The Jaccard Index (JI) measures the overlap or similarity between two shapes or polygons (A and B). It returns a value between 0 and 1, where 1 means the shapes are identical and 0 means they have no overlap:

$$JI(A, B) = \frac{A \cap B}{A \cup B} \tag{3}$$

Hausdorff Distance. The Hausdorff distance measures the degree of mismatch between two non-empty sets of points $A = \{a_1, ..., a_p\}$ and $B = \{b_1, ..., b_p\}$ by measuring the distance of the point A that is farthest from any point of B and vice versa [13]. In simpler terms, it measures how far apart two sets are from each other by finding the maximum distance between a point in one set and its closest point in the other set. Formally, it can be denoted as follows:

$$HD(A, B) = max(h(A, B), h(B, A)) \tag{4}$$

where $h(A, B)$ is the directed Hausdorff distance from shape A to shape B, and $h(B, A)$ is the directed Hausdorff distance from shape B to shape A. The directed Hausdorff distance is defined as:

$$h(A, B) = \max_{a \in A} \min_{b \in B} ||a - b|| \tag{5}$$

where a is a point in shape A, b is its closest point in shape B, and $||a - b||$ denotes the Euclidean distance between two points.

Temporal Consistency. We know that for the same fire outbreak, an area that is established as burned cannot cease to be so at a later stage. Likewise, we know that the burned area never decreases. With these considerations in mind, we define the temporal consistency TC as a complement of a geometric difference:

$$TC_{stride} = 1 - \frac{A_t - A_{t+stride}}{A_{t+stride}}, \forall t \in \{1, 2, ..., T - stride\} \tag{6}$$

where A_t and $A_{t+stride}$ represent the burned area region in separated by *stride* samples. To assess different time scales, we consider various values of stride from a geometric progression $stride_n = ar^{n-1}, \forall n \in \{1, 2, ..., N\}$, with N smaller than the total number of polygons in the sequence, a the coefficient of each term and r is the common ratio between adjacent terms. To assess the performance for each stride, we calculate the average of TC_{stride}. Finally, we can also estimate the overall Temporal Consistency by computing the mean of all TC_{stride} averages.

4 Performance Evaluation

Our study uses data from a video captured by a DJI Phantom 4 PRO UAV equipped with an RGB camera during a prescribed fire at Torre do Pinhão, in northern Portugal (41ř 23 37.56", -7ř 37 0.32"). The UAV remained in a nearly stationary stance during the data collection. The footage is approximately 15 min long, with a frame rate of 25 fps and a resolution of 720×1280, amounting to 22500 images. Prior works have employed this video, as referenced in [3,4].

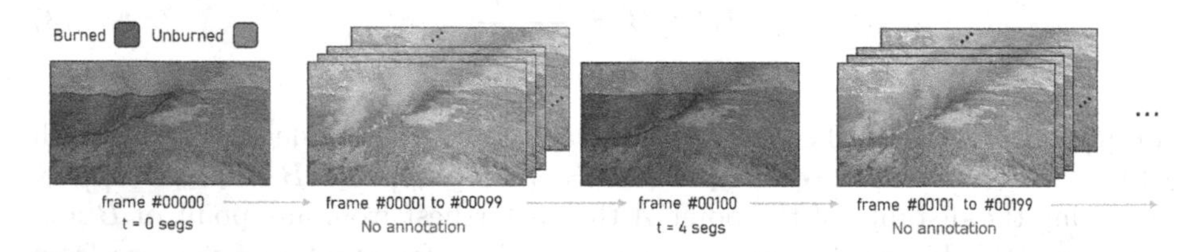

Fig. 3. ***BurnedAreaUAV*** **dataset.** The annotation was done for the entire length of the video for `burned_area` and `unburned_area` classes.

In [23], we manually annotated the burned area, as outlined in Fig. 3. The process yielded 249 annotated frames, of which 226 were used for training and 23 regions for testing. The training and testing sets were created with a periodicity of 100 frames, but having an offset of 50 frames between them. This training subset is considered a reliable representation of the polygon representing the burnt area and serves as the basis for interpolation with *Periodic Sampling*. We further subsample the resulting 226 frames set by applying Distance-Based Sampling, using Jaccard's distance as a dissimilarity measure and a tolerance threshold of $\alpha = 0, 15$, compressing the number of samples to 13.

Additionally, we trained a segmentation model based on the U-Net architecture [24] from scratch on the *BurnedAreaUAV* training set, producing segmentation masks for all video frames, which were then converted to Well-known

Text (WKT) compatible polygons. The polygons produced by the U-Net model obtained an overall Jaccard index value superior to 0.95 on the test set of the *BurnedAreaUAV* dataset, and are therefore considered good approximations of the actual progression of the burnt area. We designated this set of 22,500 polygons as *U-Net Samples.*

4.1 Evaluation Scenarios

Experiment Description. In this experiment we evaluated three different algorithms: (**I**) the Mckenney interpolation method [17], (**II**) the Shape-Based interpolation [25] and (**III**) the C-VAE-based method described in Sect. 3.1.

For each algorithm, we generate in-between samples corresponding to the frame timestamps of the original video by employing the 226 samples resulting from Periodic Sampling as well as the subset of 13 Distance-Based samples. We compare the polygons generated by the algorithms to the ones generated by the automatic segmentation (*U-Net Samples*) and validated using the test subset of the *BurnedAreaUAV* dataset and calculate the Jaccard and Hausdorff similarity metrics. We complement the evaluation by accessing the quality of the generated polygons in terms of the Temporal Consistency indicator as formulated in Sect. 3.3. To calculate the JI and HD, we discarded the samples that supported the calculation of the intermediate regions, both for Periodic Sampling and for Distance-Based sampling. That is, out of a universe of 22,500 observations corresponding to the video frames, we considered 22,274 intermediate regions for the Periodic Sampling and 22,487 for the Distance-Based sampling. All the metrics were calculated considering the resolution of the original footage (720×1280).

Experimental Setup. The experiments were conducted on a computer running Windows 10, equipped with an Intel i7-10700K processor, an Nvidia GeForce RTX 3090 GPU, and 32 GB of RAM. The code was developed in Python, almost entirely in Jupyter Notebooks. Our C-VAE model is built upon a typical convolutional-based neural network implementation for the encoder and decoder, with minor hyperparameter tuning. The code and the datasets are available in https://github.com/CIIC-C-T-Polytechnic-of-Leiria/Reconstr_CVAE_paper.

4.2 Experimental Results

Table 1 presents the achieved values for the similarity metrics and Table 2 summarizes the results in terms of the temporal consistency evaluation.

The Shape-Based and C-VAE interpolations outperformed the Mckenney interpolation method on both the periodic and distance-based sampling (as shown in Table 1 and in Fig. 4 on the top). That is particularly notable on the *BurnedAreaUAV* test set, on which the Shape-Based algorithm and the C-VAE algorithm show relatively close performance, having the former a small advantage. The Shape-Based algorithm on both datasets achieved the best performance in terms of the Hausdorff Distance metric. Comparatively low results on the U-Net dataset are due to the error inherent to the auto-generated segmentation. We can also observe that reducing the number of support samples for interpolation did not have a very pronounced impact on the Jaccard Index or Hausdorff distance values, which supports the validity of distance based compressing algorithm (from 226 to 13 samples) for these particular datasets

Table 1. Similarity Evaluation. Comparison of JI and HD for U-Net Samples and *BurnedAreaUAV* test set using periodic and distance-based sampling.

Dataset	Algorithm	Jaccard Distance			Hausdorff Distance		
		Mean	SD	min-max	Mean	SD	min-max
Periodic Sampling							
U-Net Samples	Shape-Based	**0.958**	**0.011**	**0.870-0.982**	42.460	37.503	9.849-243.994
	C-VAE	0.951	0.011	0.852-0.975	**41.866**	**26.045**	**9.849-242.745**
	Mckenney	0.892	0.048	0.519-0.982	72.195	44.284	9.659-364.660
BurnedAreaUAV Test Set	Shape-Based	**0.959**	**0.016**	**0.925-0.977**	**48.382**	**33.312**	**19.444-117.000**
	C-VAE	0.949	0.017	0.916-0.974	60.815	24.926	23.000-107.201
	Mckenney	0.822	0.073	0.493-0.864	113.161	33.832	86.279-266.303
Distance Based Sampling							
U-Net Samples	Shape-Based	**0.928**	**0.020**	**0.845-0.982**	**68.315**	**38.443**	**10.296-306.026**
	C-VAE	0.905	0.026	0.825-0.987	76.464	52.362	16.000-338.095
	Mckenney	0.876	0.040	0.763-0.978	85.426	38.922	10.471-269.194
BurnedAreaUAV Test Set	Shape-Based	0.910	0.011	0.889-0.928	**60.815**	**33.312**	**19.444-117.000**
	C-VAE	**0.930**	**0.021**	**0.887-0.964**	85.220	14.827	52.773-108.853
	Mckenney	0.850	0.038	0.799-0.960	103.068	30.744	23.014-146.521

Table 2. Temporal Consistency Comparison. Average temporal consistency across different algorithms for periodic and distance-based sampling.

Periodic Sampling				Distance Based Sampling			
Algorithm	Avg. Temp. Consistency			Algorithm	Avg. Temp. Consistency		
	Mean	SD	min-max		Mean	SD	min-max
Shape-Based	0.986	0.011	0.971-1.000	Shape-Based	0.994	0.006	0.985-1.000
C-VAE	**0.993**	**0.007**	**0.982-0.998**	C-VAE	**0.999**	**0.001**	**0.997-1.000**
Mckenney	0.970	0.019	0.951-0.995	Mckenney	0.983	0.018	0.948-0.998

The results in Table 2 indicate the average Temporal Consistency for all considered stride values (1, 10, 100, 1,000 and 10,000) and show that the C-VAE model can produce polygons with higher consistency for the burnt area evolution phenomenon in both datasets. Figure 6 corroborates that by showing the superior monotonicity and *smoother* evolution of the burned area representations generated by the C-VAE model. Analysis of Fig. 5 indicates that the C-VAE algorithm is superior for strides up to 1,000 and performs less well for strides of 10,000, which indicates less ability to retain consistency over a longer time window.

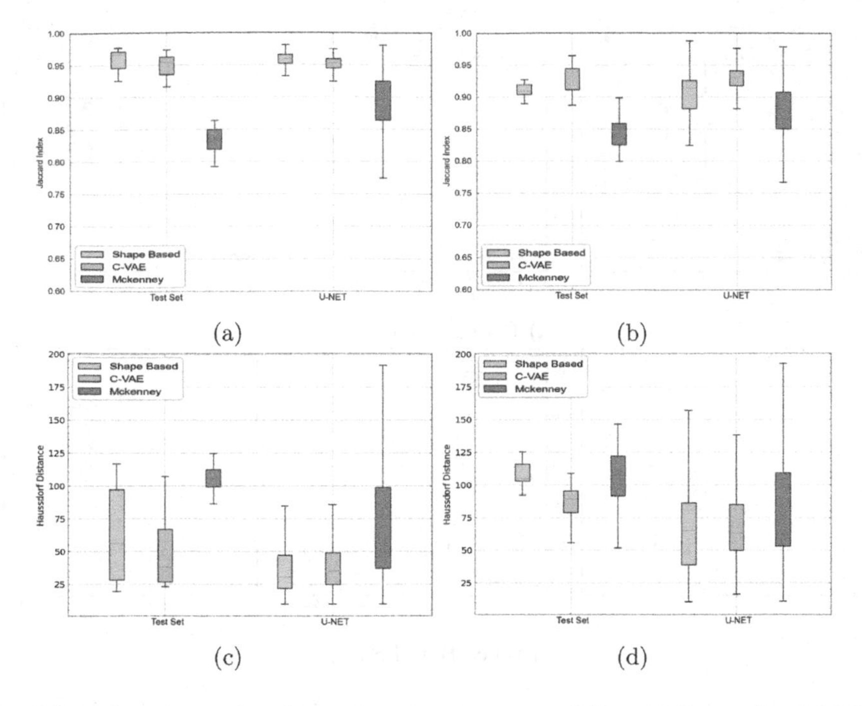

Fig. 4. Boxplots comparing the performance in terms of JI and HD metrics. (a) JI for periodic sampling, (b) JI for distance-based sampling, (c) HD for periodic sampling, (d) HD for distance-based sampling.

4.3 Discussion

A high-quality interpolation should exhibit two main features: firstly, the intermediate points should closely resemble the ground truth data; secondly, the intermediate points should enable a smooth and coherent transition between the endpoints. McKenney's algorithm, which focuses on creating interpolations with valid topology (*i.e.* without self-intersecting segments), revealed deformations and inconsistencies when applied to real-world noisy datasets like *BurnedAreaUAV*. These issues have also been noted by other authors [7,20], which is consistent with our own results

Although the Shape-Based algorithm achieved a good similarity score, it tends to produce artifacts and struggles to interpolate polygons with significantly different topologies [15]. The lower performance observed for interpolations using distance-based sampling in the *BurnedAreaUAV* test subset may reflect this issue, but additional experimentation is needed to confirm this hypothesis.

C-VAEs can learn continuous and smooth representations of complex high-dimensional data by simultaneously optimizing two loss terms: the reconstruction and the KL divergence losses. It is encouraged to describe the latent state for an observation with distributions close to the prior, but deviate when necessary to describe the input's salient features. Minimizing the KL divergence forces the encodings to be close to each other while still remaining distinct, allowing

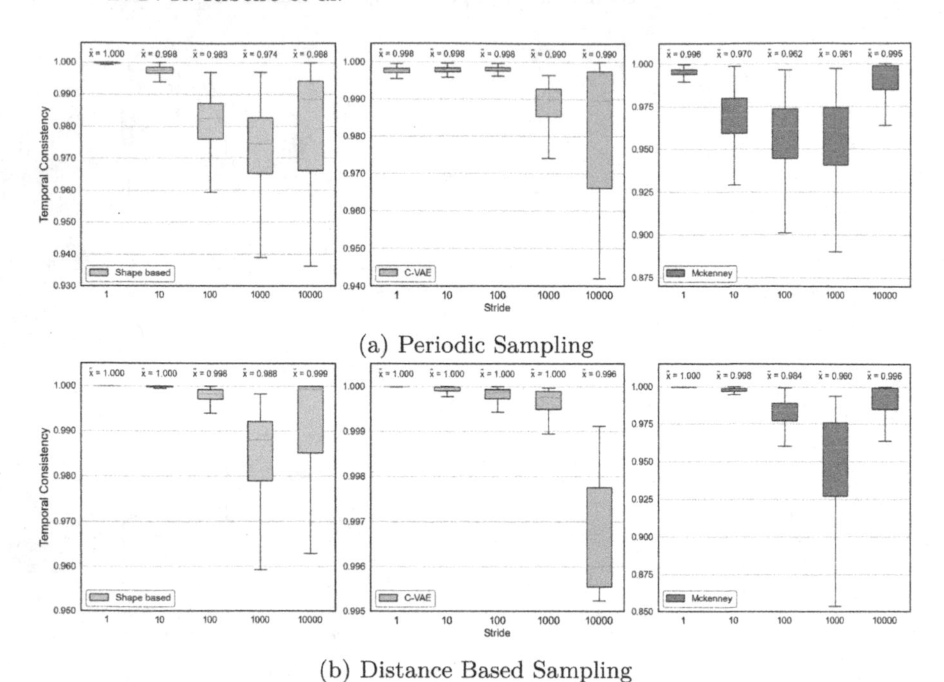

(a) Periodic Sampling

(b) Distance Based Sampling

Fig. 5. Results for Periodic Sampling and Distance Based Sampling, and different temporal strides. Different y-axis scales are used for better visibility.

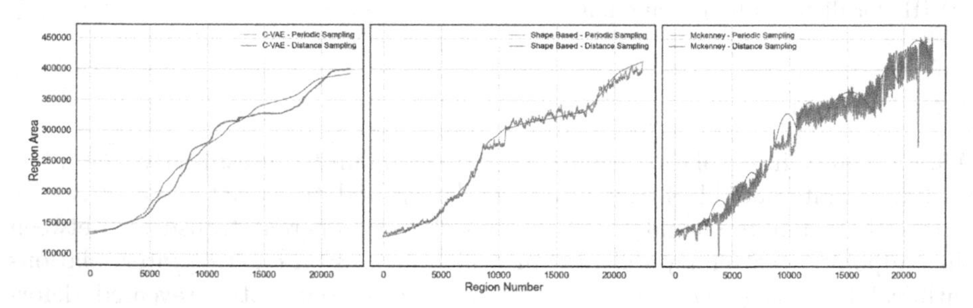

Fig. 6. Representation of the evolution burned area.

for smooth interpolation and the construction of new samples. In other words, the equilibrium reached by the cluster-forming nature of the reconstruction loss and the dense packing nature of the KL loss causes the formation of distinct clusters, enabling the decoder to interpolate smoothly and avoiding sudden gaps between clusters. This mechanism explains the superior Temporal Consistency achieved by the proposed C-VAE-based solution. However, a C-VAE also has its drawbacks. First, unlike the classical models, it has to be trained before the interpolation, which may be time-consuming. Secondly, standard VAEs tend to generate blurry outputs, translating into a poor definition of the region boundaries.

Regarding the timestamp encoding used for the C-VAE model, other methods could be considered, however, for this specific application, our approach attained competitive results, even considering the introduction of the quantization error related to the finite resolution of the floating-point numbers used to store the latent space values.

5 Conclusion

The continuous representation of spatiotemporal data requires methods for generating an entity representation for in-between observations. To implement the *moving regions* abstraction, these methods are usually region interpolation algorithms. But the recent advances in deep models show they may be a possible alternative to this problem. In this work, we compare the performance of a C-VAE model with classical algorithms for region interpolation. We use two datasets obtained from an aerial video of a prescribed fire and assess the performance of the solutions using geometric similarity and temporal consistency metrics.

The C-VAE algorithm performed competitively against the best-performing algorithm (Shape-Based) in terms of similarity metrics and also achieved superior temporal consistency. The results suggest that VAE-based models are viable options for spatiotemporal interpolation and motivate us to explore different AE variants to address the identified limitations. The C-VAE model generated a relatively realistic and smooth representation of the phenomenon evolution, a challenge faced by region interpolation methods.

In the future, we plan to test AE-based models' capabilities to generate the spatiotemporal evolution of a wider range of real-world phenomena.

Acknowledgements. This work is partially funded by FCT - Fundação para a Ciência e a Tecnologia, I.P., through projects MIT-EXPL/ACC/0057/2021 and UIDB/04524/2020, and under the Scientific Employment Stimulus - Institutional Call - CEECINST/00051/2018.

References

1. Berthelot, D., Raffel, C., Roy, A., Goodfellow, I.J.: Understanding and improving interpolation in autoencoders via an adversarial regularizer. CoRR abs/1807.07543 (2018)
2. Costa, R.L.C., Miranda, E., Dias, P., Moreira, J.: Sampling strategies to create moving regions from real world observations. In: Proceedings of the 35th Annual ACM Symposium on Applied Computing (ACM SAC), pp. 609–616 (2020). https://doi.org/10.1145/3341105.3374019
3. Costa, R.L.C., Miranda, E., Dias, P., Moreira, J.: Experience: Quality assessment and improvement on a forest fire dataset. J. Data Inform. Qual. **13**(1) (2021). https://doi.org/10.1145/3428155
4. Costa, R.L.C., Moreira, J.: Automatic quality improvement of data on the evolution of 2d regions. In: Advanced Data Mining and Applications, pp. 288–300 (2022). https://doi.org/10.1007/978-3-030-95408-6_22

5. Cristovao, P., Nakada, H., Tanimura, Y., Asoh, H.: Generating in-between images through learned latent space representation using variational autoencoders. IEEE Access **8**, 149456–149467 (2020). https://doi.org/10.1109/ACCESS.2020.3016313
6. Duarte, J., Silva, B., Moreira, J., Dias, P., Miranda, E., Costa, R.L.C.: Towards a qualitative analysis of interpolation methods for deformable moving regions. In: SIGSPATIAL '19, pp. 592–595 (2019). https://doi.org/10.1145/3347146.3359368
7. Duarte, J., Dias, P., Moreira, J.: An Evaluation of smoothing and remeshing techniques to represent the evolution of real-world phenomena. In: 13th International Symposium, ISVC 2018, Las Vegas, NV, USA, November 19–21, 2018, Proceedings, pp. 57–67 (11 2018). https://doi.org/10.1007/978-3-030-03801-4_6
8. Duarte, J., Dias, P., Moreira, J.: Approximating the evolution of rotating moving regions using bezier curves. Int. J. Geograph. Inform. Sci. **37**(4), 839–863 (2023). https://doi.org/10.1080/13658816.2022.2143504
9. Forlizzi, L., Güting, R.H., Nardelli, E., Schneider, M.: A data model and data structures for moving objects databases. In: Proceedings of the 2000 ACM SIGMOD International Conference on Management of Data, pp. 319–330 (2000)
10. Heinz, F., Güting, R.H.: A polyhedra-based model for moving regions in databases. Int. J. Geograph. Inform. Sci. **34**(1), 41–73 (2020)
11. Herman, G., Zheng, J., Bucholtz, C.: Shape-based interpolation. IEEE Comput. Graph. Appl. **12**(3), 69–79 (1992)
12. Hinton, G.E., Zemel, R.S.: Autoencoders, minimum description length and helmholtz free energy. In: Proceedings of the 6th International Conference on Neural Information Processing Systems, pp. 3–10. NIPS'93 (1993)
13. Huttenlocher, D., Klanderman, G., Rucklidge, W.: Comparing images using the hausdorff distance. IEEE Trans. Pattern Anal. Mach. Intell. **15**(9), 850–863 (1993). https://doi.org/10.1109/34.232073
14. Kingma, D.P., Welling, M.: Auto-encoding variational bayes (12 2013)
15. Lee, T.Y., Lin, C.H.: Feature-guided shape-based image interpolation. IEEE Trans. Med. Imaging **21**(12), 1479–1489 (2002)
16. Mckenney, M., Frye, R.: Generating moving regions from snapshots of complex regions. ACM Trans. Spatial Algorithms Syst. **1**(1) (2015). https://doi.org/10.1145/2774220
17. McKenney, M., Nyalakonda, N., McEvers, J., Shipton, M.: Pyspatiotemporalgeom: A python library for spatiotemporal types and operations. In: Proceedings of the 24th ACM SIGSPATIAL International Conference on Advances in Geographic Information Systems, SIGSPACIAL '16 (2016)
18. Mckennney, M., Frye, R.: Generating moving regions from snapshots of complex regions **1**(1) (2015). https://doi.org/10.1145/2774220
19. Mi, L., He, T., Park, C.F., Wang, H., Wang, Y., Shavit, N.: Revisiting latent-space interpolation via a quantitative evaluation framework. arXiv:2110.06421 (2021)
20. Moreira, J., Dias, P., Amaral, P.: Representation of continuously changing data over time and space: Modeling the shape of spatiotemporal phenomena. In: IEEE 12th International Conference on e-Science (e-Science), pp. 111–119 (2016)
21. Oring, A., Yakhini, Z., Hel-Or, Y.: Autoencoder image interpolation by shaping the latent space. In: Proceedings of the 38th International Conference on Machine Learning. vol. 139, pp. 8281–8290 (2021)
22. Rezende, D.J., Mohamed, S., Wierstra, D.: Stochastic backpropagation and approximate inference in deep generative models, pp. II-1278-II-1286 (2014)
23. Ribeiro, T.F.R., Silva, F., Moreira, J., Costa, R.L.C.: BurnedAreaUAV Dataset (v1.1) (2023). https://doi.org/10.5281/zenodo.7944963

24. Ronneberger, O., Fischer, P., Brox, T.: U-Net: convolutional networks for biomedical image segmentation. In: Navab, N., Hornegger, J., Wells, W.M., Frangi, A.F. (eds.) MICCAI 2015. LNCS, vol. 9351, pp. 234–241. Springer, Cham (2015). https://doi.org/10.1007/978-3-319-24574-4_28
25. Schenk, A., Prause, G., Peitgen, H.O.: Efficient semiautomatic segmentation of 3d objects in medical images. In: Medical Image Computing and Computer-Assisted Intervention - MICCAI 2000, pp. 186–195 (2000)
26. Sohn, K., Yan, X., Lee, H.: Learning structured output representation using deep conditional generative models. In: Proceedings of the 28th International Conference on Neural Information Processing Systems, pp. 3483–3491. NIPS'15, MIT Press (2015)
27. Tøssebro, E., Güting, R.H.: Creating representations for continuously moving regions from observations. In: Advances in Spatial and Temporal Databases, pp. 321–344 (2001)

What Happens When Two Multi-Query Optimization Paradigms Combine?

A Hybrid Shared Sub-Expression (SSE) and Materialized View Reuse (MVR) Study

Bala Gurumurthy[1(✉)], Vasudev Raghavendra Bidarkar[1], David Broneske[2], Thilo Pionteck[1], and Gunter Saake[1]

[1] Otto-Von-Guericke Universität, Magdeburg, Germany
{bala.gurumurthy,vasudevraghavendra.bidarkar,thilo.pionteck, gunter.saake}@ovgu.com

[2] German Centre for Higher Education Research and Science Studies, Hannover, Germany

Abstract. Querying in isolation lacks the potential of reusing results, that ends up wasting computational resources. Multi-Query Optimization (MQO) addresses this challenge by devising a shared execution strategy across queries, with two generally used strategies: *batched* or *cached*. These strategies are shown to improve performance, but hardly any study explores the combination of both. In this work we explore such a hybrid MQO, combining batching (Shared Sub-Expression) and caching (Materialized View Reuse) techniques. Our hybrid-MQO system merges batched query results as well as caches the intermediate results, thereby any new query is given a path within the previous plan as well as reusing the results. To study the influence of batching, we vary the factor - `derivability` - which represents the similarity of the results within a query batch. Similarly, we vary the cache sizes to study the influence of caching. Moreover, we also study the role of different database operators in the performance of our hybrid system. The results suggest that, depending on the individual operators, our hybrid method gains a speed-up between 4× to a slowdown of 2× from using MQO techniques in isolation. Furthermore, our results show that workloads with a generously sized cache that contain similar queries benefit from using our hybrid method, with an observed speed-up of 2× over sequential execution in the best case.

Keywords: Multi-Query Optimization · batched query execution · materialized view reuse

1 Introduction

In many OLAP instances, users submit multiple queries to a DBMS in a short time [GHB+20]. These queries are similar, such that processing them together might save processing time [BKSS18]. To illustrate, Fig. 1 shows the frequency of TPC-H tables fetched in the 22 benchmark queries. As we can see, multiple

A. Abelló et al. (Eds.): ADBIS 2023, LNCS 13985, pp. 74–87, 2023.
https://doi.org/10.1007/978-3-031-42914-9_6

queries access the same table (e.g., 15 queries access lineitem) that combing these queries can avoid redundant table scans. Coming up with such a common execution plan is the key goal of multi-query optimization (MQO).

MQO processes a query set commonly in two ways: shared sub-expressions (SSE) [SLZ12] and materialized view resolution (MVR) [PJ14]. The former devise a single execution plan for a batch of queries, while the latter optimizes queries individually on the fly. Though these techniques are beneficial, they also have key disadvantages: SSE does not persist results, and every batch is given a new execution plan. On the contrary, MVR persists the query results but lacks a common execution plan for a query batch. However, a hybrid of these techniques can avoid both their disadvantages.

Hybrid MQO Technique: In a gist, our hybrid MQO comes with a common query plan for a query batch and stores its intermediate results. Whenever a new query is introduced, we reuse both the existing execution plan and the stored results. Any new results are persisted replacing existing ones using caching techniques. Thus in this work, we successfully combine common query planning of shared sub-expression with results reuse from materialized views. More details about our hybrid MQO technique are discussed in Sect. 4.

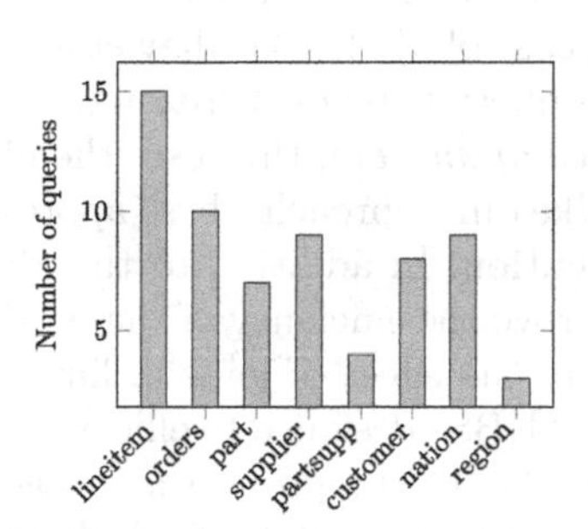

Fig. 1. No. of times TPCH tables accessed in benchmark queries

Due to its flexibility, the hybrid MQO can work in both batched and real-time contexts. Accordingly, we examine the performance of our hybrid MQO w.r.t. sequential, batched, and real-time execution of the queries under different evaluation settings. Our performance evaluation shows a consistent performance improvement which in the best case reaches up to 2× over traditional methods. Specifically, we consider the following aspects when designing and evaluating our hybrid method:

- Workload impact: The effect of various database operators on the hybrid method.
- caching impact: The effect of different cache sizes on the hybrid method.
- Batching impact: The effect of query similarities on the performance of our hybrid method.
- Baseline comparison: The viability of a hybrid method using performance comparison against SSE, MVR, and sequential isolated execution.

Our work is structured as follows: In Sect. 2, we describe related existing MQO techniques. Next, in Sect. 3, we briefly introduce background concepts for MQO. We explain the hybrid multi-query optimizer in Sect. 4 and evaluate it against existing approaches in Sect. 5. Within Sect. 5, we first describe the dataset used, followed by our performance evaluation. Finally, our findings and conclusion are summarized in Sect. 6.

2 Related Work

This section weighs the hybrid MQO with various other existing MQO approaches. A closely related work - OLTPshare by Rehrmann et al. [RBB+18] - proposes result sharing within a batch of OLTP queries. However, in our work, we analyze the sharing potential among OLAP workloads. The main goal of SSE is to identify covering expressions, hence many earlier SSE solutions used dynamic programming to identify them [MCM19]. A cloud-based solution is proposed by Silva et al. [SLZ12]. Unlike these, we statically identify SSE from a batch of queries. In the case of MVR, Bachhav et al. [BKS21] proposes a cloud-based solution to process query batches. Perez and Jermaine [PJ14] design Hawc (*History aware cost-based optimizer*), that uses the historical query plans to derive results. However, unlike our approach, these approaches don't batch upfront a set of queries before execution. In addition to the above-mentioned MQO approaches, many works improve the efficiency of these MQO approaches. Makreshanski et al. [MGAK18] study the effect of joins in hundreds of queries. Jindal et al. [JKRP18] propose BIGSUBS - designed to efficiently identify common sub-expressions. Similarly, Ge et al. [GYG+14] propose a lineage-signature method for common sub-expressions. Jonathan et al. [JCW18] study the effect of executing queries in a shared computational framework over a wide area network (WAN). All these approaches complement the hybrid MQO by enhancing its performance.

3 Background

MQO leverages the similarities across queries to avoid re-computation. Specifically, MQO approaches derive the result for a given query from the results of its predecessor. Depending on the similarity across queries, we have varying types of derivability. Likewise, we have varying MQO approaches depending on their characteristics. In this section, we give an overview of derivability as well as the MQO approaches used in our hybrid system.

Derivability: *Derivability* quantifies the amount of results that can be derived from an existing result set [DBCK17,RSSB00]. As illustrated in Fig. 2, there are four general types of derivability: *exact* - where the result is the exact copy (Fig. 2-a), *partial* - only a subset of results is present (Fig. 2-b), *subsuming* - the current result is part of the previously computed ones (Fig. 2-c) and *none* - where no results can be derived (Fig. 2-d). These types are illustrated in Fig. 2.

Please note that although the given example seems to be only applicable to selection clauses, the same applies to more complex queries such as aggregations or joins. In the next section, we will take a broader look at MQO, examining the two important types of techniques that are used in executing queries mutually.

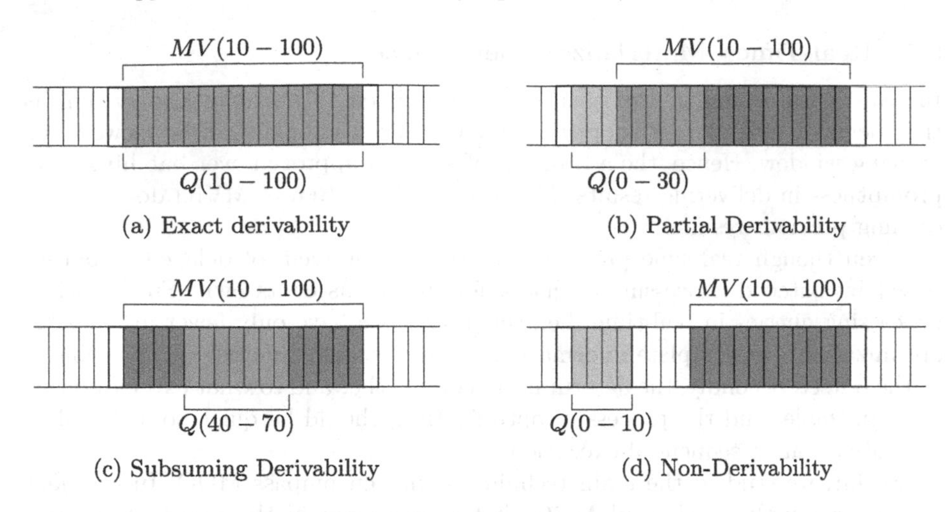

Fig. 2. Types of derivability

MQO Approaches
As we have mentioned, the goal of MQO is to produce an efficient execution plan for multiple queries, irrespective of the performance of individual ones [Sel88]. In In this section, we briefly explain the working of the two MQO approaches considered.

3.1 Batching - Shared Sub-expression

An advantage of batched processing of queries is that the information required by the optimizer – the types of queries and their operators and predicates – to make informed decisions is available upfront, making the optimization more sound [GMAK14, MCM19, SLZ12]. An obvious problem with this approach is that the time taken to create batches is inversely related to the optimization performance. Unfortunately, there is no ubiquitous solution to what the batching time should be, primarily because this is context-dependent. Alternatively, a single query can be split into many sub-queries, each of which can be considered as a part of a batch. This is advantageous in situations where there is a lot of inherent complexity in every query or if the queries contain several shared sub-expressions (SSE).

In SSE, possible common expressions are initially identified among a set of queries. The identified sub-expressions are executed in parallel, with their results being provided to the queries that require them. Specifically, a shared execution plan is created that includes covering expressions derived from combining the sub-expressions. Naturally, as the number of queries in a set increases, the possibility of finding suitable common sub-expressions also increases.

3.2 Real-Time - Materialized View Reuse

In real-time processing, the queries are submitted to the database system as they occur [PJ14]. Here, the optimized query execution plan can be viewed as a growing window. Hence, the advantage of such an approach over batching is its promptness in delivering results. Materialized View Reuse (MVR) does this by caching previous results.

Even though real-time processing solves the problem of delayed responses posed by batched processing, it has a few problems of its own. Firstly, when processing queries in real-time, i.e., separately, we have only fewer avenues for optimization – as complete information about upcoming queries is not available at the outset. Secondly, the determination of whether and to what extent a query is optimizable, and the process of optimization, should be quick so as to offset the gains against sequential execution.

So far, we studied the main techniques that encompass MQO. In the next section, we explore a hybrid MQO that places itself in the midst of the two approaches. In this way, we can leverage the benefits of both techniques while minimizing the limitations.

4 SSE-MVR Hybrid Multi-query Optimization

To support a hybrid MQO, certain extensions are needed in a traditional database system. While SSE needs batching as well as generating a shared execution plan, MVR needs access to the stored materialized views while preparing an optimal execution plan. Hence, we form the hybrid of SSE & MVR with: a query batcher, a substituter, and a cache. The overall structure of the hybrid system is shown in Fig. 3.

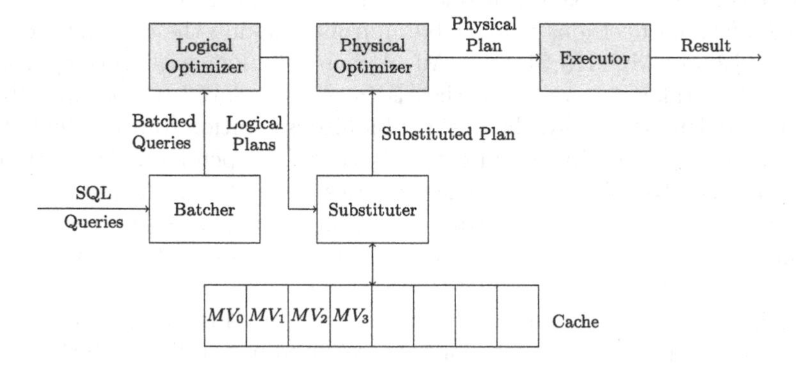

Fig. 3. Components of hybrid MQO system. The blocks colored in gray belong to a conventional database system, whereas the ones in white are exclusive to the hybrid MQO. (Color figure online)

The execution flow in the system is as follows. First, we start batching queries to develop a common query plan. We perform SSE on the given batch of queries

- pruning the redundant operators, identifying the type of derivability possible across queries, etc. Next, such a composite plan is then optimized using an existing traditional optimizer, further simplifying the query plan. The plan is then optimized for MVR based on the cached views from previous runs. If no suitable view is present, we create one from the current run and cache it for future use. In the following sections, we detail the working of the individual components of the hybrid MQO.

4.1 Query Batching

The basic necessity for creating a shared plan across two queries is to have at least a common table scan. For these queries, we then simplify their filter predicates. Similarly, in the next step, we simplify their projection. Let us consider the below queries to be batched. Then, the steps for batching are as follows.

Generating Composite Clauses: Given a batch of queries with common table scans, we start with pruning the ones with common `WHERE` clauses. However, before combining their `WHERE` clauses, we must ensure that the clauses are in their canonical form .i.e in their Conjunctive Normal Form (CNF). Next, we combine these predicates with the `OR` clause. Though such a combined `WHERE` clause might look complex and might even have redundant predicates, the logical optimizer in any traditional DBMS must be able to resolve them easily. Let us consider the queries **Q1 & Q2** below as our running example. We already see that both access lineitem table, therefore the combined predicate simplified with the `OR` clause would look as in Q_{12}. A drawback is that the composite clause may be lengthy because it may contain redundant information. These are simplified using predicate covering.

Q_1: **SELECT** l_quantity, **COUNT**(l_quantity)
FROM lineitem
WHERE l_discount > 0.06
AND l_quantity < 10
AND l_tax > 0.01
GROUP BY l_quantity

Q_2: **SELECT** l_quantity, l_discount, **COUNT**(*)
FROM lineitem
WHERE l_discount > 0.08 **OR**
(l_quantity < 15 **AND** l_tax > 0.03)
GROUP BY l_quantity, l_discount

Predicate Covering: Even after eliminating irrelevant predicates, we are often left with some predicates that can be simplified according to their access patterns. Our running example after CNF has redundant predicates over the search columns that are to be simplified. For example, `l_quantity < 10 OR l_quantity < 15` can be simplified as `l_quantity < 15` without any change in the output. Similarly, we can also simplify `l_tax < 0.03 OR l_tax > 0.01` fetching all values above 0.01. Once a composite predicate clause is generated, we have to focus on combining operations on projection clauses like aggregations.

Q_{12}: **SELECT** ...
FROM lineitem
WHERE (l_discount > 0.06 **OR**
l_discount > 0.08)
AND (l_quantity < 10 **or** l_quantity < 15)
AND (l_tax > 0.01 **or** l_tax > 0.03)
...

Generating Composite Aggregations: Similar to predicates, we can only combine queries that aggregate over a common column. Our example has Q_1 grouping over the *l_quantity* column whereas Q_2 groups on the columns *l_quantity* and *l_discount*. To combine them, we *de-aggregate* the queries, so that the columns are fetched, which are then individually aggregated. Applying this, the de-aggregated query would look like Q_{12D}. Finally, from this de-aggregated query, we can channel the results to compute aggregates based on the individual query.

```
Q12D: SELECT l_quantity, l_discount
FROM lineitem
WHERE l_discount > 0.08
AND l_quantity < 15
AND l_tax > 0.01
```

```
Q1: SELECT l_quantity, COUNT(l_quantity)
FROM Q12D
WHERE l_discount > 0.06
AND l_quantity < 10
GROUP BY l_quantity
```

```
Q2: SELECT l_quantity, l_discount, count(*)
FROM Q12D
WHERE l_tax > 0.03
GROUP BY l_quantity, l_discount
```

Though the queries seem to do redundant computations at first, we now can use the composite query to cache the results that can be used in the subsequent queries as well. The steps for MVR with such a composite query are given below.

4.2 Materialized View Reuse

At this optimization stage, the shared query plan from the batched queries is taken as input and compared with existing materialized views for result substitution. We essentially pick from the cached materialized views those that are useful in executing a batched query plan and then substitute the relevant parts of the query plan to utilize the materialized view instead of the database.

Of course, finding out whether a query is derivable requires storing materialized views. We will discuss the specifics of how the cache is designed to this end in Sect. 4.2. The process of reusing materialized views for `JOIN` operator is quite different. For SPSVERBc6s, we need the materialized views that, in addition to scanning the join of the same tables, scan on any other combination of the tables. This ensures that a subset of relations from the join could be retrieved from the view, whereas the rest are obtained from the database. Finally, since there is only a limited memory available for caching materialized views, we have to regularly evict the materialized views that are not frequently accessed.

Materialized Views Cache: Materialized view cache defines the way to access and maintain the cached views. When the cache reaches a threshold (80% of the total size in our case), a clean routine is called to remove irrelevant views from the cache. Our clean routine follows LRU eviction, such that older queries are evicted whenever a new one has to be inserted. Ultimately, using LRU we indirectly enforce a sliding window in our query batch.

We have so far looked at the individual components that constitute the hybrid MQO system: the batcher, the substituter, and the cache. In the following section, we will understand how these three components jointly form our hybrid MQO system. We will also visualize the process of batching and substitution through a relevant example.

4.3 System Integration

In this section, we briefly explain the interaction between all these components. Similar to the example above, let us assume the below batch of queries $Q_1 - Q_6$.

```
Q1: SELECT s_name, s_suppkey, s_acctbal
FROM supplier JOIN customer
ON s_nationkey = c_nationkey
WHERE s_suppkey < 6870
OR s_acctbal < 145.72
```

```
Q2: SELECT s_name, n_nationkey
FROM supplier JOIN nation
ON s_nationkey = n_nationkey
WHERE n_regionkey = 1
AND s_suppkey < 5000
```

```
Q3: SELECT s_name, sum(s_suppkey)
FROM supplier
WHERE s_acctbal < 100
GROUP BY s_name
```

```
Q4: SELECT l_quantity, AVG(l_tax)
FROM lineitem
WHERE l_quantity < 25
AND l_discount < 0.03
GROUP BY l_quantity
```

```
Q5: SELECT l_discount, l_tax, l_partkey
FROM lineitem
WHERE l_discount < 0.06
AND l_extendedprice < 10000
```

```
Q6: SELECT l_quantity, SUM(l_extendedprice)
FROM lineitem join partsupp
ON l_partkey = ps_partkey
WHERE ps_supplycost < 500
AND l_discount < 0.05
GROUP BY l_quantity
```

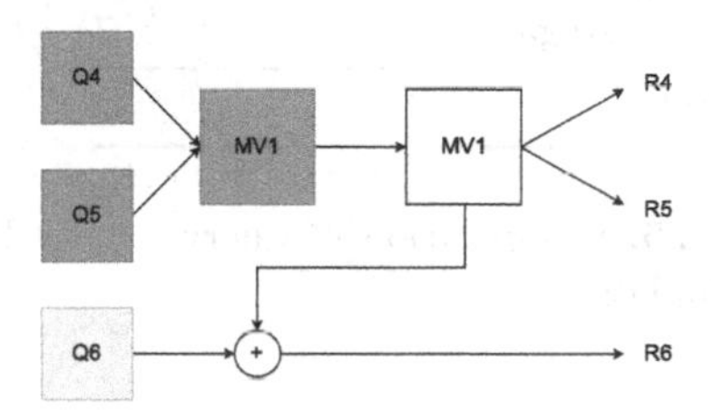

Fig. 4. Functional example of our hybrid MQO

Firstly, the batcher segregates queries that can be batched together. In our case, Q_4 and Q_5 can be batched as they both access `lineitem`. For others, we still have the possibility of achieving performance gains through materialized view substitution.

Next, the optimizer checks for any existing materialized view that can be used to execute Q_1. Since our cache is initially empty, a new view representing Q_1 is generated (MV_1 in Fig. 4). This view can be now reused for Q_2, as `s_suppkey` column is now present in MV_1. A similar case is also applicable for Q_3, Q_4 ,and Q_5.

Finally, to execute Q_6, the optimizer goes through similar steps as for the previous queries. Q_6 is a join of the relations *lineitem* and *partsupp* and the information about *lineitem* can be obtained from the materialized view MV_{4+5}. The tuples for the relation *partsupp* are fetched from the database, and then the join is performed. Now that we have seen the working of our hybrid MQO, in the following section, we evaluate the performance of our hybrid MQO.

5 Evaluation

In this section, we study the performance implications of using a hybrid multi-query optimizer. For our measurements, we extend the existing MVR in Apache calcite (using PostgreSQL plugin) with simple static batching as explained in Sect. 4.1[1]

Evaluation Setup
Due to missing OLAP benchmarks for batching, we use the TPCH dataset with scale factor 5 and custom-built queries. We have 320 such custom queries generated from 32 templates derived from the existing TPCH query set. We generate these queries based on three criteria: query type, result size, and derivability. The detail of the query split-up is shown in Fig. 5.

Type of queries	Quantity
Single column filtering	9
Multiple column filtering	4
Join	5
Aggregate	7
Join-aggregate	7
Total	32

Fig. 5. Composition of query templates

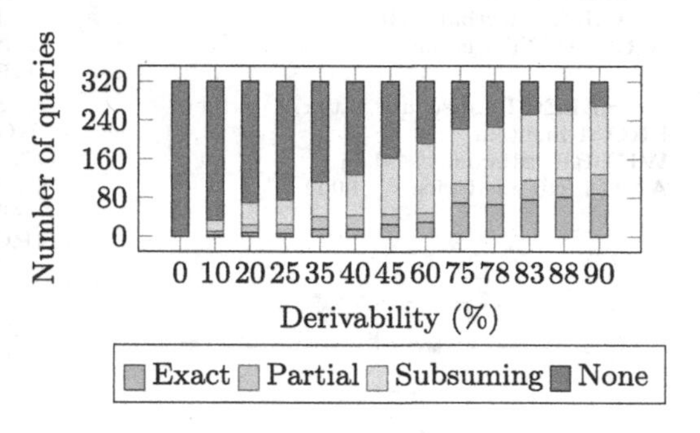

Fig. 6. Proportion of different derivabilities in generated query loads

Impact of Derivability: Furthermore, queries belonging to different derivability types (c.f. Sect. 3) also impact the overall execution. Therefore, it is necessary to ascertain the type of derivability in generated queries. To this end, we plot in Fig. 6 the split-up of different derivability types in a generated query load. As the chart shows, we keep exact derivability at a minimum, with most of the derivability from either partial or subsuming. Now using these queries, in the subsequent sections, we study the performance of hybrid optimization.

Our evaluation uses a Google Cloud - E2-Highmem - instance (with Intel Skylake) with a storage of 100 GB and main memory of 32 GB – all data is stored in main memory.

5.1 Performance Analysis of the Hybrid MQO

In this section, we present our findings in three parts. In Sect. 5.1 we study how different derivabilities and cache sizes influence the execution times of different approaches. Following this, in Sect. 5.1 we compare the performance gains

[1] Code is available here: https://github.com/vasudevrb/mqo.

and losses of different execution strategies over the whole range of tested query loads. Finally, Sect. 5.1 studies the impact of different relational operators on the performance.

The Effect of Derivability and Cache Size: Figure 7 depicts the time to execute different query loads with varying cache sizes. We vary the cache sizes from 4 MB to 2 GB to study its performance impact.

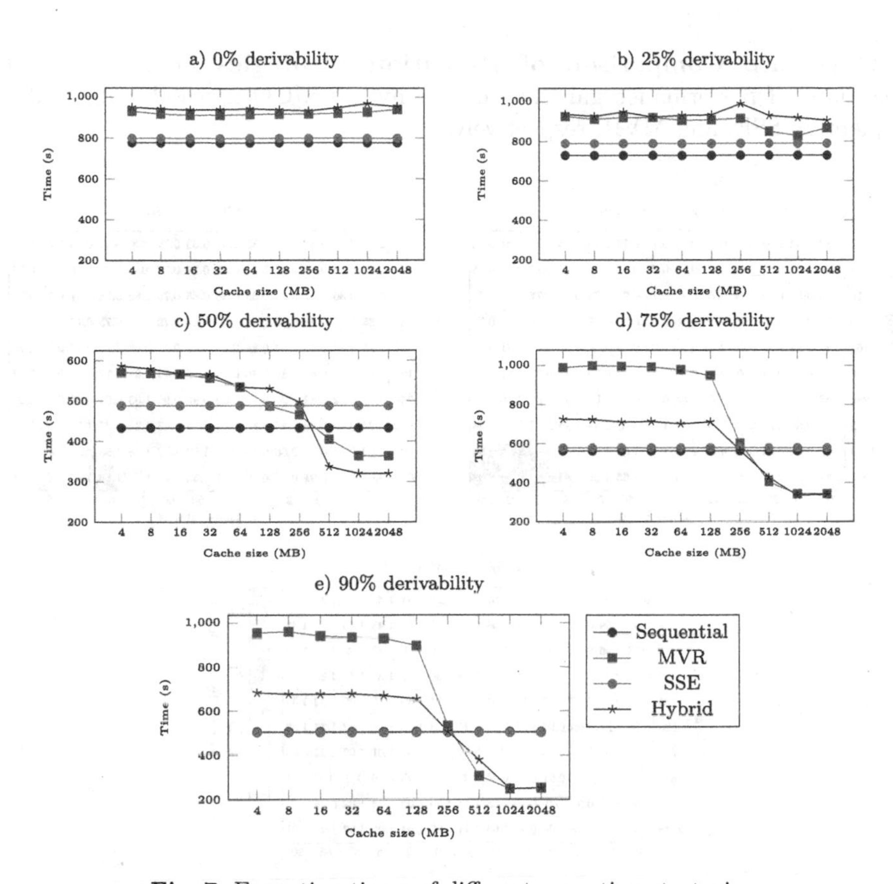

Fig. 7. Execution times of different execution strategies

Foremost, we can see that caching has no effect on sequential and SSE execution as they are independent of cache sizes. For low derivabilities (less than 25%) MVR & hybrid versions perform worse than sequential and SSE. Such poor performance is due to the time spent on creating, caching, and probing materialized views in addition to the normal execution time. However as the derivability increases there is a drastic reduction in the execution times for MVR and hybrid executions. At this point, an adequate amount of views are stored to aid the subsequent queries. This pattern of results holds for all higher derivabilities.

Finally, when derivability increases beyond 50%, we see the performance disparity between MVR and the hybrid method increase – almost a factor of 2.

In this section, we have seen how varying derivabilities and cache sizes influence execution times. Since different query loads contain fundamentally different queries, an absolute comparison of the execution times for different derivabilities is ineffective. So in the next section, we compare the performance of our hybrid system across different derivabilities.

Performance Comparison of Execution Strategies: Figure 8 depicts heatmaps of performance gain/loss of our hybrid MQO method compared to sequential, SSE, and MVR, respectively.

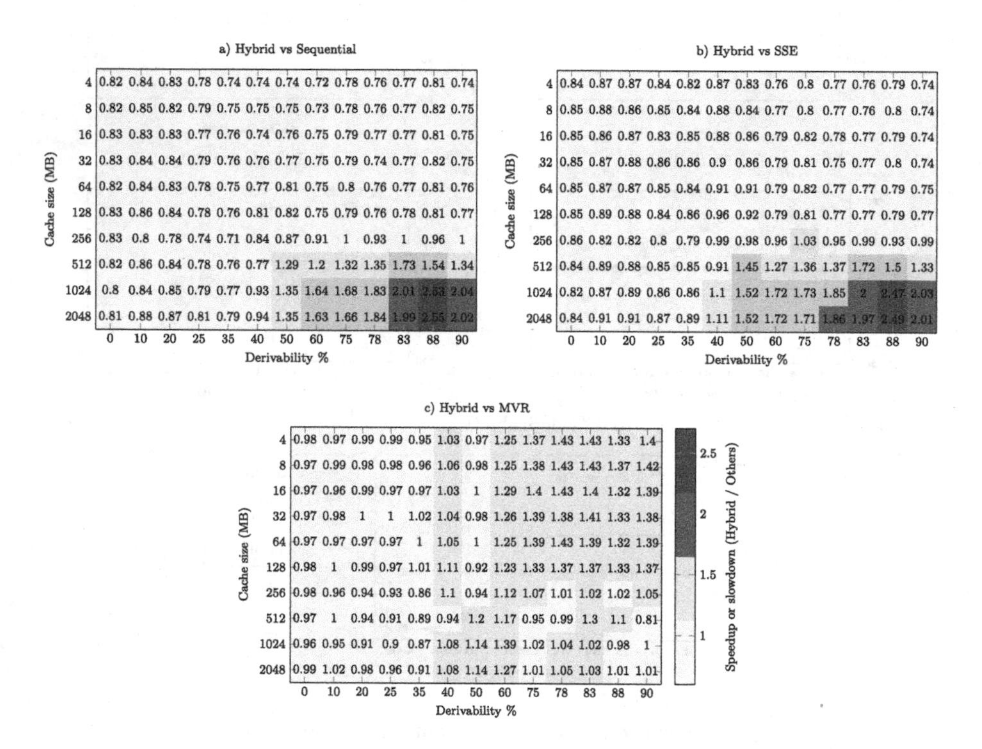

Fig. 8. Speed-up or slow-down from our hybrid mechanism in comparison with baselines

From the heatmaps, our hybrid method is approximately 25% slower for lower derivabilities regardless of cache size. However around 50% derivability, we achieve a 25%–30% gain in performance. This gain keeps increasing, such that with a 90% derivable query load, we are twice as fast as sequential execution.

In some cases, we can observe that having a larger cache harms the performance of our hybrid system. Upon closer inspection, this issue seems unique to Apache Calcite and its handling of materialized view substitutions. Specifically, calcite cycles through the entire materialized view even with 0% derivable queries, degrading performance.

Furthermore, we can observe with high derivability, the performance of our hybrid method exceeds the baseline of the basic MVR approach (Fig. 8-c) when the cache is small. This observation suggests that sharing sub-expressions contributes to this increase in performance. Similarly, for larger caches, the hybrid method is much more efficient when compared to SSE, as shown in Fig. 8-b, implying that the efficiency is a result of materialized view reuse.

Overall, we see a 2x speed-up over sequential and SSE when the queries are derivable, and the cache is large, which shows the full benefit of our approach.

Impact of Query Types: Finally in this section, we study the impact of operators with MQO techniques. The overall speed-up/slow-down is given in Fig. 9. In an overview, we see low derivabilities severely diminish the performance of our hybrid method, whereas high derivabilities enhance performance. A detailed description of the results is given below.

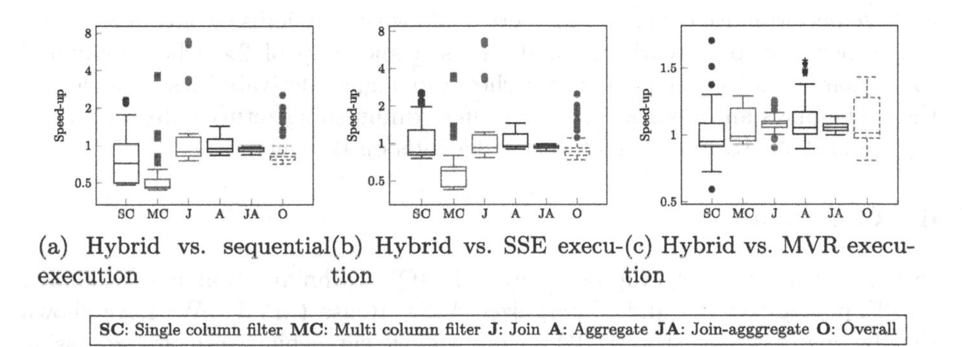

(a) Hybrid vs. sequential execution (b) Hybrid vs. SSE execution (c) Hybrid vs. MVR execution

Fig. 9. Relative performances of individual query types

We observe large fluctuations for filter queries, with speed-ups around 0.5 and 2, respectively. Multiple-column filter queries perform poorly than single-column filter queries, with most multi-column filter query loads being slower to execute with our hybrid method than with sequential ones. The performance deterioration is mainly due to the generation of predicate results from an existing materialized view. However, we observe operators such as joins and aggregates demonstrate improved performance than filters. At best, query loads containing exclusively joins show a speed-up of over 6x. As a whole, mixed query loads with filters, joins, and aggregates show insignificant speed-up. As our hybrid method performs poorly for queries that are not derivable, plotting its performance over the entire range of derivability tends to offset the gains observed for highly derivable queries. Altogether, under high derivable workloads, our hybrid method performs much better compared to sequential (2×) and SSE (2×) executions than MVR (1.4×). This mirrors our already drawn conclusions.

5.2 Discussion

Since the hybrid system has both batching and MVR, it has good performance with large cache and high derivability - which is refected in our results. As we know from our query load (c.f. Fig. 6) that exactly derivable queries are not the majority, we can conclude that the observed speedup is due to the optimizations of our hybrid method, as opposed to the relative computational efficiency of deriving exact results from materialized views. Looking at the performance with regard to the various cache sizes, a certain threshold (256 MB in our case) must be crossed to see the benefits of the hybrid execution. When the cache size and the derivability is maximum, our hybrid method executes query loads twice as fast as sequential execution and SSE. However, even with smaller caches we still get a considerable benefit from the SSE part of the system.

Finally, with different database operators, we see that filter operations have a large performance variation than joins and aggregates. Additionally, query loads containing joins and aggregates show the least variation but also show lower average performance compared to query loads with only joins or only aggregates.

As a whole, our hybrid method shows a speed-up of 2× when compared to sequential execution for larger caches and higher derivabilities. The size of the cache plays an important role but offers diminishing returns after a certain threshold (256 MB in our case), which depends on the query size.

6 Conclusion

In this paper, we have proposed a hybrid MQO technique that merges shared Sub-Expression (SSE) and Materialized View Reuse (MVR). We have shown that by composing existing MQO techniques, we can achieve a query processing system capable of halving the time taken to execute suitable query loads. Our SSE generates a composite query plan whose results are then persisted using MVR, which uses LRU as the cache eviction mechanism. We have evaluated our hybrid MQO method for different query loads, cache sizes, and compared the results with different execution strategies. from our evaluations, we see that high cache-size & derivabilities directly correspond to better performance of the hybrid system. Whereas low derivabilities can cause the hybrid system to expend additional resources managing optimization thereby increasing execution time. Additionally, comparing the hybrid MQO to MVR and SSE approaches shows that it can adapt to the workload. It reuses materialized views when the cache is larger, while with smaller caches, the performance gain is from SSE. Further, analyzing the effect of database operators suggests that complex operators such as joins and aggregates benefit more from a hybrid scheme of processing than queries that contain filters. In summary, a hybrid MQO technique combining SSE and MVR demonstrates a clear advantage of up to 2x speed-up over traditional methods.

References

[BKS21] Bachhav, A., Kharat, V., Shelar, M.: An efficient query optimizer with materialized intermediate views in distributed and cloud environment. In: Tehnički glasnik (2021)

[BKSS18] Broneske, D., Köppen, V., Saake, G., Schäler, M.: Efficient evaluation of multi-column selection predicates in main-memory. IEEE Trans. Knowl. Data Eng. **31**(7), 1296–1311 (2018)

[DBCK17] Dursun, K., Binnig, C., Cetintemel, U., Kraska, T.: Revisiting reuse in main memory database systems. In: Proceedings of ACM SIGMOD (2017)

[GHB+20] Gurumurthy, B., Hajjar, I., Broneske, D., Pionteck, T., Saake, G.: When vectorwise meets hyper, pipeline breakers become the moderator. In: ADMS@ VLDB, pp. 1–10 (2020)

[GMAK14] Giannikis, G., Makreshanski, D., Alonso, G., Kossmann, D.: Shared workload optimization. In: Proceedings of the VLDB Endowment (2014)

[GYG+14] Ge, X.: LSShare: an efficient multiple query optimization system in the cloud. In: Distributed and Parallel Databases (2014)

[JCW18] Jonathan, A., Chandra, A., Weissman, J.: Multi-query optimization in wide-area streaming analytics. In: Proceedings of ACM SIGMOD (2018)

[JKRP18] Jindal, A., Karanasos, K., Rao, S., Patel, H.: Selecting subexpressions to materialize at datacenter scale. In: Proceedings of the VLDB Endowment (2018)

[MCM19] Michiardi, P., Carra, D. Migliorini, S.: In-memory caching for multi-query optimization of data-intensive scalable computing workloads. In: Proceedings of DARLI-AP (2019)

[MGAK18] Makreshanski, D., Giannikis, G., Alonso, G., Kossmann, D.: Many-query join: efficient shared execution of relational joins on modern hardware. In: Proceedings of the VLDB Endowment (2018)

[PJ14] Perez, L., Jermaine, C.: History-aware query optimization with materialized intermediate views. In: Proceedings of the ICDE (2014)

[RBB+18] Rehrmann, R., Binnig, C., Böhm, A., Kim, K., Lehner, W., Rizk, A.: Oltpshare: the case for sharing in oltp workloads. In: Proceedings of the VLDB Endowment (2018)

[RSSB00] Roy, P., Seshadri, S., Sudarshan, S., Bhobe, S.: Efficient and extensible algorithms for multi query optimization. In: Proceedings of ACM SIGMOD (2000)

[Sel88] Sellis, T.: Multiple-query optimization. In: Proceedings of ACM SIGMOD (1988)

[SLZ12] Silva, Y., Larson, P.A., Zhou, J.: Exploiting common subexpressions for cloud query processing. In: Proceedings of the ICDE (2012)

Skyline-Based Temporal Graph Exploration

Evangelia Tsoukanara[1](✉), Georgia Koloniari[1], and Evaggelia Pitoura[2]

[1] University of Macedonia, Thessaloniki, Greece
{etsoukanara,gkoloniari}@uom.edu.gr

[2] University of Ioannina, Ioannina, Greece
pitoura@cse.uoi.gr

Abstract. An important problem in studying temporal graphs is detecting interesting events in their evolution, defined as time intervals of significant stability, growth, or shrinkage. We consider graphs whose nodes have attributes, for example in a network between individuals, the attributes may correspond to demographics, such as gender. We build aggregated graphs where nodes are grouped based on the values of their attributes, and seek for events at the aggregated level, for example, time intervals of significant growth between individuals of the same gender. We propose a novel approach based on temporal graph skylines. A temporal graph skyline considers both the significance of the event (measured by the number of graph elements that remain stable, are created, or deleted) and the length of the interval when the event appears. We also present experimental results of the efficiency and effectiveness of our approach.

1 Introduction

Graphs are often used to model relationships and interactions between real world entities. Most graphs evolve over time with nodes and edges being added, or deleted. The values of attributes associated with the nodes may also change. An interesting problem in this context is detecting important events in the evolution of the graph, such as periods of significant stability, growth or shrinkage. It is interesting to detect such events not only at the individual node and edge level, but also at a higher level. To detect events at a higher level, we build aggregated graphs where each node corresponds to sets of nodes that have the same attribute values in the original graph [16].

For example, consider a face-to-face proximity graph of students and teachers used for studying the spread of an infectious disease [5]. Nodes correspond to individuals with attributes such as class, grade, and gender, while edges to physical interactions between individuals. Detecting events in the evolution of this graph may reveal patterns in the disease spread that will facilitate the application of targeted mitigation strategies. Such patterns may emerge not only at the individual node level, but also at an aggregated one, for example, between students of the same class, or gender. For instance, a pattern may be a correlation

A. Abelló et al. (Eds.): ADBIS 2023, LNCS 13985, pp. 88–102, 2023.
https://doi.org/10.1007/978-3-031-42914-9_7

between the growth of interactions between 1st graders and an increase in the disease spread, or a correlation between the shrinkage of interactions between boys and a drop in the disease spread.

We adopt an interval-based model, where a temporal graph is modeled as a graph whose nodes, edges and attributes are annotated with the time intervals during which the corresponding elements were valid. We model the evolution of a temporal graph (including aggregated ones) between two time intervals T_1 and T_2, where T_1 precedes T_2 through three *event* graphs: (a) the *stability graph*, that includes nodes and edges that persist in both T_1 and T_2, (b) the *growth* graph that includes new nodes and edges that did not exist in T_1 but appear in T_2 and (c) the *shrinkage* graph that includes deleted nodes and edges that existed in T_1 but no longer appear in T_2. Our goal is to detect intervals in each of these graphs when significant events occur, where significance is measured by the number of graph elements that remain stable, are created, or deleted respectively.

Previous work on evolution exploration is driven either by user queries [14], or requires a threshold on the number of events [16]. Both approaches require knowledge of the underlying data to achieve appropriate parameter configuration or create a meaningful query. In this paper, we introduce a parameter-free novel approach based on temporal graph skylines. A temporal graph skyline is defined on the significance of the event and the length of the interval when the event appears. We propose a temporal skyline-based graph exploration strategy that detects all non-dominated intervals with the best length or with the largest number of events.

We use two different semantics to define the temporal graph in an interval, namely: *strict* and *loose* semantics. Strict semantics capture persistent occurrences of nodes and edges, while loose semantics capture transient changes. We are interested in the shortest intervals in the case of loose semantics so to capture abrupt changes in the graphs, and in the longest intervals in the case of strict semantics so as to capture persistence. For example, in the face-to-face proximity graph, the fact that there is a skyline point with an interval of length 10 and with significance of 30 edges between boys, for the case of stability and strict semantics, it means that: (a) there is no longer interval where 30 or more such edges remained stable, and (b) there is no interval with length 10 or more where more edges remained stable. We also extend our skyline-based strategy to detect the top-k skylines where results are ranked based on the number of tuples each result dominates. Finally, we present an experimental evaluation of our approach using three real datasets, aiming at studying both the efficiency and the effectiveness of our strategies.

The rest of the paper is structured as follows. In Sect. 2, we introduce necessary concepts and define our problem. In Sect. 3, we present the skyline computation, while in Sect. 4 we report experimental results. Section 5 summarizes related work and Sect. 6 offers conclusions.

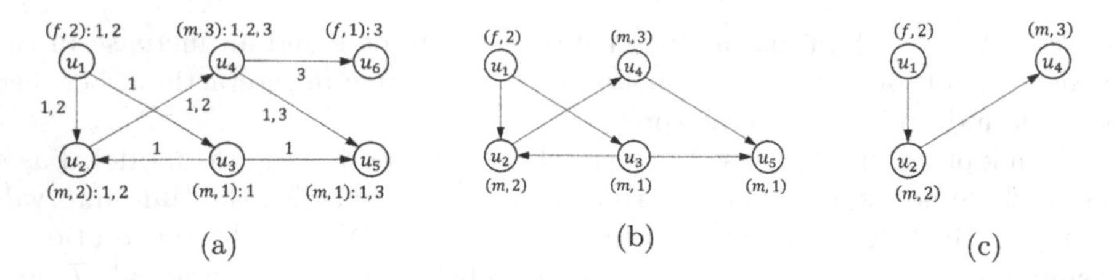

Fig. 1. (a) A temporal attributed graph $G[T]$, (b) $G[T_1 \sqcup T_2]$ with loose and (c) $G[T_1 \sqcup T_2]$ with strict semantics.

2 Temporal Graph Skylines

We adopt an interval-based model of temporal graphs, where nodes and edges are associated with intervals of validity, and each node in the graph is associated with a set of attributes that may change over time.

Specifically, let a *temporal element* be a time interval denoted as T. We define a temporal attributed graph $G(V, E, \tau u, \tau e)$ in a temporal element T, as a graph $G[T]$ where each node u and edge e is associated with timestamps $\tau u(u) \in T$ and $\tau e(e) \in T$, which are sets of temporal elements during which u and e exist in G. For each node u and time instance $t \in \tau u(u)$, there is a n-dimensional tuple, $A(u,t) = \{A_1(u,t), A_2(u,t) \ldots A_n(u,t)\}$, where $A_i(u,t)$ denotes the value of u at time $t \in \tau u(u)$ on the i-th attribute.

We collectively refer to nodes and edges of a graph as *graph elements*. An example of a temporal graph defined in $T = [1, 3]$ is shown in Fig. 1a. Nodes and edges are annotated with the time instances that they are valid. There are two attributes, *gender* and *class*, with values $\{f, m\}$ and $\{1, 2, 3\}$ respectively.

We define three different set-based operators between two temporal graphs $G[T_1]$ and $G[T_2]$. Specifically, the *union* ($\cup$) operator $G[T_1] \cup G[T_2]$ generates a new temporal attributed graph with graph elements that appear either in $G[T_1]$ or $G[T_2]$, i.e., graph elements with timestamps that include T_1 or T_2. The *intersection* ($\cap$) operator $G[T_1] \cap G[T_2]$ produces a temporal attributed graph whose elements appear in both $G[T_1]$ and $G[T_2]$. Finally, the *difference* ($-$) $G[T_1] - G[T_2]$ outputs a temporal attributed graph that includes graph elements that appear in $G[T_1]$ but not in $G[T_2]$.

Given a sequence of graphs defined at different temporal elements, we would like to combine them into a single temporal graph. To this end, we introduce *strict* and *loose* semantics. Strict semantics retain continuous occurrences of graph elements, while loose semantics retain also transient ones. Intuitively, persistence is associated with the notion of intersection which focuses on finding continuous occurrences, and therefore, consistency over a period. On the other hand, union adopts a more loose strategy and focuses on all graph elements ever appearing during a period of time.

Concretely, for two temporal elements, T_1 and T_2, let us denote the combined graph as $G[T_1 \sqcup T_2]$. Graph $G[T_1 \sqcup T_2]$ is defined with: (a) *strict semantics* as $G[T_1] \cap G[T_2]$, i.e., we want a graph element to appear in all time instances in

$T_1 \cup T_2$ and (b) *loose semantics*, as $G[T_1] \cup G[T_2]$, i.e., we want a graph element to appear in at least one time instance in $T_1 \cup T_2$. An example of $G[T_1 \sqcup T_2]$ where T_1 is time instance 1 and T_2 is time instance 2 defined with loose semantics is shown in Fig. 1b and with strict semantics in Fig. 1c.

2.1 Graph Evolution and Aggregation

In this paper, we want to capture the evolution of a graph between two consequent temporal elements T_1 and T_2 so as to detect whether graph elements remained stable in both intervals, new graph elements appeared in the most recent one, or existing ones disappeared in the most recent one. Thus, we want to study respectively *stability*, *growth* and *shrinkage*, collectively called *events*.

To model graph evolution with respect to the three types of events, we exploit the set temporal operators. For a pair of time intervals T_1, T_2 where T_1 precedes T_2: (a) the stability graph (s-graph), $G_s[(T_1, T_2)]$, is defined as $G[T_1] \cap G[T_2]$ and captures stable graph elements that appear in both T_1 and T_2, (b) the shrinkage graph (h-graph), $G_h[(T_1, T_2)]$, is defined as $G[T_1] - G[T_2]$ and captures elements that disappear from T_1 to T_2 and (c) the growth graph (g-graph), $G_g[(T_1, T_2)]$, is defined as $G[T_2] - G[T_1]$ and captures graph elements that are new to T_2 and did not appear in T_1. We use γ to denote any of the three events.

For example, $G_s[([1 \sqcup 2], 3)]$, for the graph of Fig. 1a, denotes the part of the graph that did not change between the graph in time interval $[1 \cup 2]$ and time instance 3. In case of strict semantics, the graph elements are the nodes and edges that existed in both time instances 1 and 2 and also remain in time instance 3. In case of loose semantics, the graph elements are the nodes and edges that existed in at least one of the time instances 1 and 2 and also remain in time instance 3.

Besides studying the evolution of the graph at the individual node and edge level, we want to study the evolution at a higher granularity level. For example, instead of studying stable interactions between individuals, we may want to study stable interactions between individuals of the same gender, or age. To this end, we use graph aggregation where nodes are grouped based on one or more of their attribute values, while respecting the network structure.

Given a graph $G[T]$ and a subset C of its n attributes, the graph aggregated by C in T is a weighted graph $G[T, C]$, where there is a node u' in $G[T, C]$ for each value combination of the C attributes and the weight of u', $w(u')$, is equal to the number of distinct nodes u in $G[T]$ that have the specific attribute value combination. There is an edge e' between two nodes u' and v' in $G[T, C]$ if there is an edge between the corresponding nodes in $G[T]$ and its weight $w(e')$ is equal to the number of such edges. Figure 2a shows the aggregate graph $G[T_3, \{gender\}]$, where T_3 is time instance 3.

Stability, shrinkage and growth are defined similarly for the aggregated graphs. For example, Fig. 2(b-d) shows the three event graphs for the evolution of the graph from $G[1 \sqcup 2]$ defined with loose semantics to $G[3]$ in the case of aggregation by gender, that is, of $G[1 \sqcup 2, \{gender\}]$ to $G[3, \{gender\}]$.

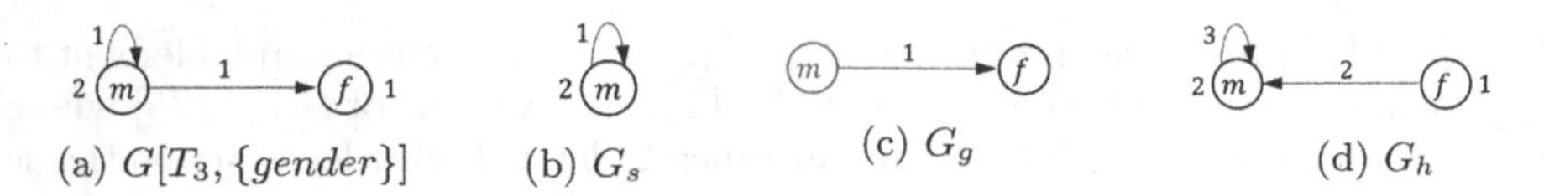

Fig. 2. (a) Aggregation on gender on T_3 for the graph of Fig. 1a, and (b-d) aggregated evolution graphs for the graph of Fig. 1b to that of Fig. 1a on T_3.

To evaluate the significance of an event, we measure the affected graph elements. We focus on events considering edges. Let c and c' be two attribute value combinations for the attributes C selected for the aggregation. For an event γ, we use $count(G_\gamma[(T_1, T_2), C], c, c')$ to denote the count of edges from nodes labeled with c to nodes labeled with c' in the γ-graph $G_\gamma[(T_1, T_2)]$ which is equal to the weight of such edges in the aggregated graph $G_\gamma[(T_1, T_2), C]$. For example in Fig. 2d, $count(G_h[(1 \sqcup 2), 3), \{gender\}], f, m) = 2$ counts the number of edges between female and male nodes that are deleted between $G[1 \sqcup 2]$ and $G[3]$.

2.2 Graph Exploration

The exploration of a temporal attributed graph is associated with the detection of intervals related with significant events in the evolution of the graph. Specifically, we would like to identify points in the history of the graph where the count of events is high, indicating interesting points in the evolution of the graph. That is, we want to locate time points, termed *reference point*, t_r, where the stability, shrinkage or growth counts, $count(G_\gamma[(T_r, t_r), C], c, c')$, are large with respect to an immediate preceding interval T_r. There are both persistent events and abrupt changes. When focusing on steep changes we are interested in finding the shortest period of an event to happen. On the other hand, when studying persistence, we aim to find the longest period for the examined event. Thus, we define minimal and maximal properties.

Given $G_\gamma[(T_r, t_r), C]$ and attribute value combinations c and c', we say that T_r is: (i) minimal if $\nexists\ T'_r$ such that $T'_r \subset T_r$ and $count(G_\gamma[(T'_r, t_r), C], c, c') \geq count(G_\gamma[(T_r, t_r), C], c, c')$, and (ii) maximal if $\nexists\ T'_r$ such that $T_r \subset T'_r$ and $count(G_\gamma[(T'_r, t_r), C], c, c') \geq count(G_\gamma[(T_r, t_r), C], c, c')$.

We say that a count for an event γ with respect to strict (loose) semantics, is *increasing* if: for any $T'_r \supseteq T_r$ defined using strict (resp. loose) semantics, $count(G_\gamma[(T'_r, t_r), C], c, c') \geq count(G_\gamma[(T_r, t_r), C], c, c')$, for any set of attributes C, any combination of attributes values c and c', and any reference time point t_r. Similarly, we say that a count for an event γ with respect to strict (loose) semantics, is *decreasing* if: for any $T'_r \subseteq T_r$ defined using strict (resp. loose) semantics, $count(G_\gamma[(T'_r, t_r), C], c, c') \leq count(G_\gamma[(T_r, t_r), C], c, c')$, for any set of attributes C, any combination of attributes values c and c', and any reference time point t_r. It holds that count for:

- stability is decreasing wrt strict and increasing wrt loose semantics,
- shrinkage is decreasing wrt strict and increasing wrt loose semantics, and

– growth is increasing wrt strict and decreasing wrt loose semantics.

Consequently for increasing counts, we are interested in minimal intervals, while for decreasing counts in maximal intervals.

2.3 Evolution Skyline

In the following, we present skyline-based exploration where based on a set of criteria we filter out the exploration results that appear to be of low interestingness. A skyline approach retrieves all items that have the "best" preferred value, in at least one of their features and are not worse in the others. In particular, for multidimensional data, we say that an item dominates another item if it is as good in all dimensions, and better in at least one dimension. A skyline query returns all non-dominated items in a dataset.

We define two criteria to evaluate the interestingness of an event γ in an aggregated graph $G_\gamma[(T_r, t_r), C]$ with respect to attribute value combinations c and c' of C: the length l_{T_r} of temporal element T_r defined as the number of time instances $t \in T_r$, and the event count $count(G_\gamma[(T_r, t_r), C], c, c')$.

Definition 1. *Evolution Skyline. Given an event γ, a graph $G[T, C]$ and sets of attribute values, c and c' of C, the evolution skyline is defined as all non-dominated tuples (t_r, T_r, w) where $t_r \in T$, $T_r \subset T$ and $w = count(G_\gamma[(T_r, t_r), C],$ $c, c')$.*

For an event γ with semantics that have increasing counts, we say that a tuple (t_r, T_r, w) dominates a tuple (t'_r, T'_r, w') if $l_{T_r} < l_{T'_r}$ and $w \geq w'$, or $w > w'$ and $l_{T_r} \leq l_{T'_r}$. Respectively, for an event γ with semantics that have decreasing counts, we say that a tuple (t_r, T_r, w) dominates a tuple (t'_r, T'_r, w') if $l_{T_r} > l_{T'_r}$ and $w \geq w'$, or $w > w'$ and $l_{T_r} \geq l_{T'_r}$. Taking advantage of the increasing and decreasing properties, we restrict the number of such tuples as follows.

Lemma 1. *Any non-dominated tuple includes an interval T which is either maximal or minimal depending on the use of strict or loose semantics.*

Proof. Without loss of generality, let us consider strict semantics. If the interval T_r of a non-dominated tuple (t_r, T_r, w) is not maximal, then there would exist a pair (t_r, T'_r, w') such that $T_r \subset T'_r$ and for lengths of T_r, T'_r, would be $l_{T'_r} > l_{T_r}$ in which $w' \geq w$ events occur. Thus, (t_r, T'_r, w') dominates (t_r, T_r, w).

As skylines may contain many results, especially when considering graphs evolving over long periods of time, we extend our definition of the evolution skyline to the *top-k evolution skyline*. To this end, we define the *domination degree* of a result tuple (t_r, T_r, w), *dod*(t_r, T_r, w), as the number of other tuples it dominates. To determine the top-k skyline, we rank all (t_r, T_r, w) tuples in the evolution skyline according to their domination degree and retrieve only the top-k ranking ones.

3 Skyline-Based Exploration

We now describe the evolution skyline computation process, that given an event γ, a temporal attributed graph $G[T]$, aggregate attributes C and attribute value sets c and c', computes all non-dominated tuples (t_r, T_r, w), $t_r \in T, T_r \subset T$ and $w = count(G_\gamma[(T_r, t_r), C], c, c')$.

A naive approach to compute the skyline would entail two phases. First, computing all possible (t_r, T_r, w), i.e., for each $t_r \in T$ enumerate all $T_r \subset T$ and computing the corresponding w, and then comparing all these tuples so as to eliminate all the dominated ones. However, this exhaustive approach can be easily improved by pruning all dominated results as they are derived. We propose incrementally evaluating the skyline for each reference point t_r, while at the same time updating the final skyline as new results are derived.

Thus, given a reference point t_r, we enumerate all possible T_r to compute our candidate tuples. The order in which we enumerate the temporal elements depends on the semantics used. In particular, when the event γ has increasing counts and the skyline consists of minimal temporal elements, then we start with the shortest temporal element and gradually extend it using either strict or loose semantics. Similarly, when γ has decreasing counts and the skyline consists of maximal temporal elements, then we start with the longest temporal element and gradually extend it using either strict or loose semantics.

The process repeats iteratively for each t_r, T_r pair. First, the event graph $G_\gamma[(T_r, t_r)]$ is computed and then aggregated on C producing $G_\gamma[(T_r, t_r), C]$ and $count(G_\gamma[(T_r, t_r), C], c, c')$ is evaluated. To compute the event graph and its aggregation, we leverage the methods introduced in the GraphTempo framework [16] that provides efficient algorithms for the implementation of all temporal operators and attribute aggregation differentiating between static and time-varying attributes to handle both more efficiently.

Every derived tuple (t_r, T_r, w) is considered as a candidate for the skyline and compared against previous candidate tuples. If it is dominated, it is ignored and we proceed with the next shorter or longer T'_r. Otherwise, (t_r, T_r, w) is added to the skyline and any previously added tuple (t'_r, T'_r, w') that it dominates is pruned from the skyline.

The process for skyline evaluation can be extended to support *top-k* skyline evaluation if we evaluate and maintain the domination degree of each (t_r, T_r, w) that is added in our skylines. Initially, $dod(t_r, T_r, w)$ is set to 0. When (t_r, T_r, w) is added to the skyline, if it substitutes other tuples already in the skyline, its $dod(t_r, T_r, w)$ is set to the sum of the *dod*s of all tuples it substitutes.

Algorithm 1 presents in detail the implementation of the evolution skyline for the event of stability ($\cap$) where we are interested in maximal temporal elements. We next prove that Algorithm 1 correctly computes the skyline, while similar proofs can be derived for other events and semantics.

Lemma 2. *If a tuple belongs to the skyline, Algorithm 1 will return it, and vice versa, any result returned by Algorithm 1 belongs to the skyline.*

Algorithm 1: SKYLINE-BASED STABILITY ($\cap$) EXPLORATION

```
   Input: A temporal attributed graph G[T], attribute C, attribute values c, c'
   Output: Skyline set S_sky, dominance set S_dom
1  Initialize hash tables S_sky : l_{T_r} → [(t_r, T_r, w)], S_dom : (t_r, T_r, w) → |d|, d : set of
   dominated items
2  for each reference point t_r do
3      T_r = longest possible interval for t_r
4      l_{T_r}, l_{T_rmax} = length(T_r)
5      while l_{T_r} ≥ 1 do
6          Compute Intersection G_s[(T_r, t_r)] = G[t_r] ∩ G[T_r]
7          Compute Aggregation G_s[(T_r, t_r), C], c, c')
8          w_curr = count(G_s[(T_r, t_r), C], c, c')
9          S_dom[(t_r, T_r, w_curr)] ← 0
10         l_prev ← l_{T_r}
11         while ∄ S_sky[l_prev] and l_prev ≤ l_{T_rmax} do
12             l_prev+ = 1
13         if l_prev > l_{T_rmax} then
14             w_prev ← 0
15         else
16             w_prev ← S_sky[l_prev][count]
17         if w_curr > w_prev then
18             if l_{T_r} = l_prev then
19                 S_dom[(t_r, T_r, w_curr)]+ = 1
20                 S_dom[(t_r, T_r, w_curr)]+ = S_dom[S_sky[l_prev]]
21             S_sky[l_{T_r}] = [(t_r, T_r, w_curr)]
22         else if w_curr = w_prev then
23             if l_{T_r} = l_prev then
24                 S_sky[l_prev] ← append[(t_r, T_r, w_curr)]
25         else
26             S_dom[S_sky[l_prev]]+ = 1
27         if l_{T_r} > 1 then
28             l'_prev ← l_{T_r} − 1
29             while ∄ S_sky[l'_prev] and l'_prev ≥ 1 do
30                 l'_prev− = 1
31             if S_sky[l'_prev][count] ≤ w_curr then
32                 S_dom[(t_r, T_r, w_curr)]+ = 1
33                 S_dom[(t_r, T_r, w_curr)]+ = S_dom[S_sky[l'_prev]]
34                 S_sky[l'_prev] ← []
35         Reduce T_r and l_{T_r} by 1
36 return S_sky, S_dom
```

Table 1. *DBLP* Graph

#TP	2000	2001	2002	2003	2004	2005	2006	2007	2008	2009	2010	2011	2012	2013	2014	2015	2016	2017	2018	2019	2020
#Nodes	1708	2165	1761	2827	3278	4466	4730	5193	5501	5363	6236	6535	6769	7457	7035	8581	8966	9660	11037	12377	12996
#Edges	2336	2949	2458	4130	4821	7145	7296	7620	8528	8740	10163	10090	11871	12989	12072	15844	16873	18470	21197	27455	28546

Table 2. *MovieLens* Graph

#TP	May	Jun	Jul	Aug	Sep	Oct
#Nodes	486	508	778	1309	575	498
#Edges	100202	85334	201800	610050	77216	48516

Table 3. *Primary School* Graph

#TP	1	2	3	4	5	6	7	8	9	10	11	12	13	14	15	16	17
#Nodes	228	231	233	220	118	217	215	232	238	235	235	236	147	119	211	175	187
#Edges	857	2124	1765	1890	1253	1560	1051	1971	1170	1230	2039	1556	1654	1336	1457	1065	1767

Proof. Algorithm 1 considers skylines with maximal temporal elements. Assume for the purposes of contradiction, that there is a tuple $s = (t_r, T_r, w)$ that belongs to the skyline but is not returned by the algorithm. Since, the algorithm considers for all reference points, all intervals that means that s was considered and it was either (a) not inserted in the skyline in lines 25–26 or (b) was pruned in lines 17–21 or in lines 31–34. If it was not inserted in lines 25–26, it means that is was dominated by a previous tuple with a largest count and same or larger length, thus s does not belong to the skyline, which is a contradiction. If it was pruned in lines 17–21, it means that the new tuple has a larger weight and the same length, thus s does not belong to the skyline. If it was pruned in lines 31–34, it means that the new tuple has at least the same weight and a longer interval, thus s does not belong to the skyline. Similarly, suppose a non-dominated tuple $s' = (t'_r, T'_r, w') \in S_{sky}$. Then, it holds that $\exists\ s = (t_r, T_r, w)$ that dominates s'. If s is examined after s', if $w > w'$ the check of lines 17–21 would prune s', or if $l_{T_r} \geq l_{T'_r}$ the check of lines 31–34 would prune s'. Otherwise, if s' is examined after s, lines 25–26 would prune s'. Thus, $s' \notin S_{sky}$.

4 Experimental Evaluation

We use three real-world datasets: (a) *DBLP*, a collaboration network where each node corresponds to an author and each edge denotes co-authorship, (b) *MovieLens* (ML) where each node represents a user and an edge between two users denotes that they have rated the same movie, and (c) *Primary School* (PS) which is a contact network that describes the contacts between students and teachers. *DBLP* and *MovieLens* are described in detail in [16], and *Primary School* in [17]. Tables 1, 2, 3 show nodes and edges per time point for each graph.

Our current implementation builds on a dataframe-based representation of the temporal graph that provides an efficient implementation of temporal operators and attribute aggregation [16]. As future work, we plan to implement skylines in a property graph database such as Neo4j to provide an integrated solution. There are a few works that implement temporal graphs in a property graph database [3,4,15]. However, the supported modeling of time-varying attributes

Table 4. Skyline runtime (s) and size

		DBLP					ML		PS	
		2005	2010	2015	2020		Oct		17	
Event	$c - c'$	Time				Size	Time	Size	Time	Size
$G_s(\cap)$	F-F	0.33	1.32	3.77	10.18	5	1.00	3	1.86	10
	F-M	0.35	1.51	3.98	8.81	8	0.83	2	2.00	11
	M-F	0.31	1.35	4.15	8.98	8	0.81	2	1.54	11
	M-M	0.31	1.35	3.65	9.35	12	0.91	2	1.85	13
$G_s(\cup)$	F-F	0.82	6.16	24.65	77.70	20	12.52	5	4.61	17
	F-M	0.87	6.07	24.80	78.25	20	12.27	5	3.96	16
	M-F	0.93	6.22	24.34	78.32	20	12.46	5	3.90	16
	M-M	0.80	5.81	24.62	78.94	20	12.44	5	4.49	16
$G_g(\cup)$	F-F	1.01	5.79	19.78	54.47	8	17.11	5	3.18	12
	F-M	0.97	6.2	20.29	53.58	14	16.91	5	3.35	15
	M-F	0.98	6.54	20.74	57.68	16	17.06	5	3.31	14
	M-M	0.93	5.42	19.83	51.97	19	17.10	5	3.15	13
$G_h(\cup)$	F-F	1.03	6.77	24.30	66.90	20	19.10	5	3.34	13
	F-M	1.12	7.35	24.38	66.89	20	19.92	5	3.80	16
	M-F	0.99	6.92	24.15	66.78	20	19.50	5	3.34	15
	M-M	1.01	6.71	24.44	66.85	20	18.72	5	3.37	15

and relationships through additional nodes or edges results in increasing the size of the graphs substantially.

Our methods are implemented in Python 3.7.9 and our experiments are conducted in a Windows 10 machine with Intel Core i5-2430, 2.40 GHz processor and 8 GB RAM. Our code and data are publicly available[1].

Performance Evaluation. In the first set of experiments, we evaluate the performance of our skyline algorithms by measuring execution times. Table 4 reports runtimes in seconds for each dataset for different value combinations of gender. For *DBLP*, gender is deduced using appropriate software[2]. We use $\cup$ to denote loose semantics, and $\cap$ strict. For *DBLP*, we measure runtime while we increase the size of the graph, considering intervals up to reference points $t_r = \{2005, 2010, 2015, 2020\}$. As expected, runtimes increase as the graph gets larger for any value combination and event. Also, events with loose semantics ($\cup$) exhibit higher runtime compared to strict ones ($\cap$), as the size of the graph gets larger when applying loose semantics which impacts the overall execution time. Concerning the different type of events, we notice a similar behavior across all datasets, as $G_s(\cap)$ is the fastest to compute and next follows $G_s(\cup)$, $G_g(\cup)$, and $G_h(\cup)$. Execution times depend on the size of the graph (with *DBLP* being the largest and slowest) and not on the specific value combinations.

[1] https://github.com/etsoukanara/skylinexplore.

[2] https://pypi.org/project/gender-guesser.

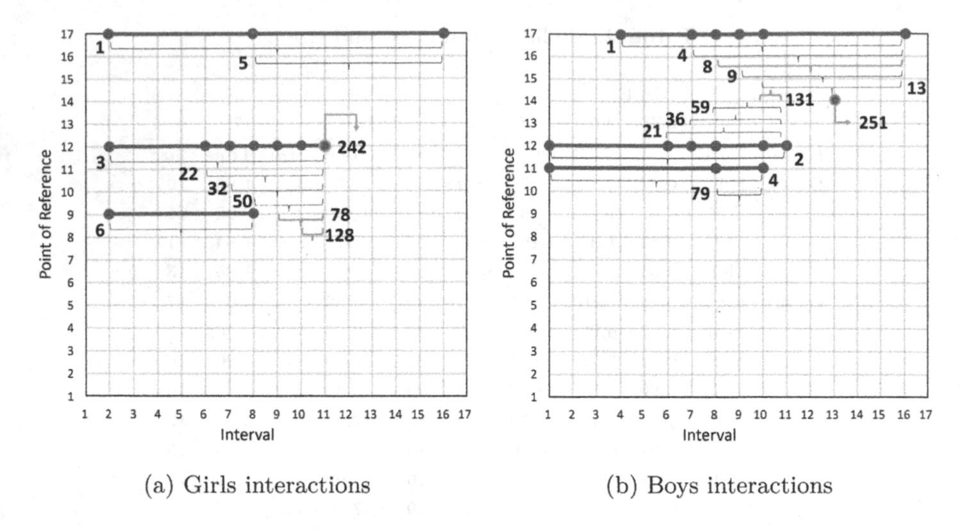

(a) Girls interactions

(b) Boys interactions

Fig. 3. Skyline for stability event for (a) F-F and (b) M-M interactions for *Primary School.*

Qualitative Evaluation. In this set of experiments, we study the produced skylines. Table 4 reports skyline sizes of each graph. As a general observation, the size of the skyline seems to depend on the number of the time points of each graph and not its size. We also observe that all value combinations seem to have similar skyline sizes for each event. However, in *DBLP*, we observe higher divergence between F-F and M-M interactions for both $G_s(\cap)$ and $G_g(\cup)$. This rather shows that the results for F-F collaborations are limited to few periods of stability that dominate all others possibly due to the overall small number of F-F interactions that also limits their growth. Regarding the semantics of the events, in most cases, loose semantics are associated with a larger skyline size with respect to strict, as loose semantics generate a higher number of candidates during the algorithm execution. This is an indication that in most cases, there are few stable interactions and that interactions exhibit high fluctuations.

Table 5 depicts the top-3 skylines for F-F interactions. We notice that events with loose semantics report longer periods and higher counts compared to strict ones. We observe that the most significant reference points for *DBLP* is 2020 for stability and shrinkage and we report the most important growth events on 2019. For *MovieLens*, *sep* seems to report a significant shrinkage while we notice a great number of new co-ratings in *aug*. Last, for *Primary School*, time instance 12 reports the highest stability overall.

We provide a more detailed illustration in Fig. 3, where we report results for F-F and M-M for *Primary School* for stability with strict semantics. The skyline for boys includes 13 results, while for girls 10. Also, there are longer stability intervals for boys than girls interactions. We observe that in the skylines for both girls and boys, intervals with length longer than 7 maintain at most 9 stable interactions, which can be seen as a bound on the duration and size of isolation bubbles. On the other hand we observe an interval of duration 4 (interval [7, 11] at point of reference 12) with stability at least 32 for both girls and boys indicating a potential lower risk zone for disease spread.

Table 5. Top-3 skyline

		DBLP		ML		PS	
Event	$c - c'$	Skyline	*dod*	Skyline	*dod*	Skyline	*dod*
$G_s(\cap)$	F-F	(['20], ['19], 62)	19	([*sep*], [*aug*], 86)	4	([12], [11], 242)	15
		(['20], ['18, '19], 18)	15	([*sep*], [*jun*, *aug*], 1)	1	([12], [10, 11], 128)	14
		(['20], ['17, '19], 7)	7			([12], [9, 11], 78)	13
$G_s(\cup)$	F-F	('20, ['15, '19], 2535)	44	([*sep*], [*aug*], 34001)	6	([3], [2], 476)	15
		('20, ['16, '19], 2091)	28	([*sep*], [*jul*, *aug*], 52599)	2	([4], [2, 3], 659)	14
		('20, ['18, '19], 1339)	21	([*sep*], [*jun*, *aug*], 60376)	1	([5], [2, 4], 857)	13
				([*sep*], [*may*, *aug*], 72027)	1		
$G_g(\cup)$	F-F	(['19], ['00, '18], 679)	14	([*aug*], [*may*, *jul*], 33952)	4	([13], [7, 12], 200)	16
		(['19], ['11, '18], 685)	2	([*aug*], [*jun*, *jul*], 33955)	3	([2], [1], 342)	15
		(['19], ['14, '18], 688)	2	([*aug*], [*jul*], 33968)	2	([13], [9, 12], 209)	14
$G_h(\cup)$	F-F	(['20], ['15, '19], 2447)	41	([*sep*], [*aug*], 33915)	6	([3], [2], 280)	15
		(['20], ['16, '19], 2003)	28	([*sep*], [*jul*, *aug*], 52500)	2	([5], [3, 4], 520)	15
		(['20], ['18, '19], 1260)	21	([*sep*], [*jun*, *aug*], 60269)	1	([5], [1, 4], 730)	13
				([*sep*], [*may*, *aug*], 71920)	1		

5 Related Work

Regarding graph evolution, most works define temporal operators and support time and attribute aggregation but provide no exploration strategies. In [4], GQL is extended to T-GQL by temporal operators to handle temporal paths and evolution is modeled through continuous paths. TGraph [12] uses temporal algebraic operators such as temporal selection for nodes and edges and traversal with temporal predicates on a temporal property graph. TGraph is extended with attribute and time aggregation that allows viewing a graph in different resolutions [1] but only stability is studied. In [13,14], GRADOOP extensions for temporal property graphs such as temporal operators for grouping and pattern matching are introduced. The system provides different graph visualizations, i.e., the temporal graph view, the grouped graph view, and the difference graph view that illustrates new, stable and deleted elements between two graph snapshots. Unlike our work, the system is driven by user queries and provides no exploration strategy. The GraphTempo framework [16] also defines temporal operators and supports both time and attribute aggregation. An evolution graph is built by overlaying an intersection and two difference graphs and interesting events in intervals are detected based on a threshold-based strategy. An interactive tool for exploring the graph, its aggregations and the evolution graph is also provided

[17]. While our approach relies on the GraphTempo model, the skyline approach we propose in this paper is novel. Unlike GraphTempo's exploration strategy that requires selecting appropriate thresholds, our proposed skyline-based approach detects all events effectively without the need for parameter configuration. Finally, works that do not utilize temporal operators mostly rely on versioning, introducing however redundancy in their models. In the EvOLAP Graph [7], versioning is used both on attributes and graph structure to enable analytics on changing graphs. In [6], explicit labeling of graph elements is designed to support analytical operations over evolving graphs, and particularly time-varying attributes, while in [2] the designed conceptual model captures changes on the topology, the set of attributes and the attributes' values.

Though skyline queries are very popular for multi-dimensional data, there is not much work on skyline queries over graphs. A domain where they were first defined is for road networks where the best detours based on a given route [8] or the best places to visit [9] are detected based on distances along with other possible criteria. In [11] skylines of routes based on multiple criteria, such as distance, and cost are defined. The network is modeled as an multi-attribute graph and a vector of different optimization criteria is stored for each edge. Beyond road networks, dynamic skylines over graphs based on shortest distances in the graph are defined in [19], where given a set of points by the user a skyline of the closest nodes to those points are retrieved. A different approach that does not rely on node distances is defined in [18], where a skyline consists of subgraphs that best match a given user query, also represented as a subgraph. Matching relies on isomorphisms and efficiency is addressed with the use of appropriate encoding schemes that capture both structural and numeric features of the graph nodes and support pruning. Finally, skylines on knowledge graphs are defined in [10]. The focus is on supporting skyline queries over entities in an RDF graph through SPARQL queries and the efficient evaluation of such queries, but while the data are modeled as a graph, skylines are defined on node attributes and do not take into account graph structure.

6 Conclusions

We introduced a skyline-based exploration strategy to detect interesting intervals when a significant number of events occur in the evolution of a temporal attributed graph. We considered aggregate graphs and explored stability, growth and shrinkage, while we defined time intervals using loose and strict semantics to capture both transient and persistent events. We experimentally evaluated the efficiency of our approach and presented results on three real datasets. We plan to explore more efficient pruning strategies and also extend the skyline definition to consider all value combinations of the aggregate attributes at the same time.

Acknowledgments. Research work supported by the Hellenic Foundation for Research and Innovation (H.F.R.I.) under the "1st Call for H.F.R.I. Research Projects to Support Faculty Members & Researchers and Procure High-Value Research Equipment" (Project Number: HFRI-FM17-1873, GraphTempo).

References

1. Aghasadeghi, A., Moffitt, V.Z., Schelter, S., Stoyanovich, J.: Zooming out on an evolving graph. In: Proceedings of the 23rd International Conference on Extending Database Technology, EDBT 2020. OpenProceedings.org (2020)
2. Andriamampianina, L., Ravat, F., Song, J., Vallès-Parlangeau, N.: Graph Data Temporal Evolutions: From Conceptual Modelling To Implementation. Data Knowl, Eng (2022)
3. Cattuto, C., Quaggiotto, M., Panisson, A., Averbuch, A.: Time-varying social networks in a graph database: a neo4j use case. In: First International Workshop on Graph Data Management Experiences and Systems, GRADES 2013, co-located with SIGMOD/PODS 2013. CWI/ACM (2013)
4. Debrouvier, A., Parodi, E., Perazzo, M., Soliani, V., Vaisman, A.A.: A model and query language for temporal graph databases. VLDB J. (2021)
5. Gemmetto, V., Barrat, A., Cattuto, C.: Mitigation of infectious disease at school: targeted class closure vs school closure. BMC Infectious Diseases (2014)
6. Ghrab, A., Skhiri, S., Jouili, S., Zimányi, E.: An analytics-aware conceptual model for evolving graphs. In: Bellatreche, L., Mohania, M.K. (eds.) DaWaK 2013. LNCS, vol. 8057, pp. 1–12. Springer, Heidelberg (2013). https://doi.org/10.1007/978-3-642-40131-2_1
7. Guminska, E., Zawadzka, T.: EvOLAP Graph – evolution and OLAP-aware graph data model. In: Kozielski, S., Mrozek, D., Kasprowski, P., Małysiak-Mrozek, B., Kostrzewa, D. (eds.) BDAS 2018. CCIS, vol. 928, pp. 75–89. Springer, Cham (2018). https://doi.org/10.1007/978-3-319-99987-6_6
8. Huang, X., Jensen, C.S.: In-route skyline querying for location-based services. In: Kwon, Y.-J., Bouju, A., Claramunt, C. (eds.) W2GIS 2004. LNCS, vol. 3428, pp. 120–135. Springer, Heidelberg (2005). https://doi.org/10.1007/11427865_10
9. Jang, S., Yoo, J.: Processing continuous skyline queries in road networks. In: International Symposium on Computer Science and its Applications (2008)
10. Keles, I., Hose, K.: Skyline queries over knowledge graphs. In: The Semantic Web - ISWC 2019. Springer International Publishing (2019)
11. Kriegel, H.P., Renz, M., Schubert, M.: Route skyline queries: A multi-preference path planning approach. In: 2010 IEEE 26th International Conference on Data Engineering (ICDE 2010) (2010)
12. Moffitt, V.Z., Stoyanovich, J.: Temporal graph algebra. In: Proceedings of The 16th International Symposium on Database Programming Languages, DBPL 2017. ACM (2017)
13. Rost, C., Gómez, K., Fritzsche, P., Thor, A., Rahm, E.: Exploration and analysis of temporal property graphs. In: Proceedings of the 24th International Conference on Extending Database Technology, EDBT 2021. OpenProceedings.org (2021)
14. Rost, C., et al.: Distributed temporal graph analytics with GRADOOP. VLDB J. (2022)
15. Semertzidis, K., Pitoura, E.: Time traveling in graphs using a graph database. In: Proceedings of the Workshops of the EDBT/ICDT 2016 Joint Conference, EDBT/ICDT Workshops 2016. CEUR-WS.org (2016)
16. Tsoukanara, E., Koloniari, G., Pitoura, E.: Graphtempo: An aggregation framework for evolving graphs. In: Proceedings 26th International Conference on Extending Database Technology, EDBT 2023. OpenProceedings.org (2023)
17. Tsoukanara, E., Koloniari, G., Pitoura, E.: TempoGRAPHer: A tool for aggregating and exploring evolving graphs. In: Proceedings 26th International Conference on Extending Database Technology, EDBT 2023. OpenProceedings.org (2023)

18. Zheng, W., Lian, X., Zou, L., Hong, L., Zhao, D.: Online subgraph skyline analysis over knowledge graphs. IEEE Transactions on Knowledge and Data Engineering (2016)
19. Zou, L., Chen, L., Özsu, M.T., Zhao, D.: Dynamic skyline queries in large graphs. In: Kitagawa, H., Ishikawa, Y., Li, Q., Watanabe, C. (eds.) Database Systems for Advanced Applications. Springer, Berlin Heidelberg (2010)

Temporal Graph Processing in Modern Memory Hierarchies

Alexander Baumstark(✉), Muhammad Attahir Jibril, and Kai-Uwe Sattler

TU Ilmenau, Ilmenau, Germany
{alexander.baumstark,muhammad-attahir.jibril,kus}@tu-ilmenau.de

Abstract. Updates in graph DBMS lead to structural changes in the graph over time with different intermediate states. These intermediate states in a DBMS and the time when the actions to the actual data take place can be processed using temporal DBMSs. Most DBMSs built their temporal features based on their non-temporal processing and storage without considering the memory hierarchy of the underlying system. This leads to slower temporal processing and poor storage utilization. In this paper, we propose a storage and processing strategy for (bi-) temporal graphs using temporal materialized views (TMV) while exploiting the memory hierarchy of a modern system. Further, we show a solution to the query containment problem for certain types of temporal graph queries. Finally, we evaluate the overhead and performance of the presented approach. The results show that using TMV reduces the runtime of temporal graph queries while using less memory.

Keywords: Bitemporal Graph Processing · Materialized View · Memory Hierarchy

1 Introduction

Graph DBMSs represent inter-connected data as a graph with nodes that are connected via relationships. Additional data can be directly assigned to the nodes and relationships as properties. In transactional graph DBMS, the underlying graphs are subject to changes resulting in a structural change of the graph over time with different intermediate states. Assuming a graph DBMS as the underlying DBMS for a social media platform reveals that the structure and properties are subject to changes, e.g., when a user follows another user, deletes a message, or likes a post of another user. Each of these operations creates a new state of the underlying graph. Analyzing these different states over time can be useful for analysis in order to make future business decisions, i.e., through gained patterns or trained models. Considering the time dimension is an already known feature in modern DBMSs which are referred to as temporal DBMSs [2,5,8,9,14]. While a usual DBMS does not preserve the old state of the data before updates, a temporal DBMS invalidates the old data and creates new valid data. The support of temporal features in SQL was introduced with the SQL:2011 standard and

A. Abelló et al. (Eds.): ADBIS 2023, LNCS 13985, pp. 103–116, 2023.
https://doi.org/10.1007/978-3-031-42914-9_8

enables the processing of evolving data using temporal query operators [9]. However, current graph (and relational) DBMS lack support for the efficient storage and processing of temporal data. Most of the current available DBMSs consider only a static data model, where changes (insert, updates, deletes) remove the previous state of the data. These DBMSs implement temporal features based on the non-temporal storage and processing models while ignoring the memory hierarchy of modern systems with different memory types like disk, flash, DRAM, or Persistent Memory (PMem). The higher the layer the less capacity is available but access is faster. Optimizing a DBMS by exploiting the memory hierarchy can lead to higher processing performance and efficient storage utilization.

In this work, we want to accelerate the performance of the execution of temporal queries by leveraging the memory hierarchy. To do this, we make use of materialized views (MVs), which is a well-known technique in databases for accelerating queries using already processed results. Further, we store the results of temporal queries in MVs, which are sub-graphs of the main graph, and place them in a higher layer of the memory hierarchy, for example in Persistent Memory (PMem). We built this work around our Poseidon Graph Database[1] which is designed to exploit the memory hierarchy of modern systems. It is optimized around the memory characteristics of PMem which are also applicable to the storage in DRAM. Further, it supports the storage of data on disk and GPU device memory. However, as this graph DBMS was designed for the storage of non-temporal data we extend the data model to support temporal processing.

Contribution. The contributions of this paper are as follows. 1. We present a data model for the processing of (bi-) temporal graph data that is built on top of the property graph model. 2. Considering the memory hierarchy of modern systems we introduce an approach for the storage and processing of temporal data in materialized views. 3. We demonstrate an approach to solve the query containment problem for queries on different time intervals. 4. We present results from an evaluation of the presented approach by using temporal queries for the SNB dataset provided by the LDBC.

2 Related Work

Processing temporal information in DBMS is a well-known research area for decades. It ranges from the representation of temporal information to the processing using temporal query languages. Temporal databases belong to a well-known field of research and are nowadays offered in many commercial systems. The SQL:2011 standard [9] introduces storage and operators in order to support temporal processing using SQL. We implement the basic operators introduced in this work for Poseidon. Based on this standard, a wide range of commercial and open-source relational databases support the processing of temporal

[1] https://dbgit.prakinf.tu-ilmenau.de/code/poseidon_core.

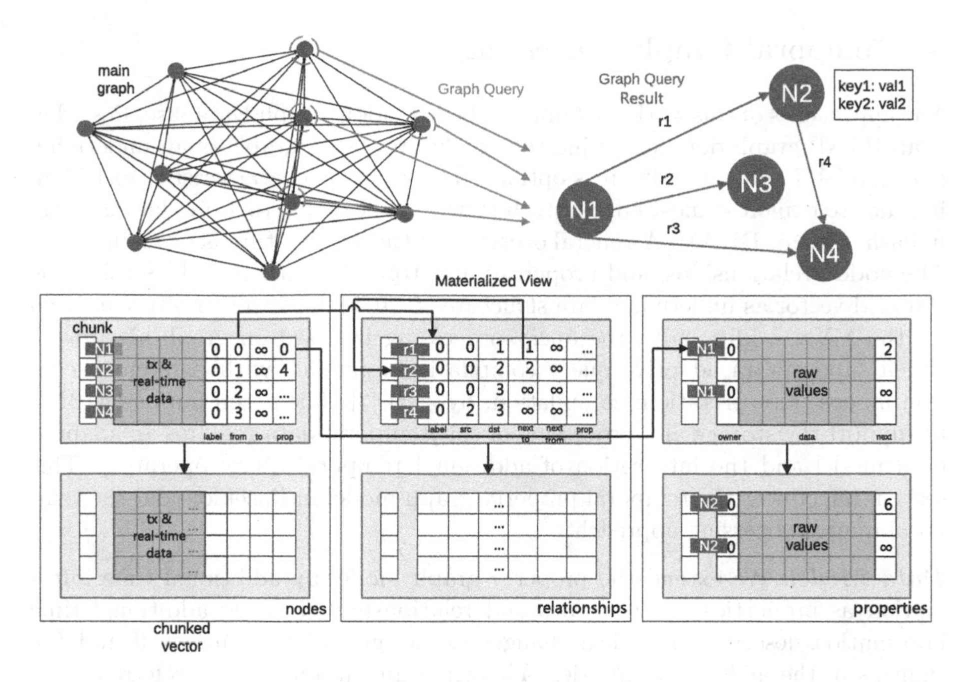

Fig. 1. Architectural overview of the graph database Poseidon.

data [5,8,14]. The SAP HANA DBMS provides wide support for temporal features, like temporal aggregation, time travel, and temporal joins [8]. The Oracle DBMS supports features of the SQL:2011 since version 12. It uses the so-called flashback technology with a different syntax for SQL queries [5]. Teradata is another DBMS that supports temporal processing according to the SQL:2011 standard [1]. For temporal graph query processing, there are already works that provide operators and algorithms. Gradoop is a system used to process and analyze temporal graph data [13]. It uses the temporal property graph model as the underlying data model and is able to process distributed graph workflows. A similar distributed system is Raphtory [12]. ArangoDB is an open-source graph database that provides additional support for processing time-travel queries [2]. The underlying data model makes use of techniques known from persistent data structures, also known as non-ephemeral data structures [4]. Regardless, while there are numerous systems that provide support for processing temporal data, there is no work on exploiting the modern memory hierarchy for the storage of temporal data. Temporal data in graphs tend to grow rapidly with the size of the graph and the number of changes to the structure. Therefore, it is necessary to optimize the storage by exploiting the memory hierarchy. T-Cypher is a temporal graph query language that extends the Cypher graph query language by temporal operators [10]. It is designed to be compliant with the Cypher syntax by providing built-in temporal support. For this, the Cypher graph query language is extended by temporal operators that process time information. We use a set of these queries to evaluate the presented approaches of this work.

3 Temporal Graph Processing

The approaches of this work are built for the Poseidon Graph Database. Poseidon is an HTAP graph database using the labeled property graph as an underlying data model. Fundamentally, it is optimized around the characteristics of PMem in order to exploit it most efficiently, but it is also capable to store data directly in flash, disk, or DRAM. A general overview of the architecture is given in Fig. 1. The nodes, relationships, and properties are stored in separate tables using the chunked vector as underlying data structure [7]. It is the same structure as used for the MVs. Additionally, the query processing is based on graph algebra [6] which supports an adaptive query compilation approach with different execution modes (interpretation, compilation, hybrid) [3]. The extension of a DBMS to support the storage and processing of (bi-)temporal data requires an adapted data model and the integration of additional temporal query operators. This section introduces the temporal property graph model in Poseidon and the temporal query processing approach.

Data Model. We extend the property graph model by additional time information as properties on the nodes and relationships. Storing additional time information describes historical changes in the graph structure itself and the changes in the assigned properties. This time dimension is often referred to as the valid or real-world time interval of a record, which indicates the start and end time point when the object was available. A user can define these periods, by adding them directly to the records. Further, we extend the model by the transaction time interval which is the interval when a transaction started processing the record and the time when the transaction committed the changes to the DBMS. Storing two different time dimensions allows processing the historical changes in a graph at a certain time point with respect to the state when the changes take place. We define the time dimensions used in our model as follows where events are usual changes (insert, updates, deletes) in a DBMS.

Definition 1 (Time Domains). *T_{tx} is the transaction-time domain. T_v is the valid time domain. An event in one of the domains $t_i \in T_{tx} \cup T_v$ is an interval (t_{begin}, t_{end}) where t_{begin} marks the beginning time of the event and t_{end} the end time of the event. Events in the time domains T_{tx} and T_v are linearly ordered so that $t_i < t_{i+1}$ where $t_i \in T_{tx} \cup T_v$.*

Based on these time domains we define the data model for a temporal property graph formally as follows.

Definition 2 (Temporal Property Graph). *A graph G consists of nodes N and directed relationships $R \in N \times N$, denoted by $G = (N, R)$. A node $n \in N$ is a tuple (id, t_{tx}, t_v) where id is a unique identifier to identify a node using $id : N \rightarrow ID$ and t_{tx} and t_v are time intervals denoting the validity of the node considering the time domains T_{tx} and T_v. A relationship $r \in R$ is a tuple (id, t_{tx}, t_v) where id is a unique identifier to identify a relationship using $id : R \rightarrow ID$ and t_{tx} and t_v are time intervals denoting the validity of the* *relationship wrt. the time domains T_{tx} and T_v. From the set of labels L, a label is assigned to each node and relationship using the label function $l : (N \cup R) \rightarrow L$. A property is a key-value pair $(k, v) \in P$. The properties P are $P = K \times D$, where K is the set of property names and D is the set of property values. Properties can be assigned using $p : (N \cup R) \rightarrow \mathcal{P}(P)$.*

Considering the structure of a graph comprised of nodes and directed relationships it should be noted that the graph is only valid in a certain time interval when all nodes and relationships are also valid. A relationship is only valid when both the source and destination node are valid in the appropriate time domain. In contrast to the model used in other systems, this model restricts multiple nodes with the same identifier in the same graph. Dropping this constraint relaxes the model definition and has the advantage of improving the query processing, as the storage of a graph contains only valid nodes and relationships. Moreover, it also enables storing non-temporal graphs and processing non-temporal queries without additional overhead. The core of temporal databases concerns the validity of tuples within a time period. In our model, we define the validity of a node or a relationship as follows:

Definition 3 (Tuple Validity). *A node $n \in N$ is valid in the time period $(t_{begin}, t_{end}) \in (T_{tx} \cup T_v)$ if $n_{begin} \geq t_{begin} \wedge n_{end} \leq t_{end}$.*

In other words, a node or relationship is valid in a given time period if the valid time (or transaction time) period lies in the given time period.

Query Processing. For query processing, we provide a number of operators for the creation of MVs and the processing of temporal data. Temporal query processing often involves the validation of data, i.e., checking if the validity of the records lies between the given real-world time t_r or transaction time t_{tx}. The TNodeScan operator iterates over the given node table and retrieves the start and time for each node from the property storage. Then, the valid time period of the nodes is checked if it is inside the given time period. For the transaction time, the same procedure is applied but instead, the interval is directly accessible from the node record. Valid nodes within the given interval are passed to the next operator in the pipeline. The TForeachRelationship operator is used to find in- and outgoing relationships of a node. Internally, it reads the identifier of the first relationship directly from the node's record, retrieved from the relationship table. Further relationships with the same direction are retrievable by accessing the next relationships identifier field of the record (see Fig. 1). TExpand retrieves the source or destination node of a relationship and checks the validity the same way. All temporal graph traversal operators work apart from that in the same way as their non-temporal equivalents NodeScan,ForeachRelationship, and Expand. For temporal processing we provide a set of operators from the interval algebra, to check the temporal information between multiple tuples (before, after, etc.). Lastly, we provide operators for the creation of (temporal) MVs. In Poseidon, it is possible to create a complete snapshot of the main graph using TSnapshot or of only a sub-graph, placing

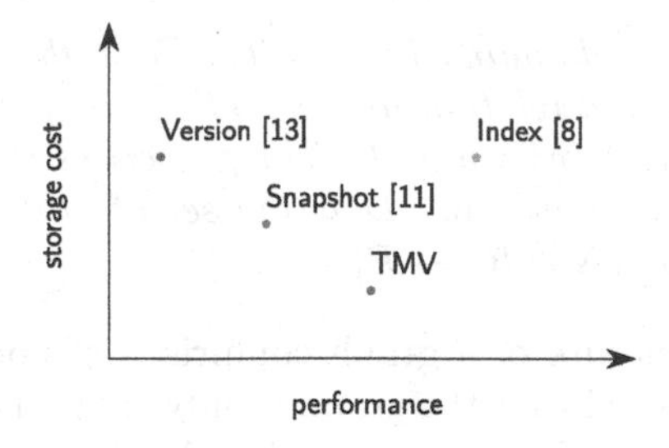

Fig. 2. Comparison of different approaches for storage and processing temporal data regarding storage costs and query expected performance

the TMVInsert operator in the query pipeline. The MVs can be scanned using the described scan operators.

4 Temporal Materialized Views

There are numerous approaches for temporal processing in DBMS. A simple and naive approach is to use the available storage, and query operators with UDFs. However, it arrives with heavy filtering for the temporal validity of records and inefficient storage utilization. Moreover, the temporal data has to be managed by the user, i.e., by creating new versions after updates. In Fig. 2 a comparison between different approaches regarding storage costs and performance is given. Another common way is to use a versioning approach, where every update creates a new version of the object which is subject to change. With this approach, there are several additional constraints that must be managed by the DBMS. First, the database must drop the uniqueness in the storage, since unique data may appear in different versions multiple times. Second, multiple copies of the records increase storage costs. Third, the processing of non-temporal queries degrades, since the additional effort when searching for objects. Especially in graph DBMS, the additional copies lead to poorer performance at graph traversing since related data may now appear in different storage locations. Using snapshots of the main graph for every relevant time period solves the uniqueness problem of the previous, but still, there are multiple copies of the whole graph which makes maintenance difficult. An indexing approach for versioned data is a fast way but comes again with the problems of the versioning approach when the index cannot be used. Using materialized views (MVs) for temporal data lies between the performance of the snapshot and indexing approach with fewer storage costs since MVs store a sub-graph of the main graph. Additionally, the performance of non-temporal queries is not affected. MVs typically store the results of a previously executed query. In relational databases, MVs are an additional queryable relation that contains the resulting tuples of a query. This concept can be transferred to a graph DBMS, by defining the MVs over a graph. Instead of creating a further relation, an MV for a graph is an additional graph that contains the relevant nodes and relationships to answer a query. We propose an approach for temporal graph processing using MVs, which we extend to temporal MVs (MVs). Furthermore, by placing the MVs in a higher layer of the memory hierarchy, the overall query processing can be improved. In this section, we show the data model, structure, placement strategy of MVs, and a solution for the containment problem for certain queries.

Data Model. Graph queries typically traverse the graph in order to find a path between a start and an end node according to a given pattern. Therefore, storing the traversed paths which are the tuple results of the graph queries is sufficient for an MV. For operators that combine two or more sub-queries, e.g., joins, paths from all sub-queries are stored in the MV. Therefore, the (sub-)graph stored in the MV may not have the same structure or may not be coherent. Calculated tuples which are the results of aggregations or projections are not inserted directly in the MV, as they can be recomputed using the graph stored in the MV. Another advantage of not storing computed results is that the MV can be reused for a wider range of further queries. We define an MV over a graph formally as follows.

Definition 4 (Materialized View (MV)). *$Q_G = o_1, \ldots, o_n, r$ is a query over a graph G, consisting of the operators o_1 to o_n which define the traversal of the graph to find a path from node 1 to node n and the remaining non-traversing operators r. A materialized view is a pair (Q_G, T_G), where Q_G denotes the defining query of the view over the graph G and T_G the contents of the materialized view over the graph G. The content of T_G are all tuples of the query $Q_G = o_1, ..o_n$, forming a graph G_{MV}. When the materialized view is up to date, then $T_G = Q(G)$, where $Q(G)$ are the results of the defining query over the graph G and every tuple in T_G is a path in G.*

Combining this definition with the temporal property graph model creates the MV, which we define as follows.

Definition 5 (Temporal Materialized View (TMV)). *A temporal materialized view (TMV) is a tuple $(MV(G), t_{tx}, t_v)$ where $MV(G)$ is a materialized view of the temporal property graph G and t_x and t_v are the transaction time interval of $MV(G)$ and the valid time respectively. The content of an TMV comprises all tuple results of a query Q_G which are valid for the given transaction and valid time intervals t_x and t_v.*

Creation. To insert only the valid nodes and relationships of a time period the real-time and transaction time properties must be checked. For this, the creation of the TMV is placed inside a query pipeline as an additional operator. The time intervals must be given as a parameter to the TMV creation operator TMVINSERT. The predicate can be dropped when the validity is checked at a previous operator, e.g., using a TNODESCAN with the appropriate predicate. However, TMVs can also be created by scanning the underlying storage directly. For this, the storage where the nodes are stored is scanned, checked for validity, and inserted into the TMV. After this, the relationships are processed using the same steps and linked with the source and destination relationships but outside the query processing pipeline which decreases the creation time because it is not required to analyze and split the tuples before storing.

Structure and Placement. As the underlying data structure, we use a chunked vector which is a data structure optimized for PMem and DRAM. Fundamentally, it is a linked list of fixed sizes arrays. For the TMV we use separate chunked vectors for nodes, relationships, and properties which enables more efficient sequential access when searching individual records. When inserting a tuple into a TMV, the tuple is split into its components of nodes and relationships and temporarily stored in separate lists. All nodes of the tuple are stored before all relationships to maintain the link between two available nodes when storing the relationships. Furthermore, it is possible to store all other (n-hop parameter) relationships of the nodes, which are not part of the actual query to increase the stored graph of the TMV, which can be helpful in other queries. However, increasing the n-hop parameter increases the processing time of the query as additional result processing is necessary. For every node and relationship, the identifier from the actual main storage is stored as an additional property. This is crucial, as the identifier of all records is unique, but the same records can be placed in multiple TMVs. Using an additional property to link to the original record helps to identify these records. The valid time interval is added to the property list on the first position to enable fast access when processing the validity. The remaining properties from the nodes and relationships are copied, stored in the storage of the MV, and linked with their equivalents in the TMV. For handling the transaction time, we store the start and end times of a transaction directly in the nodes and relationships records. Operators can access this data by searching the properties of the records. The proposed design for TMVs can be stored on any storage type. As the size of a TMV is much smaller than the size of the main graph it is suitable to place the TMV in a higher layer of the memory hierarchy where access is faster. While the main graph is stored on disk, several TMVs can be stored directly on DRAM, PMem, or any other fast storage medium. However, while access to DRAM is fast there are no persistence guarantees. Storing TMVs on DRAM requires either reprocessing the query to restore the TMV or loading it back to disk and recovering after re-instantiating the system. Therefore, we propose a TMV placement approach where we place every MV on PMem which enables near-DRAM access performance with persistence. The main graph is always stored on disk or flash memory while TMVs, which are sub-graphs of the main graph, are stored on PMem. If PMem is not available, the TMVs are stored on disk and loaded to DRAM when processing. Local data, which is data processed by a transaction, is loaded directly to DRAM and is written to persistent storage at commit. The higher layers of the memory hierarchy are more limited regarding size than disk or flash storage. In cases when there is no more space left for TMVs, the least frequently used TMVs are pushed to lower layers of the memory hierarchy (LRU strategy). The TMVs are loaded again to the higher layer when they are needed by a query.

Indexing and Selection. To manage the presented approach as dynamically as possible, we use an interval tree for indexing multiple TMVs. This has the further advantage of automatically selecting a suitable TMV in query execution. The interval tree is a suitable data structure for answering queries that apply exactly to a specific time interval. The basic idea behind the interval tree is similar to a binary search tree, but using the interval values (t_{low}, t_{high}). The order of the tree is determined by t_{low}. Furthermore, the basic assumption of overlapping intervals determines the operation of the data structure. An interval I overlaps with interval J if the following holds true.

$$I_{low} \leq J_{high} \wedge I_{high} \geq J_{low} \tag{1}$$

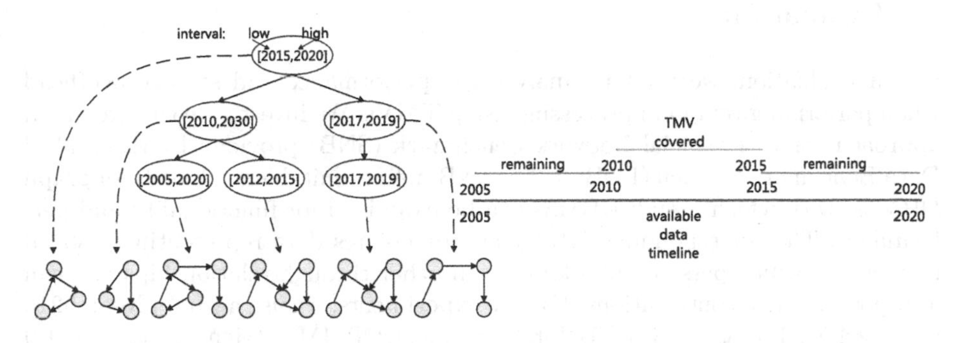

Fig. 3. Example interval tree where each node stores an interval and a pointer to the results of a query stored in an TMV.

Based on this assumption, the tree can be traversed and corresponding intervals with their associated views can be found. Figure 3 shows such an example interval tree. When searching for an applicable view that contains the values of the interval (2016, 2017), it is necessary to first traverse from the root to the right subtree, since the assumption does not apply in this case and the low key 2016 is greater than 2015. Since the following node in the right subtree is 2017 as low key and this is greater than 2016, the left subtree of the root must be searched anyway. In the same way, the key is searched in the sub-tree (2010, 2030) and found in this sub-tree, since it contains all overlapping intervals according to the condition (1).

Containment. An interval tree as an index for temporal data shows a promising data structure for answering the containment of other queries searching for the same pattern in the graph but at different time intervals. For this, we make use of condition (1) of overlapping time intervals. For temporal graphs, we define the query containment problem as follows.

Definition 6 (Temporal query containment problem). *Given two queries $Q_A(tx_A, v_A)$ and $Q_B(tx_B, v_B)$ where tx_q and v_q with $q \in \{A, B\}$ are the transaction and valid time intervals of the queries. Tuple results of the queries are* *denoted as* R_A *and* R_B *respectively. The query* Q_A *is contained in query* Q_B *if* $R_A \subseteq R_B$ *and* tx_A *or* v_A *are overlapping with* tx_B *or* v_B *respectively.*

Considering the example from Fig. 3 but storing only TMVs for the intervals 2010–2015. In order to obtain results for the time interval 2005–2030, the corresponding TMVs of the sub-trees of [2010–2015] need to be scanned. There are no validity checks for the tuples required in this TMV required which enhances the processing time. The query needs to be executed only on the main graph for the remaining time intervals (2005–2010) and (2015–2030) which are not covered by the interval tree. In the case of more available TMVs, fewer interval lengths need to be scanned from the main graph.

5 Evaluation

In our evaluation, we want to analyze the performance and storage overhead when performing temporal processing using TMVs. We investigate the presented approach using the Social Network Benchmark (SNB) provided by the Linked Data Benchmark Council (LDBC). The SNB defines suitable workloads for graph DBMSs, with which Graph DBMSs can be examined for functionality and performance. The corresponding data generator creates data representing a social network, in which person nodes know each other through relationships and can also post and like contributions. For the experiments, we use an Intel Xeon 5215 with 384 DDR4 RAM, 1.5 TB Intel Optane DCPMM (PMem), and 4 × 1.0 TB Intel SSD using Centos 7.9. For the study of query processing, we use the given example queries presented for the T-Cypher specification of the Cypher extension [10] and are also described in the Poseidon repository using our own query representation in graph algebra[2]. These queries are particularly suitable because they take the temporal aspects of the SNB dataset into account. Applicable temporal queries from the LDBC are not available at the time of this work. In the following, we focus on 3 of the presented T-Cypher example queries, which explore different aspects of our approach. Query 1 (Q1), examines the processing of intervals and is supposed to find the most recent likes of a person, which are up to a later point in time. This query has a high selectivity and stores a large graph when storing the results as a TMV with our approach. Executing the query without temporal operators requires setting a selection in order to check validity in the given time. Q2 examines the graph for a connection of two nodes that simultaneously have another connection via a common node. The resulting connections should have been created at the same time. The selectivity of this query is less than Q1 but requires more computation time, as more paths are processed. Q3 examines the SNB graph by the most recent likes to a post of a friend or friend-of-friend created before a certain date. It has the lowest selectivity but the highest computation time. The TMVs are placed directly in PMem.

[2] https://dbgit.prakinf.tu-ilmenau.de/code/poseidon_core/-/tree/mv_v2/queries.

Experiment 1: Storage and Overhead. Storing bitemporal information requires additional memory. The following evaluation is intended to show the overhead of the approach proposed in this paper. If we compare the storage of a usual non-temporal graph with the storage of a bitemporal graph, we need for each node and each relation additionally the valid time as well as the transaction time. The real-time for our approach, which is usually passed by the user, is stored in the properties of the corresponding records. The additional overhead associated with the storage of bitemporal information is only related to the storage of real-time and transaction time properties. Real-world time properties are stored in the property storage of each entity. These values only require a maximum 8-byte integer value. The corresponding key, which is compressed in Poseidon using a dictionary, requires another 8-byte integer value. Consequently, at least 32 bytes are required for the storage of real-time values (start/end time) per record. In our approach, the transaction time is stored directly in the records of the entities and requires 8-byte integer values for the start and end time. The total size of a node record is 80 bytes and 64 bytes for relationship records.

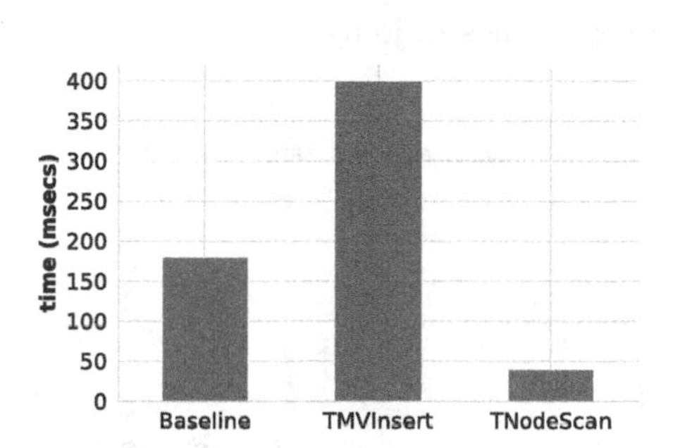

Fig. 4. Query execution baseline compared with the materialization time and scan of the created MV.

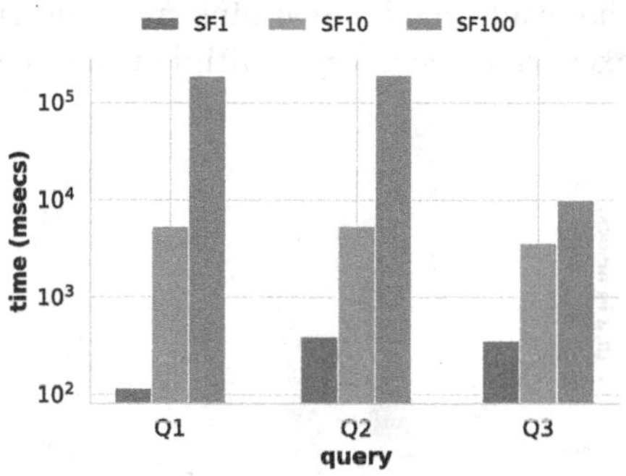

Fig. 5. Overhead of TMV creation for the queries Q1–Q3 on scaling factors 1–100.

Figure 5 shows the creation overhead for Q1–Q3 on the different scaling factors. The creation of TMV comes with additional overhead since the results must be copied and written to the appropriate storage. As the number of results increases, the overhead for creating TMVs increases as well. Still, when placing TMV in a higher layer of the memory hierarchy, writing to storage is more efficient than directly writing to disk, since access is faster (Table 1).

The storage sizes of the TMVs for the queries on different scale factors are shown in Sect. 1. Q1 touches the greatest amount of the dataset since it searches for 1-hop connections between nodes. Finding such a pattern in the graph is fast, as shown in the execution times. More complex queries like Q3 touch a small part of the graph. Therefore, the size of the corresponding TMV is small. Most of the computation is saved when using the TMV in future query execution.

Table 1. TMV sizes with nodes/relationships for scale factor 1–100 in thousand and total sizes for the scale factor 1–100.

query	SF 1	SF 10	SF 100
Q1	6/9	60/90	600/900
Q2	6/6	68/12	680/18
Q3	1/1	6/10	60/100

Scale Factor	Size (GB)
1	4
10	45
100	451

The same observation, with a different complexity, is shown with the size of the TMV in Q2.

Experiment 2: Temporal Query Processing. TMVs are only beneficial for query processing when scanning the view is faster than obtaining the results again. The first benchmark in Fig. 4 shows the execution of a query on a graph with 10.000 Person nodes and 10.000 Book nodes, connected via 1-hop relationships. The results show that creating the view is costly compared to direct execution. However, scanning from the view is four times faster than obtaining the results again. Scanning from the view can enhance the processing time when executing the query multiple times, e.g., sub-pipelines in joins.

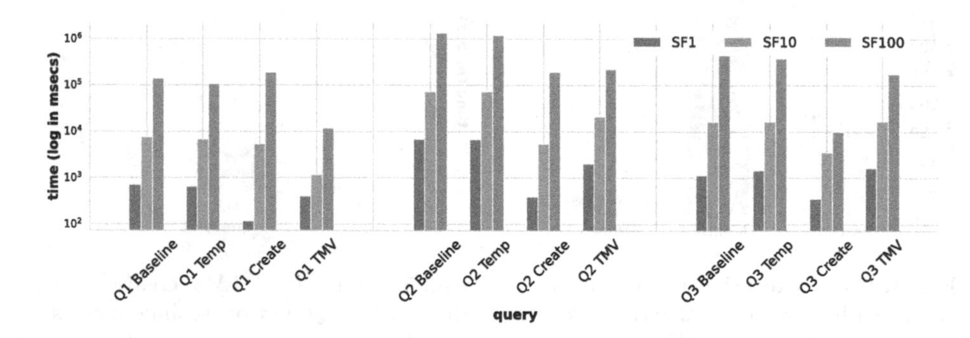

Fig. 6. Q1–Q3 on scale factors 10–100 with different approaches.

Experiment 3: Execution. Execution times for the queries on different scaling factors of the LDBC-SNB dataset are given in Fig. 6. We compare the baseline with non-temporal operators and UDFs (Baseline), with execution using temporal operators (Temp), the TMV creation (Create), and with the execution based on previously created TMVs. When the size of the original graph is similar to the size of the TMV, the runtimes of the baseline and the execution of the queries on TMVs are in a similar range, partially the TMV approach is marginally slower, which can be seen in Q3 for SF1 and SF10. This can be attributed to the different access speeds of DRAM and PMem. TMVs are placed in PMem and the original graph is buffered from disk to DRAM via the buffer manager. The significant difference can then be seen with a very large graph, as is the case with SF100.

There is a huge improvement in runtime when using TMVs, especially in larger graphs. The baseline execution is limited to the processing of filter expressions and UDFs. The execution of corresponding operators for temporal queries leads to an additional improvement of the runtime, which again increases with the size of the graph. Furthermore, temporal validity constraints can be processed at an earlier position in the query pipeline, leading to an improved runtime. The creation of TMVs (Create) takes only a small part of the actual query processing.

Experiment 4: Temporal Query Containment. In this experiment, we evaluate the query containment using the interval tree as the underlying index data structure. We create TMVs for the queries Q1–Q3 which contain the results up to a certain percentage. The content can be obtained directly for the results while the remaining results (i.e. remaining time intervals) need to be processed from the original storage. With more pre-processed results the remaining processing time decreases. However, when the TMVs cover only a small percentage of the desired time interval, the processing time increases as there is more overhead scanning the original graph for multiple time intervals. In general, the processing time decreases with the results covered in the TMV, i.e., when more TMVs with overlapping intervals are available (Fig. 7).

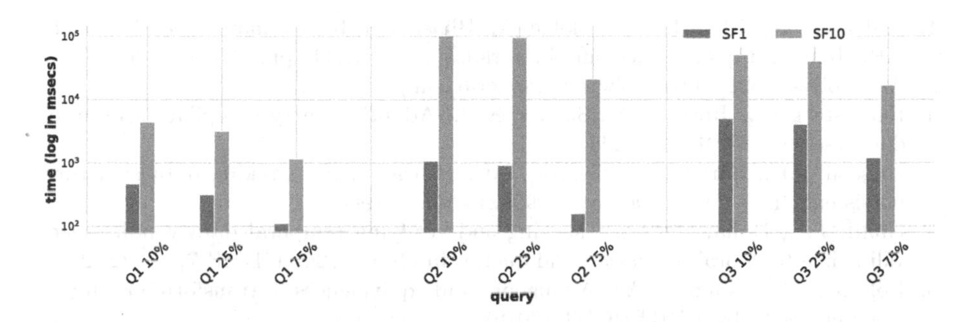

Fig. 7. Q1–Q3 on scale factors 1–10 using different TMVs to obtain remaining results.

6 Conclusion and Future Work

Storing temporal information in a graph DBMS increases the storage and processing overhead. Data must be stored multiple times for several time periods. By considering the memory hierarchy of modern systems, temporal information can be stored efficiently by providing fast access for the answering of queries. This work has shown an approach with MV that improve the storage and processing of temporal information in a graph database. The benefit of the proposed approach increases with the size of the stored main graph. Further improvements are achieved by the placement of the TMVs in a higher layer of the memory hierarchy, like PMem. Exploiting the memory hierarchy of modern systems is a reliable approach to speed up temporal query processing, as temporal workloads store multiple versions of data which creates conceivably a large amount of data.

Placing relevant sub-graphs, i.e., recent versions of the graph, as MVs in higher layers of the memory hierarchy improves the performance of future queries. This work has shown that exploiting the memory hierarchy is an important aspect when trying to improve temporal query processing. This work was focused on the optimization of answering temporal queries using the results of previous execution (TMVs). For future work, it is beneficial to investigate the selection of TMVs for queries, that share only a subset of the results. Furthermore, the suitable placement of the TMVs depends on the query workload and the resulting size. Therefore, a cost model that evaluates the suitability of the placement on the respective medium would be useful to further improve query execution times.

Acknowledgements. This work was partially funded by the German Research Foundation (DFG) in the context of the project "Hybrid Transactional/Analytical Graph Processing in Modern Memory Hierarchies (#TAG)" and "Processing-In-Memory Primitives for Data Management (PIMPMe)" as part of the priority programs "Scalable Data Management for Future Hardware" (SPP 2037) (SA 782/28-2) and "Disruptive Memory Technologies" (SPP 2377) (SA 782/31).

References

1. Al-Kateb, M., Ghazal, A., Crolotte, A., Bhashyam, R., Chimanchode, J., Pakala, S.P.: Temporal query processing in Teradata. In: EDBT, pp. 573–578 (2013)
2. ArangoDB (2022). https://www.arangodb.com/
3. Baumstark, A., Jibril, M.A., Sattler, K.U.: Adaptive query compilation in graph databases. In: ICDEW (2021)
4. Driscoll, J.R., Sarnak, N., Sleator, D.D., Tarjan, R.E.: Making data structures persistent. In: ACM STOC, pp. 109–121. ACM (1986)
5. Gohil, J.A., Dolia, D.P.M.: Checking and verifying temporal data validity using valid time temporal dimension and queries in Oracle 12C. IJDMS **7**, 11–22 (2015)
6. Hölsch, J., Grossniklaus, M.: An algebra and equivalences to transform graph patterns in Neo4j. In: EDBT/ICDT (2016)
7. Jibril, M.A., Baumstark, A., Götze, P., Sattler, K.: JIT happens: transactional graph processing in persistent memory meets just-in-time compilation. In: EDBT 2021, Nicosia, Cyprus, pp. 37–48 (2021)
8. Kaufmann, M., Manjili, A.A., Hildenbrand, S., Kossmann, D., Tonder, A.: Time travel in column stores. In: ICDE, pp. 110–121. IEEE (2013)
9. Kulkarni, K., Michels, J.E.: Temporal features in SQL: 2011. ACM SIGMOD Rec. **41**(3), 34–43 (2012)
10. Massri, M., Miklos, Z., Raipin, P., Mey, P.: T-cypher: a temporal graph query language (2023). https://project.inria.fr/tcypher/
11. Oracle: Mav introduction. https://www.oracle.com/webfolder/technetwork/de/community/dbadmin/tipps/mav_introduction/. Accessed 17 Nov 2021
12. Raphtory, January 2023. https://github.com/Raphtory/Raphtory
13. Rost, C., et al.: Distributed temporal graph analytics with GRADOOP. VLDB J. **31**(2), 375–401 (2022)
14. Saracco, C.M., Nicola, M., Gandhi, L.: A Matter of Time: Temporal Data Management in DB2 for z. IBM Corporation, New York (2010)

Data Science and Fairness

Comparing and Improving Active Learning Uncertainty Measures for Transformer Models

Julius Gonsior[1(✉)], Christian Falkenberg[1], Silvio Magino[1], Anja Reusch[1], Claudio Hartmann[1], Maik Thiele[2], and Wolfgang Lehner[1]

[1] Technische Universität Dresden, Dresden, Germany
{julius.gonsior,Christian.Falkenberg,Silvio.Magino,Anja.Reusch,claudio.hartmann,wolfgang.lehner}@tu-dresden.de
[2] Hochschule für Technik und Wirtschaft Dresden, Dresden, Germany
maik.thiele@htw-dresden.de

Abstract. Despite achieving state-of-the-art results in nearly all Natural Language Processing applications, fine-tuning Transformer-encoder based language models still requires a significant amount of labeled data to achieve satisfying work. A well known technique to reduce the amount of human effort in acquiring a labeled dataset is *Active Learning* (AL): an iterative process in which only the minimal amount of samples is labeled. AL strategies require access to a quantified confidence measure of the model predictions. A common choice is the softmax activation function for the final Neural Network layer. In this paper we compare eight alternatives on seven datasets and show that the softmax function provides misleading probabilities. Our finding is that most of the methods primarily identify hard-to-learn-from samples (outliers), resulting in worse than random performance, instead of samples, which reduce the uncertainty of the learned language model. As a solution this paper proposes a heuristic to systematically exclude samples, which results in improvements of various methods compared to the softmax function.

Keywords: Active Learning · Transformer · Softmax · Uncertainty · Calibration · Deep Neural Networks

1 Introduction

The most common use case of Machine Learning (ML) is supervised learning, which inherently requires a labeled dataset to demonstrate the desired outcome to the ML model. This initial step of acquiring a labeled dataset can only be accomplished by often rare-to-get and costly human domain experts; automation is not possible as the automation via ML is exactly the task which should be learned. For example, the average cost for the common label task of segmenting a single image reliably is 6,40 USD[1]. At the same time, recent advances

[1] According to scale.ai as of December 2021 (as of 2023 the cost is not publicly visible anymore): https://web.archive.org/web/20210112234705/https://scale.com/pricing.

A. Abelló et al. (Eds.): ADBIS 2023, LNCS 13985, pp. 119–132, 2023.
https://doi.org/10.1007/978-3-031-42914-9_9

in the field of Neural Network (NN) such as Transformer-encoder [33] models for Natural Language Processing (NLP) (with BERT [4] being the most prominent example) or Convolutional Neural Networks (CNN) [16] for computer vision resulted in huge Deep NN which require even more labeled training data. Reducing the amount of labeled data is therefore a primary objective of making ML more applicable in real-world scenarios. The focus of this paper is on the NLP domain and Transformer-encoder NN respectively, but the proposed methods can also be applied without any further work to other deep NN models and domains.

Noting that deep NN require a large amount of labeled data, Active Learning (AL) is a popular method to reduce the required human effort in creating a labeled dataset, by reducing labeling of nearly identical, and therefore redundant samples. AL is an iterative process, which step-by-step decides, which samples to label first, based on the existing knowledge in the form of the currently labeled samples. The challenge of applying AL is the almost paradoxical problem: how to decide which samples are most beneficial to the ML model, without knowing the label of the samples, since this is exactly the task to be learned by the to-be-trained ML model.

Despite successful application in a variety of domains [6,9,19], AL fails to work for very deep NN such as Transformer-encoder models, rarely beating pure random sampling. The common explanation [3,8,14,23] is that AL methods favor hard-to-learn samples, often simply called *outliers*, which therefore neglects the potential benefits from AL. Another potential explanation – so far to our knowledge not covered in the AL literature – could be the calculation of the uncertainty of the NN, which is a core component of most AL strategies, and directly influencing in the selection of the outliers. For NN the softmax function is used to calculate the uncertainty, but as pointed out before [22], simply interpreting the softmax function as the true model confidence is a fallacy.

Our main contributions are: a) an empirical comparison of eight alternative methods to the vanilla softmax function as Uncertainty-measures in the context of AL, applied for fine-tuning Transformer-encoder models, and b) proposing the novel and easy to implement method *Uncertainty Clipping (UC)* of mitigating the negative effect of uncertainty based AL methods of favoring outliers.

The remainder of this paper is structured as follows: In Sect. 2, we briefly explain AL and give an overview of Related Works about AL strategies. Sect. 3 presents alternative Uncertainty-measures, Sect. 4 describes our experimental setup. Results are discussed in Sect. 5 and we conclude in Sect. 6.

2 Active Learning 101

This section introduces in Sect. 2.1 the standard AL cycle, gives an overview about AL strategies and explains the focus of this paper on uncertainty-based AL (Sect. 2.2, and ends with an introduction to uncertainty-based AL in Sect. 2.3.

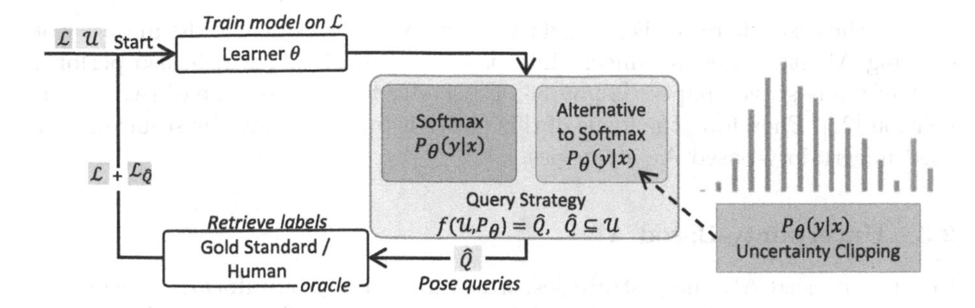

Fig. 1. Standard Active Learning Cycle including our proposed *Uncertainty Clipping (UC)* to influence the uncertainty based ranking by ignoring the top-k results

2.1 Active Learning Cycle

AL is a well-known technique for saving human effort by iteratively selecting exactly those unlabeled samples for expert labeling that are the most useful ones for the overall classification task [29]. The goal is to train a classification model θ which maps samples $x \in \mathcal{X}$ to a respective label $y \in \mathcal{Y}$; for the training, the labels $\mathcal{Y}$ have to be provided by an *oracle*, often one or multiple human annotators. Figure 1 shows a standard pool-based AL cycle: Given a small initial labeled dataset $\mathcal{L} = \{(x_i, y_i)\}_{i=0}^{n}$ of n samples $x_i \in \mathcal{X}$ and the respective label $y_i \in \mathcal{Y}$ and a large unlabeled pool $\mathcal{U} = \{x_i\}, x_i \notin \mathcal{L}$, a ML model called *learner* $\theta : \mathcal{X} \mapsto \mathcal{Y}$ is trained on the labeled set. A *query strategy* $f : \mathcal{U} \longrightarrow Q$ then subsequently chooses a batch of b unlabeled samples Q, which will be labeled by the oracle (human expert) and added to the set of labeled data $\mathcal{L}$. This AL cycle repeats τ times until a stopping criterion is met.

2.2 Overview of AL Strategies

AL strategies rely on their core all on the same set of two simple heuristics [36]: *informativeness/uncertainty* [17,24,31] and *representativeness/diversity* [2]. The first selects samples that reduce the uncertainty of the learner model, whereas the latter prefers samples that represent the overall sample distribution in the feature vector space. Most recent AL strategies [10,12] aim to combine both heuristics.

The most apparent problem of uncertainty based strategies are the inability to discover unknown clusters behind the existing classification boundary, most visible for chessboard-like datasets [1] due to the – sometimes – wrong assumption that samples, which lie far behind the boundary, belong certainly all to the same class. Diversity based strategies on the other hand solve this problem, but only function well if the underlying sample vector space contains well-formed clusters and require a larger amount of labeled data before functioning well [36]. One common neglected aspect of combined and advanced AL strategies is the very high runtime, which is often especially bad for large datasets [10], which are exactly the case where AL is applied the most. As a consequence, the majority of existing AL strategies are uncertainty-based [14,26,27,36], which also perform best in many survey papers [27,35,36], especially for our use-case of text classification [27]. Therefore, the focus of this paper is on enhancing the so-far mostly used uncertainty-based AL strategies.

2.3 Uncertainty-Based AL

Commonly used AL query strategies, which are relying on informativeness, use the uncertainty of the learner model θ to select the AL query. The uncertainty is defined as the inverse confidence/probability of the learner $P_\theta(y|x)$ in classifying a sample x with the label y. The idea behind is to label samples in those regions, where the learner model is most uncertain about, and thereby intentionally decreasing the models overall uncertainty. The most simple informativeness AL strategy is *Uncertainty Least Confidence* (LC) [17], which selects those samples, where the learner model is most uncertain about, i.e. where the probability $P_\theta(\hat{y}|x)$ of the most probable label $\hat{y}$ is the lowest: $f_{LC}(\mathcal{U}) = \text{argmax}_{x \in \mathcal{U}} (1 - P_\theta(\hat{y}|x))$.

3 Uncertainty-measures

In the following we will first discuss the suitability and problems of the softmax activation function as an Uncertainty-measure (Sect. 3.1), present alternative methods in Sect. 3.2, and end with our novel meta-strategy *Uncertainty Clipping*, which significantly enhances several Uncertainty-measures (Sect. 3.3).

3.1 Softmax as Uncertainty-measure

All uncertainty-based AL strategies have in common that they need the *uncertainty* of the learner model in a quantified form to rank the unlabeled samples. In this paper, this will be called *Uncertainty-measure*. The inverse of an Uncertainty-measure, the confidence probability of an ML model, should reflect, how probable it is that the models own predictions are true. For example, a confidence of 70% should mean a correct prediction in 70 out of 100 cases. Due to the property of having 1 as the sum for all components the *softmax* function is often used as a makeshift probability measure for the certainty of NNs: $\sigma(\boldsymbol{z})_i = \frac{exp(z_i)}{\sum_{j=1}^{K} exp(z_j)}$, for $i = 1, \ldots, K$. The output of the last neurons i before entering the activation functions is called *logit*, and denoted as z_i, K denotes the amount of neurons in the last layer. But as has been mentioned in the past by other researchers [3,8,15,23,34], the training objective in training NNs is purely to maximize the value of the correct output neuron, not to create a true confidence probability. An inherent limitation of the softmax function is its inability to have – in the theoretical case – zero confidence in its prediction, as the sum of all possible outcomes always equals 1. Previous works have indicated that softmax based confidence is often overconfident [5]. Especially NNs using a typical ReLU activation function for the inner layers can be easily tricked into being overly confident in any prediction by simply scaling the input x with an arbitrarily large value $\alpha > 1$ to $\tilde{x} = \alpha x$ [11,22].

3.2 Alternative Uncertainty-measures

Even though an Uncertainty-measure is a crucial component for uncertainty-based AL, few research has been done on solely comparing Uncertainty-measures, with the goal of AL for Transformer-encoder models in mind. We selected seven methods from the literature suitable as alternative Uncertainty-measures of Deep NNs such as Transformer-encoder models. They can be divided into five categories [7]: a) single network deterministic methods, which deterministically produce the same result for each NN forward pass (*Inhibited Softmax (IS)* [21], *TrustScore (TrSc)* [13] and *Evidential Neural Networks (Evi)* [28]), b) Bayesian methods, which sample from a distribution and result therefore in non-deterministic results (*Monte-Carlo Dropout (MC)* [5]), c) ensemble methods, which combine multiple deterministic models into a single decision (*Softmax Ensemble* [30]), d) calibration methods, which calibrate the softmax function (*Label Smoothing (LS)* [32] and *Temperature Scaling (TeSc)* [37]) and e) test-time augmentation methods, which, similarly to the ensemble methods, augment the input samples, and return the combined prediction for the augmented samples. The last category is a subject of future research as we could not find a subset of data augmentation techniques which reliably worked well for our use case among different datasets.

In the following the core ideas of the individual methods are briefly explained, more details, reasonings, and the exact formulas can be found in the original papers.

Inhibited Softmax (IS). The Inhibited Softmax method [21] is a simple extension of the vanilla softmax function by an additional constant factor $\alpha \in \mathbb{R}$, which enhances the effect of the absolute magnitude of the single logit value z_i on the softmax output: $\sigma(\boldsymbol{z})_i = \frac{exp(z_i)}{\sum_{j=1}^{K} exp(z_j)+exp(\alpha)}$.

TrustScore (TrSc). The TrustScore [13] method uses the set of available labeled data to calculate a *TrustScore*, independent on the NN model. In a first step the available labeled data is clustered into a single high density region for each class. The TrustScore ts of a sample x is then calculated as the ratio between the distance from x to the cluster of the nearest class $c_{closest}$, and the distance to the cluster of the predicted class $\hat{y}$: $ts_x = \frac{dist(x,c_{closest})}{dist(x,c_{\hat{y}})}$ The distance metric as well as the calculation of the clusters is based on the k-nearest neighbors algorithm.

Evidential Neural Networks (Evi). Evidential Neural Networks [28] treat the vanilla softmax outputs as a parameter set over a Dirichlet distribution. The prediction acts as *evidence* supporting the given parameter set out of the distribution, and the confidence probability of the NN reflects the Dirichlet probability density function over the possible softmax outputs.

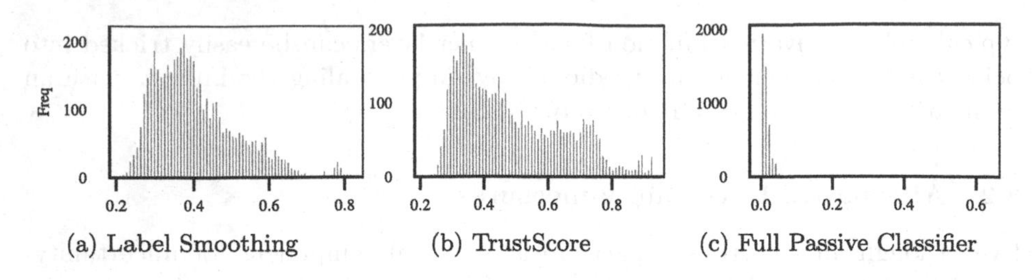

(a) Label Smoothing (b) TrustScore (c) Full Passive Classifier

Fig. 2. Histogram distribution of exemplary uncertainty values for a single AL iteration of TREC-6 dataset before Uncertainty Clipping. The x-axis ranges from 0, full certainty, to 1, full uncertainty.

Monte-Carlo Dropout (MC). Monte-Carlo Dropout [5] is a Bayesian method that uses the NN dropout regularization method to construct for "free" an ensemble of the same trained model. For Monte-Carlo Dropout, the dropout method is applied during the prediction phase. As neurons are disabled randomly, this results in a large Gaussian sample space of different models. Therefore, each model, with differently dropped out neurons, results in a potentially different prediction. Combining the vanilla softmax prediction using the arithmetic mean produces a combined Uncertainty-measure.

Softmax Ensemble. The softmax ensemble approach uses an ensemble of NN models, similar to Monte-Carlo Dropout. The predictions of the ensemble can be interpreted as a vote upon the prediction. The disagreement among the votees acts then as the Uncertainty-measure and can be calculated in two ways, either as Vote Entropy (VE), or as Kullback-Leibler Divergence (KLD) [20]:$VE(x) = -\sum_i \frac{V(\hat{y_i},x)}{K} log \frac{V(\hat{y_i},x)}{K}$ with K being the number of ensemble models, and $V(x, \hat{y_i})$ denoting the number of ensemble models assigning the class $\hat{y_i}$ to the sample x. The complete equation to calculate the KLD is omitted for brevity.

Temperature Scaling (TeSc). Temperature Scaling [37] is a model calibration method that is applied after the training and changes the calculation of the softmax function by inducing a temperature $T > 0$: $\sigma(z_i) = \frac{exp(z_i/T)}{\sum_{j=1}^{K} exp(z_j/T)}$ This has a dampening effect on the overall confidence. The value for the temperature T is computed empirically using the existent labeled set of samples during application time. The parameter is therefore different for each AL iteration.

Label Smoothing (LS). Label Smoothing [32] removes a fraction α of the loss function per predicted class and distributes it uniformly among the other classes by adding $\frac{\alpha}{K-1}$ to the other loss outputs, with K being the number of classification classes. In contrast to Temperature Scaling, Label Smoothing is not applied after a network has been trained, but directly during the training process due to a modified loss function.

3.3 Uncertainty Clipping (UC)

The aforementioned Uncertainty-measures can directly be used in AL strategies to sort the pool of unlabeled samples to select exactly those samples for labeling that have the lowest confidence/highest uncertainty. As repeatedly reported by others [3,8,14,23], AL is rarely able to outperform pure random sampling when used to fine-tune Transformer-encoder models. Purely labeling based on the uncertainty score results in labeling many outliers/hard-to-learn-from samples, which results in an often bad classification performance.

Figure 2 displays exemplarily the histograms of the prediction probabilities/uncertainty values for the two methods Label Smoothing and TrustScore for a single AL iteration for the TREC-6 dataset. Both distributions have a characteristic small peak of uncertainty to the far right, and we theorize, that these are the outliers which are labeled first. For comparison, the uncertainty values of a *passive* classifier are shown – an NN model trained on the full available training dataset –, such a model is obviously very confident in its predictions.

To prevent this issue we propose the simple, yet effective method *Uncertainty Clipping*, where the top-k% of the most uncertain samples are ignored by the AL strategy, resulting in ignoring the displayed small peaks to the far right. Therefore, not the most uncertain samples – potentially many outliers – but the second-most uncertain samples are labeled. This method can be used in combination with any AL strategy using uncertainty for ranking the pool of unlabeled samples.

4 Experimental Setup

Details to reproduce our evaluation are provided in this Section. We are open to join the reproducability initiative and to offer other researchers the re-use of our work by making our source code fully publicly available on GitHub[2].

4.1 Setup

We extended the AL framework *small-text* [25], tailored for the use-case of applying AL to Transformer-encoder networks. Some of the aforementioned methods such as Label Smoothing can be applied directly during the training of the network and have a positive effect on the training outcome, whereas others like Monte-Carlo Dropout are applied after the training is complete. As we are only interested in evaluating the effect of alternative Uncertainty-measures instead of the potentially positive influence on the classification quality, we are effectively training two Transformer-encoder models simultaneously. One is the original vanilla Transformer-encoder model with a linear classifier and a softmax on top of the CLS embedding. This one is used for the class predictions and for calculating the classification quality. The other one is solely used for the AL selection process, and includes the implemented alternative Uncertainty-measures. Even

[2] https://github.com/jgonsior/active-learning-softmax-uncertainty-clipping.

though this adds computational overhead to our experiments, it enables us to evaluate the effect of the Uncertainty-measure for AL independently of potentially other positive – or negative – effects on the classification quality.

Parameters for the compared methods were selected empirically using hyper parameter tuning. The constant factor α for Inhibited Softmax was set to 1.0, for TrustScore the k for the k-nearest neighbor calculation to 10. We used an ensemble of 50 softmax-based Transformer-encoder models for Monte-Carlo Dropout, each one with a different seed for the random-number-generator. For the softmax ensemble we could only use 5 different Transformer-encoder models with the vanilla softmax activation function, as the runtime was drastically higher compared to Monte-Carlo Dropout (MC). For Label Smoothing the fraction α was set empirically fixed to 0.2. For Temperature Scaling 1.000 different values for T between 0 and 10 were tested; the temperature resulting in the smallest cross-entropy was finally used. Whenever possible, we used the original implementations of the methods with slight adaptations to make them work together with Transformer-encoder models and the small-text framework. We use the original $BERT_{base}$ [4] model and the updated version $RoBERTa_{base}$ [18] as Transformer-encoder models.

Active Learning Simulation. To evaluate the effectiveness of the Uncertainty-measure we are simulating AL in the following way: For a given labeled dataset, we start with an initially labeled set of 25 samples, and perform afterwards 20 iterations of AL, ignoring the known labels for the other samples, following the procedure of [27]. In each iteration, a batch of 25 samples is selected by the AL strategy for labeling, which is simulated using the known ground-truth labels. Each simulation was repeated 10 times using different initially labeled samples to ensure statistical significance.

As baselines, we decided to deliberately only include uncertainty-based AL strategies. Firstly, the scope of this work is to evaluate improvements for the uncertainty calculation part of AL strategies. More advanced strategies have the potential, to overshadow the influence of the changed Uncertainty-measures. Secondly, according to recent AL surveys such as [27,36] the vast majority of AL strategies uses uncertainty. And thirdly, other works such as [14,27] already evaluated those advanced strategies on nearly the same datasets.

We used the most simple uncertainty-based AL strategy, directly using the Uncertainty-measure without further pre-processing for ranking, Uncertainty Least Confidence (see Sect. 2) in combination with the softmax function replaced with the alternative Uncertainty-measures. Two additional uncertainty-based baselines were included: Uncertainty Entropy [31], and Uncertainty Max-Margin [24]. Another baseline was the passively trained model on all available training data, and pure random sampling.

4.2 Datasets, Hardware, and Metrics

We used seven common NLP datasets in our experiments to *fine-tune* the pretrained Transformer-encoder models in the AL simulations: AG's News (AG),

CoLA (Co), IMDB (IM), Rotten Tomatoes (RT), Subjectivity (SU), SST2 (S2), and TREC-6 (TR). The datasets were selected as a diverse set of popular NLP datasets, including binary and multi-class classification of different domains of varying difficulty. All experiments were conducted on a cluster consisting of NVIDIA A100-SXM4 GPUs, AMD EPIC CPU 7352 (2.3GHZ) CPUs and NVMEe disks. Each experiment was run on a single graphic card, with 120GB memory, and 16 CPU cores.

The AL experiments can be evaluated in multitude of ways. At its core, after each AL iteration, a standard ML metric is being measured for the labeled and withheld dedicated test set. We decided upon the *accuracy (acc)* metric, calculated on a withheld test dataset. It is possible to compare the test accuracy values of the last iteration, the mean of the last five iterations (acc_{last5}), and the mean of all iterations. The last one equals to the area-under-the-curve when plotting the so-called AL learning curve. As an effective AL strategy should select the most valuable samples for labeling first, those metrics that include the accuracy of multiple iterations are often closer to real use-cases. Nevertheless, at the beginning of the labeling process the fluctuation of the test accuracy for most strategies is very high and contains often surprisingly few information about which strategy is better. The influence of the initially labeled samples is simply so high, that a better strategy, with a bad starting point, has no chance to be better than a bad strategy, which has a good starting point. But after a couple of AL iterations, the results stabilize, and good strategies can be reliably distinguished from bad strategies, as each strategy tends to approach its own characteristic threshold, regardless of the starting point. Therefore, we decided to use the accuracy of the last five iterations, deliberately ignoring the first iterations with the highly fluctuating results.

Additionally, we measured the runtime. AL, applied in real-life scenarios, is an interactive process. Decisions of the AL strategy should be made in the magnitude of single-digit seconds, longer calculation times render the annotation process impractical.

As the experiments were each repeated for 10 times, we report in the following the arithmetic mean for the 10 repetitions, independently per each dataset.

5 Results

To compare the Uncertainty-measures we are looking at the test accuracies, with and without our proposed Uncertainty Clipping in Sect. 5.1. Additionally, we performed an analysis of the runtimes in Sect. 5.2.

5.1 Test Accuracies

Figure 3 displays the distributions of the average test accuracy of the last five AL iterations acc_{last5} per method and per dataset. Each result is display using two boxplots: The light gray one displays the original values, the dark gray one the ones using Uncertainty Clipping. The underlying displayed distribution

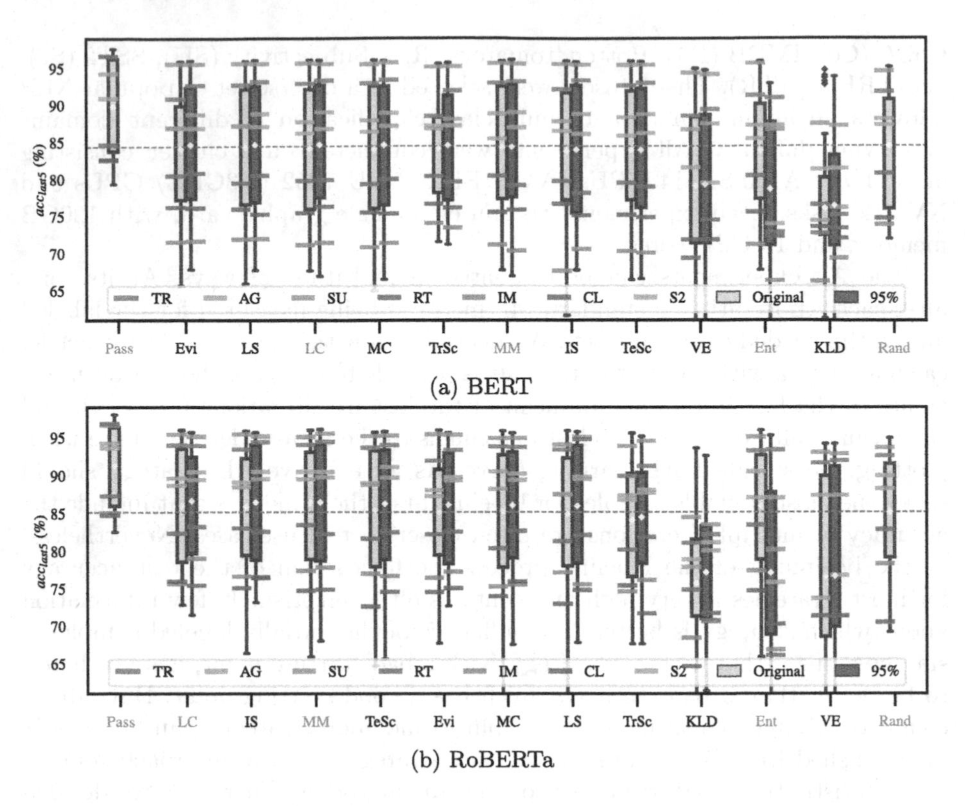

(a) BERT

(b) RoBERTa

Fig. 3. Distribution of acc_{last5} including the *Uncertainty Clipping* variants and the average acc_{last5} values per dataset as colorful line, ordered by acc_{last5} after Uncertainty Clipping. The arithmetic mean of the runs per method are included as a yellow diamond in the middle of the plots. The vanilla softmax based baselines Ent, MM und LC are marked in blue, and the baselines Random Selection as well as the Passive classifier are marked red.

consists of the acc_{last5} values combined for all datasets. As each method was evaluated using 10 different starting points, we additionally display the arithmetic mean of the 10 repetitions per dataset as an additional colorful stick. The methods are ordered after the mean acc_{last5} value using Uncertainty Clipping. As Uncertainty Clipping does not work on the baselines of a full passive classifier (Pass) and random selection (Rand), only one boxplot is shown for these two.

General Remarks First, it is obvious that the difference between the individual methods differs per dataset, but averaged over all datasets (the yellow diamond in the middle of the boxplots indicates the arithmetic mean) the differences become marginally small. Still, it can be safely stated already that the two Softmax Ensemble techniques (KLD and VE) perform far worse than any other technique. This confirms the finding of [8].

Uncertainty Clipping We discarded the top-5% of the most *uncertain* values when selecting samples for labeling, displayed in the dark gray boxplots. The clipping threshold of 5% was found empirically using hyperparameter tuning, other values up to 10% also worked well. The reasoning behind this is that a good Uncertainty-measure in an AL strategy selects mostly outliers, as these are indeed the most uncertain values. But solely labeling outliers results in worse performance than Random sampling, as can be seen in Fig. 3 when looking only on the light gray boxplots. Our paradoxical finding is therefore that the Softmax function is, as expected, a bad method for Uncertainty-measure, but purely uncertainty-based AL strategies actually *benefit* from a *slightly less than perfect* Uncertainty-measure.

Comparing the light and dark gray boxplots per method, Uncertainty Clipping often seems to reduce mostly the lower ends of the distribution. This is expected, as applying Uncertainty Clipping can only result in the ignoration of results, and thereby prevention of bad AL choices. As no new results are added, improvement of already good choices is not to be expected. The effect of Uncertainty Clipping becomes more clear if one compares the colorful lines per dataset in each boxplot from the left and the right side. These lines indicate the arithmetic mean per dataset. Especially for the dataset AG-News and Trec-6 the clipping improves the final test accuracy drastically, as the right lines are often higher than the left ones. These datasets, which are more influenced by Uncertainty Clipping, appear therefore to contain a higher percentage of outliers. Except for the ensemble methods and Uncertainty Entropy, all methods benefit from Uncertainty Clipping.

Baselines In addition to pure Random Sampling selection, and Uncertainty Least Confidence with the vanilla softmax function, Uncertainty Entropy and Uncertainty Max-Margin – both also using the vanilla softmax function – were included in our evaluation. Uncertainty Entropy is the only strategy which is highly negatively influenced by Uncertainty Clipping. This is explainable: this strategy calculates the entropy of all uncertainty values, which transforms the shape of the distribution of the original uncertainty values into a right-skewed distribution with many high values. Using a fixed threshold of 5% for Uncertainty Clipping, is therefore not directly transferable to the Entropy-based strategy. The baseline strategies Uncertainty Least Confidence and Uncertainty Max-Margin perform better on RoBERTa than on BERT, indicating that the softmax function is better calibrated to true probabilities for RoBERTa than for the original BERT model.

Best overall performing method The results differ based on the used Transformer-encoder model. For BERT, no method was able to beat random sampling without our proposed Uncertainty Clipping, whereas with it, the majority was able to justify the usage of AL. For RoBERTa many methods were better than Random sampling even without Uncertainty Clipping, but benefited even further from using it. Otherwise, the question of whether to use the softmax activation function or an alternative Uncertainty-measure can be concluded with a confirmation for the status-quo to continue using the softmax function. The pure softmax function-based method Uncertainty Least Confidence (LC) perform still comparably good, with only marginally small differ-

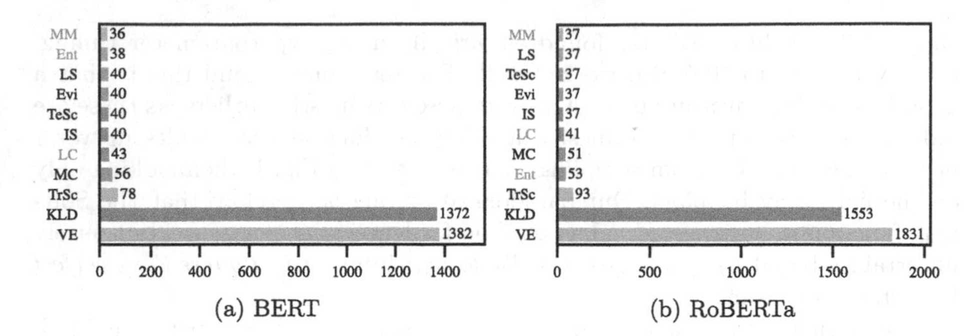

Fig. 4. Runtime Comparison of all methods averaged over the datasets in seconds

ences to most alternative methods. This confirms the conclusion of the deep investigation of [22] about the softmax function, stating that it *might have a sounder basis than widely believed.* Still, from the alternative methods, Evidential Neural Networks (Evi) seems promising on both Transformer-encoder models, with Label Smoothing (LS), Monte-Carlo Dropout (MC), Temperature Scaling (TeSc), or TrustScore (TrSc) close behind. Also, the baseline AL strategy Uncertainty Max-Margin (MM) seems promising.

5.2 Runtime Comparison

AL is in practice a human-in-the-loop process. Annotators want a responsive systems, which tells them immediately what to label next, without long waiting times. Therefore, a fast runtime of the AL query selection is a crucial factor in making AL real-world usable. In Fig. 4, we display the runtimes averaged over all datasets for our methods in seconds per complete AL loop, measuring only the time of the AL computations, not the Transformer-encoder model fine-tuning time[3]. First, it becomes clear that the overhead of the two Ensemble methods (KLD and VE) of training multiple Transformer-encoder models in parallel is simply too high. Apart from that, almost all other methods seem to perform equally fast.

6 Conclusion

Noting the importance of NN Uncertainty-measures for AL, and the potential shortcomings of simply using vanilla softmax as such a measure, we experimentally compared eight alternative methods over seven datasets using both the original BERT Transformer-encoder model as well as the improved RoBERTa variant. After discovering that better Uncertainty-measures result in selecting

[3] The average training time of the Transformer-encoder models is around 600 s combined for a single AL experiment.

only outliers for labeling, we proposed Uncertainty Clipping, improving all methods, including vanilla softmax, which showed to be the altogether best performing technique.

Acknowledgments. The authors are grateful to the Center for Information Services and High Performance Computing ZIH at TU Dresden for providing its facilities for high throughput calculations. This research was funded by the German Federal Ministry of Education and Research (BMBF) through grants 01IS17044 Software Campus 2.0 (TU Dresden).

References

1. Baram, Y., Yaniv, R.E., Luz, K.: Online choice of active learning algorithms. J. Mach. Learn. Res. **5**(Mar), 255–291 (2004)
2. Coleman, C., et al.: Selection via proxy: Efficient data selection for deep learning. ICLR (2020)
3. D'Arcy, M., Downey, D.: Limitations of active learning with deep transformer language models (2022)
4. Devlin, J., Chang, M.W., Lee, K., Toutanova, K.: Bert: Pre-training of deep bidirectional transformers for language understanding. In: NAACL, pp. 4171–4186. Association for Computational Linguistics (2019)
5. Gal, Y., Ghahramani, Z.: Dropout as a bayesian approximation: Representing model uncertainty in deep learning. In: ICML, pp. 1050–1059. PMLR (2016)
6. Gal, Y., Islam, R., Ghahramani, Z.: Deep bayesian active learning with image data. In: International Conference on Machine Learning, pp. 1183–1192. PMLR (2017)
7. Gawlikowski, J., et al.: A survey of uncertainty in deep neural networks. arXiv preprint arXiv:2107.03342 (2021)
8. Gleave, A., Irving, G.: Uncertainty estimation for language reward models. arXiv preprint arXiv:2203.07472 (2022)
9. Gonsior, J., Rehak, J., Thiele, M., Koci, E., Günther, M., Lehner, W.: Active learning for spreadsheet cell classification. In: EDBT/ICDT Workshops (2020)
10. Gonsior, J., Thiele, M., Lehner, W.: Imital: learned active learning strategy on synthetic data. In: Pascal, P., Ienco, D. (eds.) Discovery Science, pp. 47–56. Springer Nature Switzerland, Cham (2022)
11. Hein, M., Andriushchenko, M., Bitterwolf, J.: Why relu networks yield high-confidence predictions far away from the training data and how to mitigate the problem. In: CVPR, pp. 41–50. IEEE (2019)
12. Hsu, W.N., Lin, H.T.: Active learning by learning. In: Proceedings of the Twenty-Ninth AAAI Conference on Artificial Intelligence, pp. 2659–2665. AAAI'15, AAAI Press (2015)
13. Jiang, H., Kim, B., Guan, M., Gupta, M.: To trust or not to trust a classifier. In: NeurIPS 31 (2018)
14. Karamcheti, S., Krishna, R., Fei-Fei, L., Manning, C.: Mind your outliers! investigating the negative impact of outliers on active learning for visual question answering. In: ACL-IJCNLP, pp. 7265–7281. Association for Computational Linguistics (2021)
15. Lakshminarayanan, B., Pritzel, A., Blundell, C.: Simple and scalable predictive uncertainty estimation using deep ensembles. In: NeurIPS 30 (2017)

16. LeCun, Y., Bengio, Y., Hinton, G.: Deep learning. Nature **521**(7553), 436–444 (2015)
17. Lewis, D.D., Gale, W.A.: A sequential algorithm for training text classifiers. In: SIGIR '94, pp. 3–12. Springer, London (1994)
18. Liu, Y., et al.: Roberta: A robustly optimized bert pretraining approach. arXiv preprint arXiv:1907.11692 (2019)
19. Lowell, D., Lipton, Z.C., Wallace, B.C.: Practical obstacles to deploying active learning. In: EMNLP-IJCNLP, pp. 21–30 (2019)
20. McCallumzy, A.K., Nigamy, K.: Employing em and pool-based active learning for text classification. In: ICML, pp. 359–367. Citeseer (1998)
21. Możejko, M., Susik, M., Karczewski, R.: Inhibited softmax for uncertainty estimation in neural networks. arXiv preprint arXiv:1810.01861 (2018)
22. Pearce, T., Brintrup, A., Zhu, J.: Understanding softmax confidence and uncertainty. arXiv preprint arXiv:2106.04972 (2021)
23. Sankararaman, K.A., Wang, S., Fang, H.: Bayesformer: Transformer with uncertainty estimation. arXiv preprint arXiv:2206.00826 (2022)
24. Scheffer, T., Decomain, C., Wrobel, S.: Mining the web with active hidden markov models. In: Hoffmann, F., Hand, D.J., Adams, N., Fisher, D., Guimaraes, G. (eds.) ICDM, pp. 309–318. IEEE Comput. Soc (2001)
25. Schröder, C., Müller, L., Niekler, A., Potthast, M.: Small-text: Active learning for text classification in python. arXiv preprint arXiv:2107.10314 (2021)
26. Schröder, C., Niekler, A.: A survey of active learning for text classification using deep neural networks. arXiv preprint arXiv:2008.07267 (2020)
27. Schröder, C., Niekler, A., Potthast, M.: Revisiting uncertainty-based query strategies for active learning with transformers. In: ACL, pp. 2194–2203. Association for Computational Linguistics (2022)
28. Sensoy, M., Kaplan, L., Kandemir, M.: Evidential deep learning to quantify classification uncertainty. In: NeurIPS 31 (2018)
29. Settles, B.: Active learning. Synth. Lect. Artif. Intell. Mach. Learn. **6**(1), 1–114 (2012)
30. Seung, H.S., Opper, M., Sompolinsky, H.: Query by committee. In: Proceedings of the Fifth Annual Workshop On Computational Learning Theory, pp. 287–294. COLT '92, ACM (1992)
31. Shannon, C.E.: A mathematical theory of communication. Bell Syst. Tech. J. **27**(3), 379–423 (1948)
32. Szegedy, C., Vanhoucke, V., Ioffe, S., Shlens, J., Wojna, Z.: Rethinking the inception architecture for computer vision. In: CVPR, pp. 2818–2826. IEEE (2016)
33. Vaswani, A., Shazeer, N., Parmar, N., Uszkoreit, J., Jones, L., Gomez, A.N., Kaiser, L., Polosukhin, I.: Attention is all you need. In: NeurIPS 30 (2017)
34. Weiss, M., Tonella, P.: Simple techniques work surprisingly well for neural network test prioritization and active learning (replicability study). arXiv preprint arXiv:2205.00664 (2022)
35. Yoo, D., Kweon, I.S.: Learning loss for active learning. In: Proceedings of the IEEE/CVF Conference On Computer Vision and Pattern Recognition, pp. 93–102 (2019)
36. Zhan, X., Liu, H., Li, Q., Chan, A.B.: A comparative survey: Benchmarking for pool-based active learning. In: IJCAI, pp. 4679–4686 (2021), survey Track
37. Zhang, J., Kailkhura, B., Han, T.Y.J.: Mix-n-match: Ensemble and compositional methods for uncertainty calibration in deep learning. In: ICML, pp. 11117–11128. PMLR (2020)

FARMUR: Fair Adversarial Retraining to Mitigate Unfairness in Robustness

Seyed Ali Mousavi[1], Hamid Mousavi[2(✉)], and Masoud Daneshtalab[2,3]

[1] Shahid Bahonar University, Kerman, Iran
mousavi.sayedali@math.uk.ac.ir
[2] Mälardalen University, Universitetsplan 1, 722 20 Västerås, Sweden
{seyedhamidreza.mousavi,masoud.daneshtalab}@mdu.se
[3] Tallinn University of Technology (Taltech), Tallinn, Akadeemia tee 15A, Estonia
masoud.daneshtalab@taltech.ee

Abstract. Deep Neural Networks (DNNs) have been deployed in safety-critical real-world applications, including automated decision-making systems. There are often concerns about two aspects of these systems: the fairness of the predictions and their robustness against adversarial attacks. In recent years, extensive studies have been devoted to addressing these issues independently through adversarial training and unfairness mitigation techniques. To consider fairness and robustness simultaneously, the robustness-bias concept is introduced, which means an attacker can more easily target a particular sub-partition in the dataset. However, there is no unified and mathematical definition for measuring fairness in the robustness of DNNs independent of the type of adversarial attacks. In this paper, we first provide a unified, precise, and mathematical theory and measurement for fairness in robustness independent of adversarial attacks for a DNN model. Finally, we proposed a fair adversarial retraining method (FARMUR) to mitigate unfairness in robustness that retrains the DNN models based on vulnerable and robust sub-partitions. In particular, FARMUR leverages different objective functions for vulnerable and robust sub-partitions to retrain the DNN. Experimental results demonstrate the effectiveness of FARMUR in mitigating the unfairness in robustness during adversarial training without significantly degrading robustness. FARMUR improves fairness in robustness by 19.18% with only 2.22% reduction in robustness in comparison with adversarial training on the UTKFace dataset, which is partitioned based on race attributes.

Keywords: Deep Neural Networks · Robustness · Robustness-bias · Fairness · Fairness-In-Robustness

1 Introduction

Deep Neural Networks (DNNs) have been used dramatically in real-world applications over the past decade. These applications cover a wide range of automated

A. Abelló et al. (Eds.): ADBIS 2023, LNCS 13985, pp. 133–145, 2023.
https://doi.org/10.1007/978-3-031-42914-9_10

decision-making systems, such as self-driving cars [15,31], social media [45], healthcare [27], targeted ads [35], and hiring [34]. Defense against adversarial attacks and the fairness of the predictions are two of the main concerns of these applications. Adversarial attacks can easily fool DNNs into predicting the wrong results by adding small, imperceptible noise into the input data (adversarial data). On the other side, unfair DNNs exacerbate societal stereotypes by using sensitive attributes, such as race and gender, in their predictions. These issues are addressed independently in the various research studies [6,10,38]. Adversarial training is a well-known and widely adopted defense method to increase the robustness of DNNs against adversarial attacks [25,38,43]. The goal of the AT methods is to train an adversarially robust DNN such that its predictions are locally invariant to a small neighborhood of the inputs. In terms of the fairness of DNNs, many studies have defined it based on the output accuracy for different sub-partitions of the test data. The accuracy of the model on some sub-partitions may be significantly lower than the average accuracy of the model on all test data. This is interpreted as the existence of a kind of unfairness in the DNN. This can happen for a variety of reasons, including the existence of some spurious correlations [24] and the data-driven training algorithm with unfair data used in DNNs [9]. A lot of works have been proposed to quantify, measure, and mitigate unfairness in DNNs [1,7,11,14,33]. To consider fairness and robustness jointly, the robustness-bias concept is defined such that DNNs may have similar accuracy across the sub-partitions in the dataset, but be more vulnerable to adversarial attacks on certain sub-partitions [30]. Based on this definition, different sub-partitions of the dataset can have different levels of robustness, which leads to unfairness in the robustness. However, there is no unified and clear mathematical definition for fairness in robustness, and previous metrics measured it based on adversarial attacks or computed it for only one sub-partition of data. Therefore, previous measures are highly dependent on the type of adversarial attacks and do not consider the whole DNN to measure fairness in robustness. In addition, the robustness-bias problem is investigated in the case of natural training of DNNs [30], but this issue also needs to be investigated for the adversarial training methods.

In this paper, we first introduce a unified, precise, and mathematical theory and measurement for fairness in robustness of a DNN model. Based on this metric, we empirically show that adversarial training increases the robustness-bias problem in the DNNs compared to natural training. To solve the robustness-bias problem in adversarial training, we propose the fair adversarial retraining (FARMUR) method to mitigate unfairness in robustness. It finds the vulnerable and robust sub-partitions of a partitioned dataset by analyzing their robustness and retrains the DNN on this data based on a new objective function. The new objective function consists of the robust and natural loss functions for vulnerable and robust sub-partitions. We evaluate FARMUR by comparing the fairness in robustness metric for the DNNs before and after fair adversarial retraining on various datasets.

Key contributions: The main contributions are as follows:

1. We introduce a unified, precise, and mathematical theory and measurement for fairness in robustness of the DNN models (Definition 4) (Sect. 3 and Sect. 4).
2. We analyze the fairness in robustness for natural and adversarial training methods and demonstrate that adversarial training increases unfairness in robustness.
3. We propose the fair adversarial retraining method (FARMUR) to mitigate unfairness in the robustness of adversarial training methods (Sect. 5).
4. We evaluate the FARMUR method on MNIST, Cifar-10, and UTKFace datasets. Experimentally we show that our method improves the fairness of the DNNs in terms of robustness for adversarial training methods.

Throughout this paper, we will denote the dataset, partition, and sub-partition by D, $\mathcal{P}$, and P, respectively. Also, we demonstrate the DNN by f.

2 Related Works

2.1 Fairness in DNNs

DNNs are generally regarded as high-performance black-box models. They used data-driven learning algorithms to learn useful representations from the raw data. Because the data might have biases, this data-driven algorithm in DNNs causes to amplify the biases in the data and make unfairness in the DNN model [9]. When we talk about fairness in DNNs, we need to be clear about what we are talking about in terms of fairness. In recent research works, the concept of fairness is considered as *fair predictions* [26]. It means that the DNNs predictions should be the same for different sub-partitions of data. There are different metrics to measure the fairness of a DNN. They can be categorized into individual and group fairness measurements. Individual fairness metrics consider the rule that similar inputs should have similar predictions [10]. Group fairness partitions the data based on a sensitive attribute and computes the statistic measures for each sub-partition and compares it across all sub-partitions. [3]. Moreover, Some mitigation techniques have been proposed for removing bias from DNNs [6,21,33,42]. These methods can be divided into pre-processing, in-processing, and post-processing methods. Pre-processing methods try to remove the biases from training data [4]. In-processing approaches regularize the loss function by adding fairness metric to the overall objective function [8]. Post-processing techniques try to calibrate the predictions of trained DNN models [16]. The comprehensive overview of fairness in DNNs in terms of definition, measurement, and mitigation methods can be found in [9,26]. In this paper, we look at the fairness of DNNs from the robustness perspective. Therefore, we introduce a theory, measurement, and mitigation method for fairness in robustness.

2.2 Robustness and Adversarial Training

Despite the high accuracy of DNNs, the presence of adversarial examples [13] has raised concerns about the use of these models for sensitive tasks. This sensitivity can be related to various aspects such as safety (e.g. autonomous driving vehicles [29]) and fairness (e.g. hiring [34]). A growing body of research shows that neural networks are vulnerable to adversarial examples generated with adversarial attacks [12,13,18,32,39,41]. Therefore, it requires first to define the robustness of DNNs and find the method to defend against adversarial attacks. The robustness of DNNs to has been defined in different but related ways. A commonly used definition is *robust accuracy*, which monitors the accuracy of the model on adversarial examples [5,37,44]. This definition is highly dependent on the type of adversarial attacks. Another definition, introduced by [28], is the average of the normalized distance of each point to the decision boundary. In terms of defense against adversarial attacks, adversarial training is a well-known and widely adopted defense method to increase the robustness of classifiers [25,38,43] Adversarial training is formulated as a min-max optimization problem, and the model is trained exclusively on adversarial images [25]. To find a trade-off between accuracy and robustness, TRADES [44] regularizes the loss function for clean data by incorporating a robust loss term and making a trade-off between them. Authors in [40] have shown that adversarial training can cause a serious disparity in both standard accuracy and adversarial robustness between different classes of data. The main goal of this paper is to find a trade-off between robustness and the fairness in robustness of the DNN model.

2.3 Fairness in Robustness

Recently, the concept of fairness has been integrated with robustness by using robust accuracy measurement [2,36,40]. These researches show that robustness may be at odds with fairness [2] and adversarial training algorithms. In addition, they demonstrate that adversarial training tends to introduce severe disparity of accuracy and robustness between different sub-partitions of data [40]. In terms of robustness measurement based on the distance of data points to the decision boundary, [30] show unfairness in the robustness for the DNNs that are naturally trained on clean data. It discusses fairness through robustness and has introduced the concept of robustness-bias based on the geometry of the decision boundary. However, it considers the natural training method and computes the robustness-bias for one sub-partition, not for the entire DNN. In this paper, we introduce a new unified, prices and mathematical theory and measurement for fairness in robustness based on the geometry of decision boundary for evaluating the fairness of the DNN models. Our new metric also shows that adversarial training decreases fairness in robustness. In addition, we improve the fairness in robustness by fair adversarial retraining (FARMUR) without compromising robustness significantly.

3 Fairness in Robustness: Theory

Unfairness in terms of robustness means that there are data sub-partitions that are most vulnerable to adversarial attacks. Therefore, the data on those sub-partitions is closer to the decision boundary of the DNN and can easily convert to an adversarial example. A partition ($\mathcal{P}$) consists of the non-empty sub-partitions of the dataset in which each data point is contained in exactly one sub-partition. In a classification problem, data can be partitioned in different ways. The simple way is to leverage the original labels as the sensitive attributes to partition data. However, data can be partitioned based on other sensitive attributes apart from the original labels. For the UTKFace dataset, data can partition based on age (original label) or other sensitive attributes such as race, gender, or ethnicity. Let D, $\mathcal{P}$, and f demonstrate the dataset, the partition of the dataset, and the DNN classifier, respectively. To define a new unified and precise metric for fairness in robustness, we first define the robustness of each sub-partition of $\mathcal{P}$ as follows.

Definition 1. *For every sub-partition P of a partition $\mathcal{P}$, we consider $Corr_f(P)$ as all data in P that are classified correctly by DNN classifier f and define $I_P : [0, \infty] \rightarrow [0, 1]$ as:*

$$I_P(\tau) = \frac{|\{(x, y) \in Corr_f(P) : d_\theta(x) > \tau\}|}{|P|} \tag{1}$$

where $|.|$ indicates the number of data of a sub-partition and $d_\theta(x)$ is the distance of x to decision boundary of DNN classifier. τ is a threshold value for the distance.

To make the comparison between the robustness of the sub-partitions in a partition, we define the *AUC* (Area Under Curve) metric that is independent of variable τ as follows.

Definition 2. *(robustness of a sub-partition) For every I_P in definition 1 we define:*

$$AUC(I_P) = \int_0^\infty I_P(\tau)d\tau \tag{2}$$

$$Rob(P) = AUC(I_P) \tag{3}$$

The *AUC* (Area Under Curve) metric is a well-defined function and converges to a precise value due to the following properties of I_P. For every $P \in \mathcal{P}$: 1) the I_P is a non-increasing function and 2) As $\tau \rightarrow \infty$ then $I_P(\tau) \rightarrow 0$. Based on this definition, we can sort the sub-partitions according to their robustness $Rob(P)$. To evaluate the robustness of each DNN classifier, we define $Rob(f)$ as follows:

Definition 3. *(robustness of the DNN classifier) Suppose $\mathcal{P} = \{P_1, P_2, \ldots, P_n\}$ is a partition of dataset and f be our DNN classifier. We consider $I_f : [0, \infty] \longrightarrow [0, 1]$ as*

$$I_f(\tau) = mean\{I_{P_i}(\tau) : i = 1, 2, \ldots, n\} \tag{4}$$

and define the robustness of f as

$$Rob(f) = AUC(I_f) \tag{5}$$

Where *mean* indicates the average function on $I_{P_i}(\tau)$.

This definition differs from the other papers that used robust accuracy metrics, which evaluate the proportion of correctly classified attacked data relative to the total data. The robust accuracy metric depends on the type and features of the adversarial attacks. However, our definition is based on the distance of data to the decision boundary and therefore depends only on the DNN classifier. To evaluate the fairness in robustness of the DNN classifier f, we define a metric for **fairness in robustness** as follows.

Definition 4. *(Metric for fairness in robustness) Let $\mathcal{P}$ be a partition of data set D and f be a DNN classifier on D. The metric defines as:*

$$V_f(\mathcal{P}) = Var\{Rob(P) : P \in \mathcal{P}\}. \tag{6}$$

where $Var\{.\}$ shows the variance of the robustness values. The smaller value of the $V_f(\mathcal{P})$, shows better "fairness in robustness" of the DNN classifier relative to the partition $\mathcal{P}$. Because the smaller values indicate that the distance between each sub-partition and the decision boundary is close to the average distance between all sub-partitions.

4 Fairness in Robustness: Calculation

The main challenge to evaluate the robustness of sub-partitions and the robustness of a DNN classifier is the calculation of $d_\theta(x)$; that is the distance of data point x to the decision boundary of the DNN classifier. For a linear classifier $f(x) = W^T x + b$ we can exactly compute this distance as:

$$d_\theta(x) = \frac{|f_{\hat{l}(x)}(x) - f_{\hat{k}(x)}(x)|}{||w_{\hat{l}(x)} - w_{\hat{k}(x)}||_2^2} \tag{7}$$

where $\hat{k}(x)$ is the ground label of x and

$$\hat{l}(x) = \underset{k \neq \hat{k}(x)}{\arg\min} \frac{|f_k(x) - f_{\hat{k}(x)}(x)|}{||w_k - w_{\hat{k}(x)}||_2^2}$$

The non-linear DNN classifiers have non-convex decision boundaries. Therefore, we do not have an exact formula for calculating $d_\theta(x)$. In this case, we approximate this distance by using iterative linearization of the decision boundary [28]. This algorithm modifies data values until the classification label is changed. In each iteration of the approximation algorithm, we compute the distance between the generated data point and the decision boundary. To compute the accurate distance, we consider r_{max} and r_{tot} as the distance to the decision boundary for the two last iterations (before changing the classification label and after that). We approximate $d_\theta(x)$ by averaging these two distances as:

$$d_\theta(x) = \frac{||r_{max}||_2 + ||r_{tot}||_2}{2} \tag{8}$$

5 Fair Adversarial Retraining to Mitigate Unfairness in Robustness (FARMUR)

We analyze the quantity of $V_f(\mathcal{P})$ to investigate the fairness in robustness for natural and adversarial training. Table 1 demonstrates that the $V_f(\mathcal{P})$ in natural training is smaller than in adversarial training. It shows that if we want to increase the robustness of the network through adversarial training, the fairness metric for robustness is reduced. More precisely, for a DNN classifier f and a partition $\mathcal{P}$ on the dataset, if we apply adversarial training on the model and obtain a new DNN classifier f', then we have $V_{f'}(\mathcal{P}) > V_f(\mathcal{P})$ (lower $V_f(\mathcal{P})$ shows better fairness in robustness) (see Table 1).

To improve fairness in robustness in the case of adversarial training, we propose the"Fair Adversarial Retraining" method. To this end, we first find the vulnerable and robust sub-partitions in the partition $\mathcal{P}$ of the dataset as follows.

Definition 5. *Let $\mathcal{P}$ and f be the partition of the dataset and the DNN classifier respectively. A sub-partition $P \in \mathcal{P}$ is vulnerable if $Rob(P) < Rob(f)$. The other sub-partitions are robust.*

After identifying the vulnerable and robust sub-partitions, we split the dataset D into two subsets, $D = D^{vul} \cup D^{rob}$. Then we apply "fair adversarial retraining" on these subsets with the following loss functions. For vulnerable subset (D^{vul}) we use TRADES [44] loss function as:

$$L_{TRADES} = \mathrm{E}_{(X,Y)\sim D^{vul}}[\mathcal{L}(f(X),Y) + \beta \max_{X' \in \mathcal{B}(X,\epsilon)} \mathcal{L}(f(X), f(X')] \quad (9)$$

where $\mathcal{B}(X,\epsilon)$ and β denote the neighborhood of the decision boundary of f and regularization hyperparameter. For robust subset (D^{rob}) we leverage natural loss function as:

$$L_{Natural} = \mathrm{E}_{(X,Y)\sim D^{rob}}[\mathcal{L}(f(X),Y)] \quad (10)$$

In our fair adversarial retraining method, we use the sum of these two loss functions to retrain the model.

$$L_{FARMUR} = L_{TRADES} + L_{Natural} \quad (11)$$

Figure 1 and Algorithm 1 demonstrate the main steps of our fair adversarial retraining method. The main aim of our method is to generate a DNN with high robustness and fairness in the robustness.

6 Experiments

6.1 Experimental Setup

Datasets and Partition To evaluate our method, we use MNIST [22], CIFAR-10 [20], and UTKFace [46] datasets. The UTKFace dataset is a large-scale face

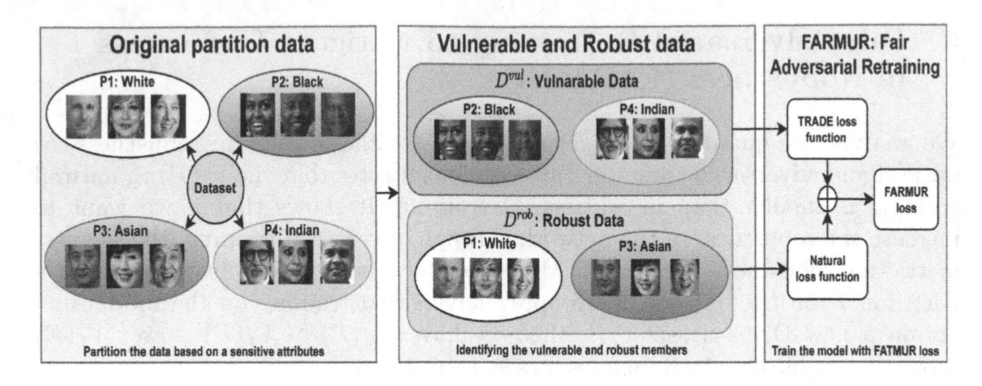

Fig. 1. The overview of FARMUR. It first finds the vulnerable and robust sub-partitions D^{vul}, D^{rob} and retrains the DNN based on the new loss function.

dataset with a long age span (ranging from 0 to 116 years old). It consists of 20,000 face images with annotations of age, gender, and race/ethnicity. Partitions in MNIST and CIFAR-10 are exactly the same as the labels in the datasets. However, in UTKFace we select the partition differently from the labels of the dataset. In this case, although the classifier has trained on the age labels, we select the race attribute for partitioning the data.

Models We use LeNet-5 [23] and AlexNet [19] to evaluate our method on the MNIST dataset. For CIFAR-10 and UTK-Face, we use ResNet18 [17] and UTK-Face classifier [30], respectively. For natural training, we use Stochastic Gradient Decent (SGD) for 120 epochs with a batch size of 64. For adversarial training, we use TRADES [44] algorithm. We set perturbation size $\epsilon = 0.031$, perturbation steps 0.007, learning rate 0.1, batch size 64, and run 120 epochs on the training dataset. For fair adversarial retraining, we first adversarially train the DNN model for 100 epochs. Then, we use the proposed FARMUR loss function on vulnerable and robust sub-partitions and we continue the training for 20 epochs with the same hyperparameters of adversarial training.

6.2 Experimental Results

Table 1 compares the performance of FARMUR against natural and adversarial training. In terms of fairness in robustness, we evaluate the natural, adversarial, and fair retraining (FARMUR) methods with the proposed metric which is defined in 4. Although adversarial training increases the robustness of the DNN in comparison with natural training ($Rob(f)$), it decreases the fairness in robustness (increase $V_f(\mathcal{P})$. It decreases the fairness in robustness by 1.46×, 6.67×, and 10.43× For MNIST (AlexNet), Cifar-10, and UTKFace datasets. FARMUR improves the fairness in robustness by 29.17% and 15.39%, and 19.18% In comparison with adversarial training for MNIST (AlexNet) and CIFAR-10 and UTK-Face datasets. The robustness of the DNN classifiers remains almost the same as

Algorithm 1 Fair Adversarial Retraining

Input: Adversarially trained DNN model f, Partition dataset $\mathcal{P}$, number of epochs T for retraining, learning rate γ, training dataset D.

Find vulnerable and robust data

Step 1: `Find vulnerable subset:` $D^{vul} = \{P \in \mathcal{P} : Rob(P) < Rob(f)\}$

Step 2: `Find robust subset:` $D^{rob} = \{P \in \mathcal{P} : Rob(P) > Rob(f)\}$

Step 3: `Split the dataset` D `into two subsets,` $D = D^{vul} \cup D^{rob}$.

Retrain DNN with new loss function

for `1:T` **do**

  for `mini-batches` $(b^{vul}, y^{batch}) \subseteq D^{vul}, (b^{rob}, y^{batch}) \subseteq D^{rob}$ **do**

    Step 1: `Calculate TRADES_loss for` (b^{vul}, y^{batch}) `by equation (9)`

    Step 2: `Calculate Natural loss for` (b^{rob}, y^{batch}) `by equation (10)`

    Step 3: `Calculate FARMUR loss based on equation (11)`

    Step 4: `Retrain model with loss in Step 3.`

  end for

end for

Output: Retrained model

the adversarial training in FARMUR retraining. It denotes that FARMUR not only improves the fairness in the robustness but also does not change significantly the robustness level of the DNN. To illustrate the effectiveness of FARMUR, we conduct some case studies on three different models and datasets. Figure 2 show the variation of $I_{\mathcal{P}}(\tau)$ with respect to τ for the DNNs trained with natural, adversarial, and FARMUR retraining methods. All the figures show that our method has a smaller variance in the robustness (lower $V_f(\mathcal{P})$) in comparison to the adversarial training method. For the UTKFace dataset, the last row of the figure determines that our proposed method can increase fairness in terms of robustness for different races. It means the attacker cannot attack some specific race more simply than others. FARMUR consists of two separate phases in terms of time complexity: finding vulnerable and robust data and retraining the DNN. On a single NVIDIA ®RTX A4000, identifying vulnerable and robust data for the UTKFace dataset requires approximately 6 GPU hours, and retraining the DNN requires approximately 4 GPU hours.

Table 1. The accuracy, robustness ($Rob(f)$), and Fairness in Robustness ($V_f(\mathcal{P})$) of different DNN classifiers on different datasets. The sensitive attributes for MNIST and Cifar-10 are same as the classification labels. For UTKFace dataset, the race attributes use for partition data.

		Accuracy (%) ↑			**Robustness ($Rob(f)$) ↑**			**Fairness in Robustness ($V_f(\mathcal{P})$) ↓**		
Model (f)	Dataset	Natural	Adversarial	FARMUR	Natural	Adversarial	FARMUR	Natural	Adversarial	FARMUR
Lenet-5	MNIST	99.08	98.74	98.91	0.79	1.26	1.24	0.014	0.019	0.009
AlexNet	MNIST	99.30	99.41	99.38	1.47	1.96	1.95	0.048	0.070	0.034
ResNet-18	Cifra-10	93.77	93.58	90.98	0.18	0.48	0.45	0.003	0.02	0.011
UTKClassifier	UTKFace	66.70	65.60	65.66	0.33	1.34	1.31	0.007	0.073	0.059

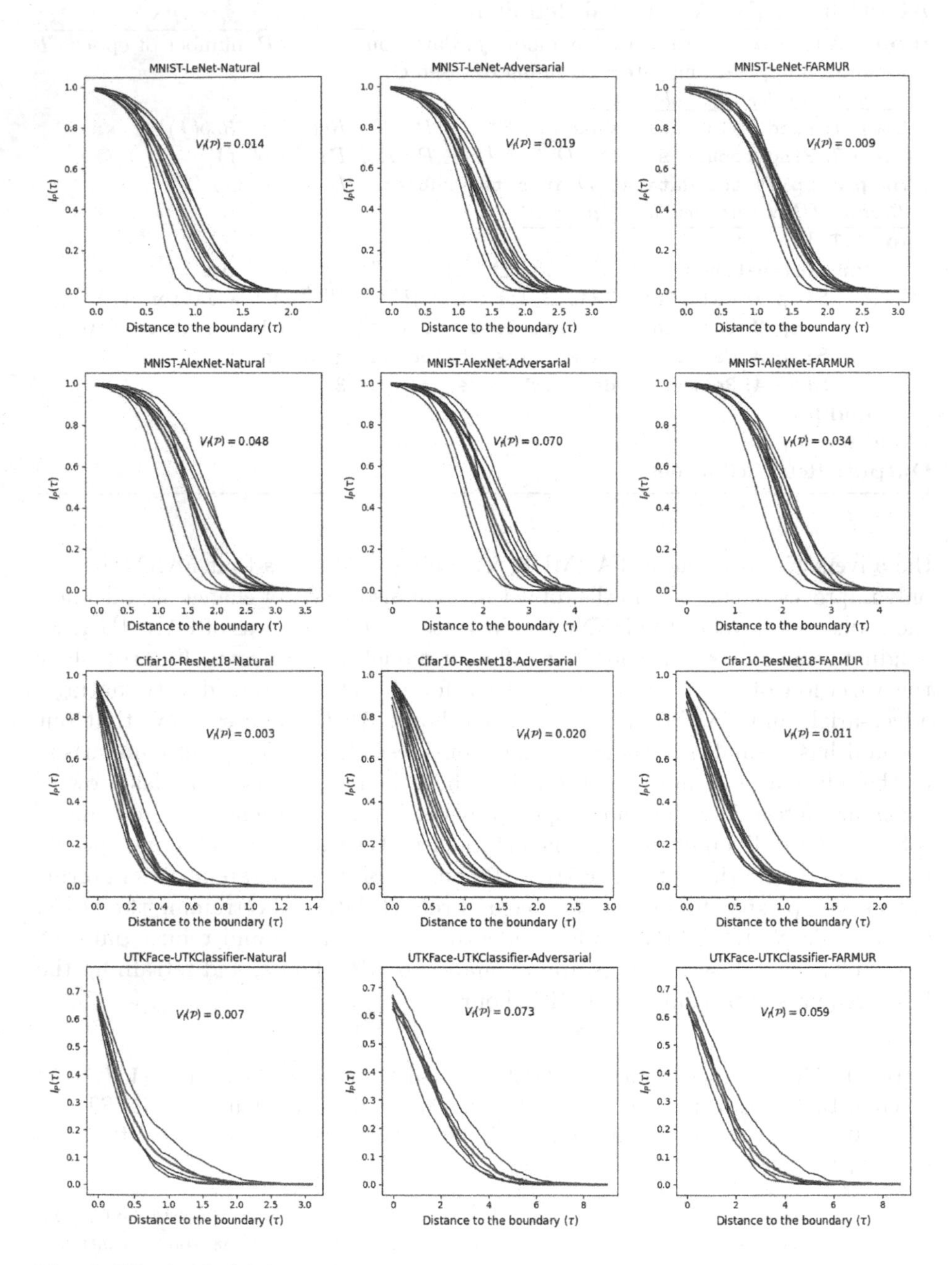

Fig. 2. We plot $I_P(\tau)$ for each sub-partition P in each dataset. Each blue line represents one sub-partition. The red line represents the mean of the blue lines. (Color figure online)

7 Conclusion and Future Work

In this paper, we introduce a new theory and metric to measure fairness in robustness based on the distance of the data to the decision boundary. Unlike other metrics, our metric is independent of the type of adversarial attacks and evaluates the fairness for the entire DNN model. Based on this metric, we demonstrate that the adversarial training methods reduce the fairness in robustness of the DNNs. Then we proposed FARMUR as a fair adversarial retraining method to mitigate unfairness in robustness of the adversarial training method. FARMUR tackles the unfairness issue in robustness by splitting the partitions into vulnerable and robust sub-partitions and retraining the model. Experimental results confirm the efficiency of our method to improve the fairness in the robustness of DNNs. For future work, we try to develop a new data evaluation mechanism to find appropriate data for adversarial training. In the other direction, we try to leverage neural architecture search to find DNNs with higher fairness in robustness.

Acknowledgement. This work was supported in part by the European Union through European Social Fund in the frames of the "Information and Communication Technologies (ICT) program" and by the Swedish Innovation Agency VINNOVA project "AutoDeep" and "SafeDeep"." The computations were enabled by the supercomputing resource Berzelius provided by National Supercomputer Centre at Linköping University and the Knut and Alice Wallenberg foundation.

References

1. Adel, T., Valera, I., Ghahramani, Z., Weller, A.: One-network adversarial fairness. In: Proceedings of the AAAI Conference on Artificial Intelligence, vol. 33, pp. 2412–2420 (2019)
2. Benz, P., Zhang, C., Karjauv, A., Kweon, I.S.: Robustness may be at odds with fairness: An empirical study on class-wise accuracy. In: NeurIPS 2020 Workshop on Pre-registration in Machine Learning, pp. 325–342. PMLR (2021)
3. Beutel, A., et al.: Putting fairness principles into practice: challenges, metrics, and improvements. In: Proceedings of the 2019 AAAI/ACM Conference on AI, Ethics, and Society, pp. 453–459 (2019)
4. Calmon, F., Wei, D., Vinzamuri, B., Natesan Ramamurthy, K., Varshney, K.R.: Optimized pre-processing for discrimination prevention. Adv. Neural Inf. Process. Syst. **30** (2017)
5. Carlini, N., Wagner, D.: Towards evaluating the robustness of neural networks. In: 2017 IEEE Symposium on Security and Privacy (SP), pp. 39–57. IEEE (2017)
6. Chouldechova, A.: Fair prediction with disparate impact: a study of bias in recidivism prediction instruments. Big Data **5**(2), 153–163 (2017)
7. Donini, M., Oneto, L., Ben-David, S., Shawe-Taylor, J., Pontil, M.: Empirical risk minimization under fairness constraints. arXiv preprint arXiv:1802.08626 (2018)
8. Du, M., Liu, N., Yang, F., Hu, X.: Learning credible deep neural networks with rationale regularization. In: 2019 IEEE International Conference on Data Mining (ICDM), pp. 150–159. IEEE (2019)

9. Du, M., Yang, F., Zou, N., Hu, X.: Fairness in deep learning: a computational perspective. IEEE Intell. Syst. **36**(4), 25–34 (2020)
10. Dwork, C., Hardt, M., Pitassi, T., Reingold, O., Zemel, R.: Fairness through awareness. In: Proceedings of the 3rd Innovations in Theoretical Computer Science Conference, pp. 214–226 (2012)
11. Dwork, C., Ilvento, C.: Fairness under composition. arXiv preprint arXiv:1806.06122 (2018)
12. Geraeinejad, V., Sinaei, S., Modarressi, M., Daneshtalab, M.: Roco-nas: robust and compact neural architecture search. In: 2021 International Joint Conference on Neural Networks (IJCNN), pp. 1–8. IEEE (2021)
13. Goodfellow, I.J., Shlens, J., Szegedy, C.: Explaining and harnessing adversarial examples. arXiv preprint arXiv:1412.6572 (2014)
14. Grgić-Hlača, N., Zafar, M.B., Gummadi, K.P., Weller, A.: Beyond distributive fairness in algorithmic decision making: feature selection for procedurally fair learning. In: Proceedings of the AAAI Conference on Artificial Intelligence, vol. 32 (2018)
15. Grigorescu, S., Trasnea, B., Cocias, T., Macesanu, G.: A survey of deep learning techniques for autonomous driving. J. Field Robot. **37**(3), 362–386 (2020)
16. Hardt, M., Price, E., Srebro, N.: Equality of opportunity in supervised learning. Adv. Neural Inf. Process. Syst. **29** (2016)
17. He, K., Zhang, X., Ren, S., Sun, J.: Deep residual learning for image recognition. In: Proceedings of the IEEE Conference on Computer Vision and Pattern Recognition, pp. 770–778 (2016)
18. Ilyas, A., Santurkar, S., Tsipras, D., Engstrom, L., Tran, B., Madry, A.: Adversarial examples are not bugs, they are features. arXiv preprint arXiv:1905.02175 (2019)
19. Krizhevsky, A.: One weird trick for parallelizing convolutional neural networks. arXiv preprint arXiv:1404.5997 (2014)
20. Krizhevsky, A., Hinton, G., et al.: Learning multiple layers of features from tiny images (2009)
21. Leben, D.: Normative principles for evaluating fairness in machine learning. In: Proceedings of the AAAI/ACM Conference on AI, Ethics, and Society, pp. 86–92 (2020)
22. LeCun, Y., Bottou, L., Bengio, Y., Haffner, P.: Gradient-based learning applied to document recognition. Proc. IEEE **86**(11), 2278–2324 (1998)
23. LeCun, Y., Haffner, P., Bottou, L., Bengio, Y.: Object recognition with gradient-based learning. In: Shape, Contour and Grouping in Computer Vision. LNCS, vol. 1681, pp. 319–345. Springer, Heidelberg (1999). https://doi.org/10.1007/3-540-46805-6_19
24. Liu, E.Z., et al.: Just train twice: improving group robustness without training group information. In: International Conference on Machine Learning, pp. 6781–6792. PMLR (2021)
25. Madry, A., Makelov, A., Schmidt, L., Tsipras, D., Vladu, A.: Towards deep learning models resistant to adversarial attacks. arXiv preprint arXiv:1706.06083 (2017)
26. Mehrabi, N., Morstatter, F., Saxena, N., Lerman, K., Galstyan, A.: A survey on bias and fairness in machine learning. ACM Comput. Surv. (CSUR) **54**(6), 1–35 (2021)
27. Miotto, R., Wang, F., Wang, S., Jiang, X., Dudley, J.T.: Deep learning for healthcare: review, opportunities and challenges. Brief. Bioinform. **19**(6), 1236–1246 (2018)
28. Moosavi-Dezfooli, S.M., Fawzi, A., Frossard, P.: Deepfool: a simple and accurate method to fool deep neural networks. In: Proceedings of the IEEE Conference on Computer Vision and Pattern Recognition, pp. 2574–2582 (2016)

29. Morgulis, N., Kreines, A., Mendelowitz, S., Weisglass, Y.: Fooling a real car with adversarial traffic signs. arXiv preprint arXiv:1907.00374 (2019)
30. Nanda, V., Dooley, S., Singla, S., Feizi, S., Dickerson, J.P.: Fairness through robustness: investigating robustness disparity in deep learning. In: Proceedings of the 2021 ACM Conference on Fairness, Accountability, and Transparency, pp. 466–477 (2021)
31. Paden, B., Čáp, M., Yong, S.Z., Yershov, D., Frazzoli, E.: A survey of motion planning and control techniques for self-driving urban vehicles. IEEE Trans. Intell. Vehicles **1**(1), 33–55 (2016)
32. Papernot, N., McDaniel, P., Wu, X., Jha, S., Swami, A.: Distillation as a defense to adversarial perturbations against deep neural networks. In: 2016 IEEE Symposium on Security and Privacy (SP), pp. 582–597. IEEE (2016)
33. Saha, D., Schumann, C., Mcelfresh, D., Dickerson, J., Mazurek, M., Tschantz, M.: Measuring non-expert comprehension of machine learning fairness metrics. In: International Conference on Machine Learning, pp. 8377–8387. PMLR (2020)
34. Schumann, C., Foster, J., Mattei, N., Dickerson, J.: We need fairness and explainability in algorithmic hiring. In: International Conference on Autonomous Agents and Multi-Agent Systems (AAMAS) (2020)
35. Speicher, T., et al.: Potential for discrimination in online targeted advertising. In: Conference on Fairness, Accountability and Transparency, pp. 5–19. PMLR (2018)
36. Tian, Q., Kuang, K., Jiang, K., Wu, F., Wang, Y.: Analysis and applications of class-wise robustness in adversarial training. arXiv preprint arXiv:2105.14240 (2021)
37. Tsipras, D., Santurkar, S., Engstrom, L., Turner, A., Madry, A.: Robustness may be at odds with accuracy. arXiv preprint arXiv:1805.12152 (2018)
38. Wang, J., Zhang, H.: Bilateral adversarial training: towards fast training of more robust models against adversarial attacks. In: Proceedings of the IEEE/CVF International Conference on Computer Vision, pp. 6629–6638 (2019)
39. Xie, C., Wang, J., Zhang, Z., Ren, Z., Yuille, A.: Mitigating adversarial effects through randomization. arXiv preprint arXiv:1711.01991 (2017)
40. Xu, H., Liu, X., Li, Y., Jain, A., Tang, J.: To be robust or to be fair: towards fairness in adversarial training. In: International Conference on Machine Learning, pp. 11492–11501. PMLR (2021)
41. Yuan, X., He, P., Zhu, Q., Li, X.: Adversarial examples: attacks and defenses for deep learning. IEEE Trans. Neural Netw. Learn. Syst. **30**(9), 2805–2824 (2019)
42. Zemel, R., Wu, Y., Swersky, K., Pitassi, T., Dwork, C.: Learning fair representations. In: International Conference on Machine Learning, pp. 325–333. PMLR (2013)
43. Zhang, D., Zhang, T., Lu, Y., Zhu, Z., Dong, B.: You only propagate once: accelerating adversarial training via maximal principle. arXiv preprint arXiv:1905.00877 (2019)
44. Zhang, H., Yu, Y., Jiao, J., Xing, E., El Ghaoui, L., Jordan, M.: Theoretically principled trade-off between robustness and accuracy. In: International Conference on Machine Learning, pp. 7472–7482. PMLR (2019)
45. Zhang, Z., He, Q., Gao, J., Ni, M.: A deep learning approach for detecting traffic accidents from social media data. Transport. Res. Part C: Emerg. Technol. **86**, 580–596 (2018)
46. Zhang, Z., Song, Y., Qi, H.: Age progression/regression by conditional adversarial autoencoder. In: Proceedings of the IEEE Conference on Computer Vision and Pattern Recognition, pp. 5810–5818 (2017)

How to Make the Most of Local Explanations: Effective Clustering Based on Influences

Elodie Escriva[1,2(✉)], Julien Aligon[2], Jean-Baptiste Excoffier[1], Paul Monsarrat[3,4,5], and Chantal Soulé-Dupuy[2]

[1] Kaduceo, Toulouse, France
[2] Université de Toulouse-Capitole, IRIT, (CNRS/UMR 5505), Toulouse, France
elodie.escriva@kaduceo.com
[3] RESTORE Research Center, Toulouse, France
[4] Artificial and Natural Intelligence Toulouse Institute ANITI, Toulouse, France
[5] Oral Medicine Department, Toulouse, France

Abstract. Machine Learning is now commonly used to model complex phenomena, providing robust predictions and data exploration analysis. However, the lack of explanations for predictions leads to a black box effect which the domain called Explainability (XAI) attempts to overcome. In particular, XAI local attribution methods quantify the contribution of each attribute on each instance prediction, named influences. This type of explanation is the most precise as it focuses on each instance of the dataset and allows the detection of individual differences. Moreover, all local explanations can be aggregated to get further analysis of the underlying data. In this context, influences can be seen as new data space to understand and reveal complex data patterns. We then hypothesise that influences obtained through ML modelling are more informative than the original raw data, particularly in identifying homogeneous groups. The most efficient way to identify such groups is to consider a clustering approach. We thus compare clusters based on raw data against those based on influences (computed through several XAI local attribution methods). Our results indicate that clusters based on influences perform better than those based on raw data, even with low-accuracy models.

Keywords: Explainable Artificial Intelligence (XAI) · Instance clustering · Prediction explanation · Machine learning explanation

1 Introduction

Analytic and predictive tools are now commonly based on Machine Learning (ML) methods and used in sensitive domains such as healthcare, finance, insurance, banking and chemistry. These methods give a prediction for a single instance based on its data, which often creates a black-box effect as methods do not inherently explain their decision process [11]. Explanation methods have therefore been perfected, providing global insights about the model's general

A. Abelló et al. (Eds.): ADBIS 2023, LNCS 13985, pp. 146–160, 2023.
https://doi.org/10.1007/978-3-031-42914-9_11

behaviour or a local one about a single situation [9] (XAI for eXplainable Artificial Intelligence). Local explanations are increasingly used in AI-assisted tools to offer more information than a single prediction [1]. Among the most popular methods are XAI local attribution methods that produce influences, especially LIME [13] and approximation of Shapley Value such as SHAP [12], the K-depth [7] and Coalitional approaches [8]. Their popularity is due to the instance-level accuracy of these explanations, which links the impact of each attribute to the prediction made for each instance and allows finer differences to be detected between all instances. Yet, providing only local influences seems insufficient to improve decision-making efficiency. Indeed [16,17] show that displaying influences along with an individual prediction did not significantly enhance the utility and understanding for the user as opposed to prediction alone. Moreover, knowing all the local explanations of a dataset does not guarantee a complete data understanding since there are as many explanations as instances in the original raw dataset, with the difficulty of finding explainability patterns in this new dataset.

In this context, we hypothesise that influences can be seen as a new data space that can be explored and used as a basis for further analysis. Indeed, influences provide new information thanks to ML modelling, which considers complex phenomena and interactions. Influence analysis can thus help identify the main trends of explanations, i.e. the characteristic relationships between the attributes. Also, it can be interesting to provide a global view of the explanations to determine whether instances are typical or atypical cases of the data. In this direction, an influence-based clustering approach is a good candidate since it can be the most straightforward approach to detect more homogeneous subgroups of influences and understand the behaviour of the modelling and the underlying dataset. Thus, in this paper, we want to propose a framework for analysing influences through a clustering approach. To the best of our knowledge, this is the first work that studies in a general framework the benefits of using local influences as a new input for clustering to identify more informative and homogeneous groups. We also explored the robustness of this framework regarding low-accuracy models or misclassified instances.

The paper is organised as follows. Section 2 gives an overview of the current local explanation methods used in experiments and how explanations are used to detect subgroups of instances. Then, Sect. 3 details our clustering framework for detecting subgroups based on local influences. Section 4 describes the experiments performed on 104 datasets to compare the use of raw data and influences from multiple local XAI methods. We study the K-medoid clusters quality to show the efficiency of using influences. We detail the metrics used to evaluate the clusters and the different approaches based on the model prediction. Globally, our results demonstrate that local influences produce better-quality clusters than raw data, even with low-accuracy models. Separating instances well classified and misclassified by the model also allows a more precise clustering. Section 5 discusses the advantages of our approach in a broader context, linking results from clustering with knowledge from modelling and explanation methods. Section 6 concludes this paper and gives short and long-term perspectives of works.

2 Related Works

In the field of local explanations, one of the first methods was based on the Shapley values, a local attribution XAI method [18], to explain machine learning predictions. With these methods, the influence of each attribute over a prediction is computed as the difference in prediction from the model with and without the attribute. Influences then represent the impact of each attribute over a prediction for each instance of the dataset. Local influences facilitate the prediction understanding without expert data science knowledge as they are easy to interpret and represent graphically. Other methods have emerged with LIME [13] that uses linear surrogate models trained with sampled data to approximate the black box model locally. The Coalitional approaches [8] approximate the Shapley value by precomputing relevant groups of instances and reducing complexity. Finally, SHAP [12] mixes Shapley values with LIME and other methods to simulate the absence of attributes by sampling, find a linear model that explains the black-box model locally and approximate the Shapley values. Nowadays, SHAP is one of the most well-known methods in the literature, easy to use and provides both agnostic and specific methods with KernelSHAP or TreeSHAP.

With the rise of explainability, ML research looks beyond simply explaining the machine learning model. Several papers in the last year have covered use cases combining machine learning explainability and clustering to find relationships between instances [3,10]. Based on a COVID-19 dataset, [3] tries to identify better clusters based on KernelSHAP values. Rather than clustering the original dataset, called raw data, they trained a classification model, computed the KernelSHAP values for each instance and performed DB-SCAN clustering on these influences. They show better identification of clusters with influences than raw data and graphically display the cluster differences using UMAP, a well-known reduction dimension technique. Other papers also used clustering to determine groups and to recommend instances based on the influences on a single dataset [5,6]. [5] was a use case on a urinary disease that explores healthcare risk stratification based on influences from TreeSHAP. Clustering patients by SHAP values allows the selection of representative patients and investigation of the risk factors for each cluster, where only raw data are insufficient to perform the same analysis. The same kind of analysis was performed on a COVID-19 dataset concerning the identification of subgroups of patients during the first lockdown in France [6]. These four papers explored the idea of using influences and clustering to find more knowledge about the data on specific medical examples. However, no paper formally evaluates the contribution of explanation clustering in general. Although their positive conclusions, these papers only use one single dataset with one XAI method, without generalizing the approach or comparing findings with other XAI local attribution methods, in opposition to what we propose in this paper. Finally, none uses prediction to differentiate subgroups of data for clustering.

3 Influence-Based Clustering Framework

In this section, we detail our influence-based clustering framework. Figure 1 shows the step-by-step process to cluster instances based on their influences:

1. A machine learning model is trained with raw data and predicts classes of all the instances from the raw dataset.
2. A local attribution XAI method explains the trained model. Users can choose the data used as input for the method. Influences are computed to explain why the ML model made such predictions.
3. A clustering algorithm is used on influences to create homogeneous groups of instances to detect their important attributes based on the modelling. Users can define the number of clusters they want to compute.

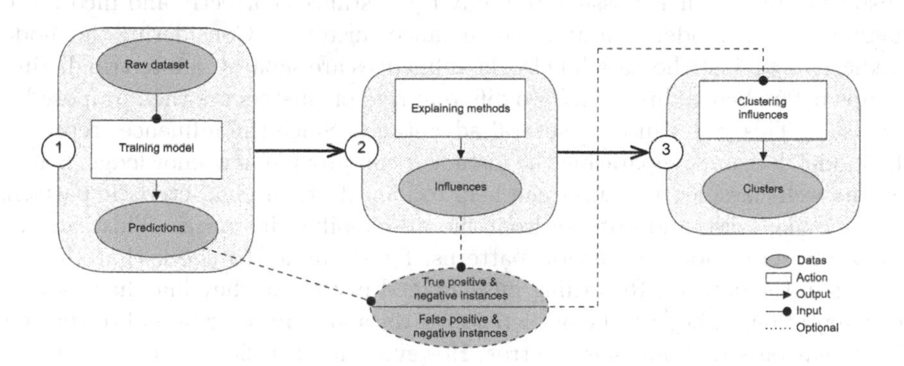

Fig. 1. Our proposal Framework.

In this framework, various elements can be modified according to user preferences. Any classification model can be used in Step 1, as they are all designed to compute predictions, and Step 3 allows any clustering method.

In step 2, the framework is designed to accept XAI local attribution methods. These influences are represented as tabular data, where each instance has a value associated with each attribute. We directly use these influences data as input for the clustering step. Influences are valuable because they provide additional information that the raw data does not: the link between the modelling predictions and the dataset attributes. Compared to raw data, explanations produced by XAI local attribution methods have the same unit across all attributes, thus avoiding any problem of value ranges. Another advantage is that influence values are less noisy since the ML model mainly focuses on attributes relevant to the underlying predictive task and excludes information not explained by the complex attributes interaction, hence the relevance of carrying out clustering. For supervised tasks, XAI local attribution methods usually generate a dataset for each class with identical dimensions as the raw data. For example, if the raw data consists of n instances and m features and the supervised task is a multi-class problem with c classes, the generated dataset (also called the influence dataset) has a $n \times m \times c$ dimension. To have an influence dataset with the same dimension as raw data ($n \times m$) one can only select a single class and its associated influences. For example, regarding binary classification, the positive class is often chosen as the class of interest for influences. Finally, XAI local attribution methods allow the selection of different data as input than the ones used for training the machine learning model. These inputs are used by XAI methods to explain the model, by creating perturbations in SHAP or LIME, for example.

An additional and optional step is to select a particular subset of the data for clustering. Indeed, it is possible to study the instances correctly and incorrectly classified by the model separately via instance clustering. Considering the model predictions against the data labels, the influences are separated into two distinct groups before being clustered. Two different sets of clusters are then proposed to the users. This step can have several advantages. Since the influences represent the model decisions, separating the instances can provide new knowledge. Studying the well-classified instances can help to identify their characteristic patterns by removing noise and outliers from the misclassified instances. This can give a more accurate idea of general patterns, for example, to check that there is no bias in the dataset. Regarding misclassified instances, they may have several representations. They can be outliers in the data and not correspond to the general behaviours without bias or error. However, misclassified instances can also be a particular subgroup of the data relevant to study. For example, this would be the case of children with cancers usually associated with older people. Due to age, the model may misunderstand this subgroup, as there are few children with non-pediatric cancers, or the input variables may be insufficient to identify this subgroup. However, it is necessary to study this subgroup to understand whether there is any specific behaviour in this subgroup and ultimately understand the overall dataset. Separating the instances can therefore allow the exploration of new patterns that can be invisible if all the data were kept. This may be even more important for influences because of their direct link to the model. Indeed, when the model prediction is incorrect, the influences reflect this error and are directly impacted by the wrong prediction of the model.

The full implementation of our proposal is available here: https://github.com/kaduceo/XAI-based-instance-selection. The source code will evolve with future works. Additional materials are also available.

Table 1. Statistics of the experimental datasets based on the number of attributes.

# of attributes	4	5	6	7	8	9
# of datasets	14	25	17	16	15	17
Mean # of insts	465	1197	654	554	650	503
Min-Max # of insts	125–1372	100–7129	100–3107	108–4052	130–4177	100–1473

4 Experiments

4.1 Experimental Protocol

For our evaluations, we use 104 datasets from an Open ML collection[1] [14] that meet the following criteria: binary classification, more than 100 instances, more than four attributes and at most nine attributes due to the computational cost of producing influences. Table 1 details statistics about the datasets used.

Binary classification is chosen to facilitate the interpretation of influences. We consider that all influences are based on class 1. In this case, influences represent the impact of each attribute on the probability of the instance being in class 1. We train a Random Forest model (RF) with a Grid Search Cross-Validation to optimise hyperparameters. This model was chosen to test tree-specific explanation methods while keeping a limited number of hyperparameters to avoid overfitting (compared to boosted trees). Only to evaluate the performances of the modelling, each dataset is divided into train and test sets according to the 75%/25% ratio. Table 2 shows the performances of all the models trained in our experiments. Models are trained adequately to capture most information of the dataset. The mean and median balanced accuracy are respectively 0.79 and 0.85, meaning most models can accurately classify test instances. Some models also have very low accuracy, the minimum being 0.42. We choose to separate models based on a threshold set to 0.8 to evaluate the behaviour of our framework on models with high and low accuracy. Thus, high-accuracy models have a median balanced accuracy of 0.92, whereas low-accuracy models have a median of 0.6.

We also study the number of instances well classified and misclassified by the ML modelling in Table 2. In all experiments, we call *true instances* well-classified instances, referring to True positive and True negative terms. *False instances* is then related to False positive and False negatives instances, so misclassified instances. We use three different separations of data: all instances together, only true instances and only false instances. As we separate true and false instances, we choose not to evaluate high-accuracy models on false instances as there are not enough instances in most datasets to create clusters and properly evaluate them and compare the results. Then, when studying false instances, we only work with models with low accuracy as the number of false instances is higher and sufficient. Also, the number of true instances is adequate to perform clustering for all models.

[1] Available in https://www.openml.org/s/107/tasks.

Table 2. Statistics of models trained. Balanced accuracy and percentages of true and false instances are presented for the 104 datasets, and separataly based on the 0.8 accuracy treshold. For true and falses instances, the median number of instances is presented along the percentage.

	Balanced Accuracy			% of True instances			% of False instances		
Models (#)	Median	Min	Max	Median	Min	Max	Median	Min	Max
All (104)	0.85	0.42	1.0	94% (307)	61%	100%	6% (21)	0%	39%
Acc > 0.8 (60)	0.92	0.81	1.0	97% (404)	85%	100%	3% (11)	0%	15%
Acc < 0.8 (44)	0.60	0.42	0.79	82% (252)	61%	98%	18% (62)	2%	39%

For exhaustive purposes, we choose three different XAI local attribution methods to compute influences: KernelSHAP, TreeSHAP, LIME and Spearman coalitional. As explained in [4], each XAI method provides influences with different strengths and disadvantages. Thus, we want to study the relevance of using local influence clustering compared to raw clustering in a global way.

Once influences are computed, instances are clustered by the influence-based approach with K-medoids as the clustering method. This method has the advantage of always selecting actual instances as centroid from the dataset, unlike other clustering methods like *k-means* where centres are not necessarily existing instances. Metrics to compute distances, so clusters, can be selected arbitrarily, with the Euclidean distance being the usually chosen distance. As both raw data and influences data are tabular data of the same dimensions, the distance metrics can be easily applied to both datasets to compute clusters without adapting the clustering method to a specific input. We use ten different percentages to choose the number of clusters: 1%, 2%, 3%, 4%, 5%, 10%, 20%, 30%, 40% and 50%. We define the number of clusters based on the percentage as $n_{cluster} = p * n_{instances}$ with p the selection percentage between 0 and 1. We set a minimum number of two to avoid too few clusters. As the size of the datasets varies greatly as shown in Table 1, we prefer to select a percentage rather than fixed numbers of instances to take into account the diversity of the datasets. As we aim to show how clustering on influences exhaustively performs against the raw data, multiple percentages per dataset can show how cluster quality evolves without looking for the optimal number of clusters (which may also be different for each method).

Finally, we evaluate if clusters are well defined and manage to group similar instances and separate dissimilar instances based on their *a-priori* labels. We select two external clustering metrics, *Entropy* and *Purity*. With external metrics, class labels are needed as metrics assess the distribution of labels within clusters to evaluate how clusters and labels are related and how clusters manage to group similar instances. Entropy measures the distribution of labels in a cluster, i.e. the ability of the algorithm to differentiate between data that do not have the same "real" class. A perfect entropy means all instances from the same class are in the same clusters. In addition, Purity measures the relative size of the majority class in a cluster to evaluate its dominance over other classes. Perfect purity describes that each cluster has only one class. These two metrics give values between 0 and 1. A perfect clustering will usually have an entropy equal to 0 and a purity equal to 1. These metrics are defined as follows [2]:

$$Entropy = \sum_{k=1}^{K} \frac{n_k}{n} \Big(- \frac{1}{\log q} \sum_{i=1}^{q} \frac{n_k^i}{n_k} \log \frac{n_k^i}{n_k} \Big) \qquad Purity = \sum_{k=1}^{K} \frac{1}{n} \max_i (n_k^i)$$

where C_k is a particular cluster of size n_k, q is the number of class in the dataset, K the number of clusters and n_k^i is the number of instances of the ith class assigned to the kth cluster.

4.2 Results

In this section, we describe the results of the experiments by first comparing the clusters based on the influences from XAI methods with the ones made with raw instances. We then study the impact of each data subgroup on the cluster quality for KernelSHAP and Spearman coalitional. For all experiments, results are presented separately based on the machine learning model's accuracy to differentiate the impact of the model performance on the influences and clustering.

Comparing Raw Data and XAI Influences Clustering. When comparing raw data clusters to the influence ones, for all instances, Fig. 2 shows raw data clusters have lower purity and greater entropy than other clusters, regardless of the percentages, the XAI methods or the model performance. Differences in entropy between clusters from raw data and influences are even greater when the model has an accuracy greater than 80%. Clusters from raw data have poorer quality than clusters from influences, indicating that clustering instances based on their influences from XAI methods gives better results than clustering the raw data. Also, as expected, when models have a lower accuracy, clusters have a lower purity and entropy, whatever the data or cluster percentages. Indeed, when the model's performance is poor while the model is adequately trained, this may indicate that the data is less generalisable or of lower quality. This hypothesis seems to be reflected in the quality of the clusters created.

When taking into account only the true instances (the instances well predicted by the model), Fig. 3a shows similar results as Fig. 2: clusters based on influences have better quality than the ones based on raw data (for all XAI methods, percentages of selection and model accuracy). The purity and entropy are almost perfect, even with low selection percentages. Clusters have also better quality with only the true instances than with all the instances of the dataset. Figure 3b only considers the false instances (instances misclassified by the model). Globally, the cluster quality is degraded, especially the purity. Since purity checks the proportion of the majority class in each cluster, grouping instances misclassified by the model logically lower the cluster purity. No XAI method seems to have a good result on small percentages, even if they all have

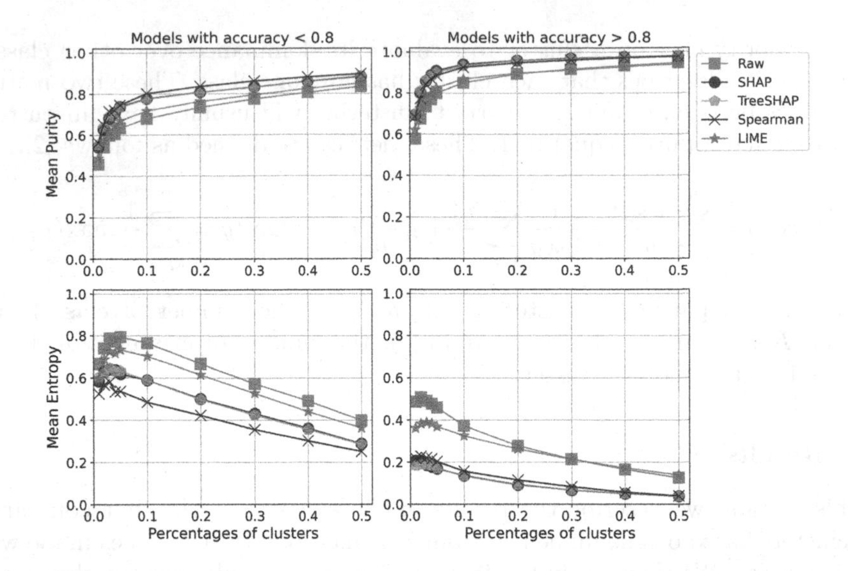

Fig. 2. Comparison of clustering for XAI methods trained on all instances.

better results than raw clustering. With false instances, we analyse cases where the model fails to generalise or describe the data correctly. As influences represent the model decision, influences of misclassified instances may have lower quality than true instances. They may, however, be representative of why the model does not generalise and understand these data. Thus, these clusters can indicate where the problems lie in the data or the model.

Moreover, even if this is not the aim of this paper, we can briefly compare XAI methods between them based on cluster quality. Although all are attribution methods with similar global behaviour, the calculated influences appear to be sufficiently different to produce dissimilar cluster results, especially in entropy. Spearman coalitional seems slightly better at clustering false instances on models with low accuracy. Clusters based on LIME have purity and entropy close to the clusters based on raw data, making this XAI method the one with the worst results. Based on the subgroup of data studied, one method may be preferable to another depending on the context. This seems consistent with the findings of [4], where depending on the dataset, the interdependence of attributes, the dimensionality or the model, one XAI method can be more efficient than others. The same reasoning seems to apply here, where according to the subgroup studied, one XAI method can be better than the others.

In the next sub-section, we select only KernelSHAP and Spearman coalitional to study the impact of using only specific subgroups of data, as TreeSHAP is almost identical as KernelSHAP and LIME have the worst results.

Comparing the Impact of Using Different Data Subgroups. In this subsection, we aim to show in which circumstances well classified or misclassified

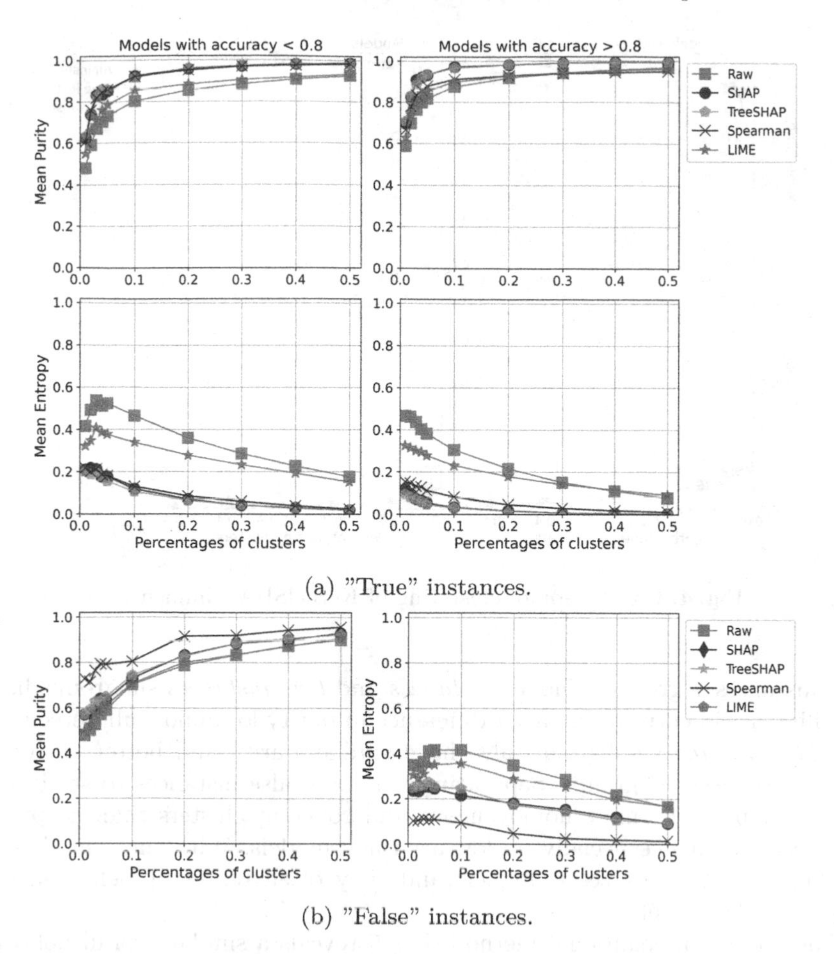

(a) "True" instances.

(b) "False" instances.

Fig. 3. Comparison of clustering for XAI methods trained on (a) only "true" instances and (b) only "false" instances for models with an accuracy below 0.8.

instances can be used to produce clusters of good quality (or not), notably in the worst case (degraded accuracy on a set of misclassified instances). Figure 4 and 5 show the cluster quality for the three data modalities, with influences respectively from KernelSHAP and Spearman coalitional.

Figure 4 shows little difference in cluster quality between *all instances* and *true instances* subgroups for models with high accuracy. Purity metric is high and almost equal for both modalities, and the *all instances* subgroups have slightly higher entropy. Influences from true instances produce almost perfect clusters even with low cluster percentages and are little affected by the model accuracy. As models with high accuracy have fewer false instances, their influences may only produce noises for the clustering. Removing them give slightly better global results, as clusters have better entropy. For models with low accuracy, there are more differences between the subgroups, presumably because the proportion of

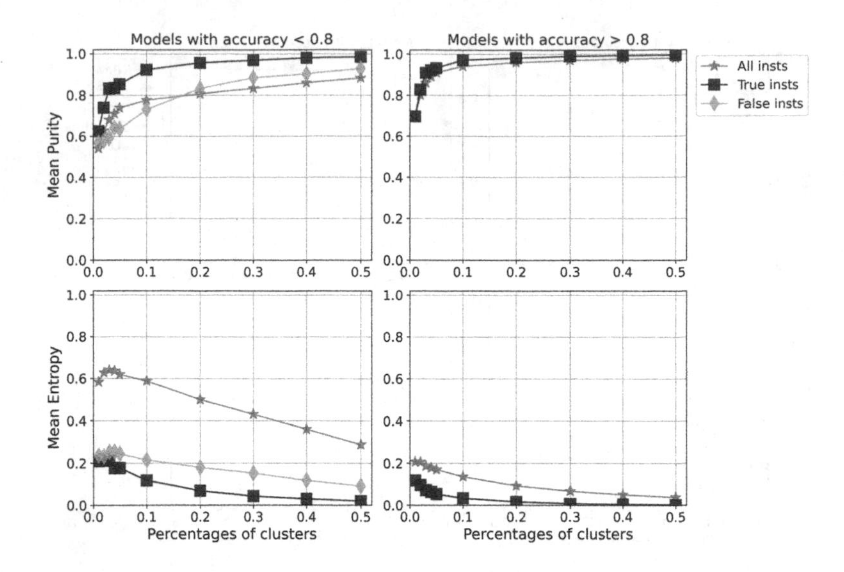

Fig. 4. Comparison of clustering of KernelSHAP influences.

false instances is greater. The *all instances* and *true instances* subgroups have a 0.4 difference in entropy and a 0.1 difference in purity for almost all percentages. The *false instances* subgroups also have similar purity and better entropy as the *all instances* subgroups. Separating true and false instances to study them separately produces more homogeneous and coherent clusters than keeping all instances together, especially on low-accuracy models. With these models, the number of false instances is higher, and they often represent behaviours not caught by the model.

For Spearman coalitional method, Fig. 5 reveals a similar overall behaviour to KernelSHAP regarding the cluster quality depending on the subgroups, especially on high-accuracy models and on the *true instances* subgroups. However, for low accuracy models and unlike KernelSHAP, there are some differences when using only false instances. The *false instances* subgroups have slightly higher purity and lower entropy, especially on low percentages. The different use of input data by both methods can explain this behaviour. KernelSHAP use the input to produce perturbations for the model, creating new instances and studying a larger area of the data space than just the input data (here, the false instances). In contrast, Spearman coalitional does not produce any perturbations and uses the input data as is to explain the model. The data space is then smaller, therefore, less exhaustive. Using only false instances may lead to influences more precise for this subgroup, compared to using all instances or instances with perturbations, hence the difference between the two subgroups for Spearman coalitional and the difference with KernelSHAP. Moreover, for low-accuracy models, clusters from *true instances* and *false instances* subgroups are better than the clusters from all instances.

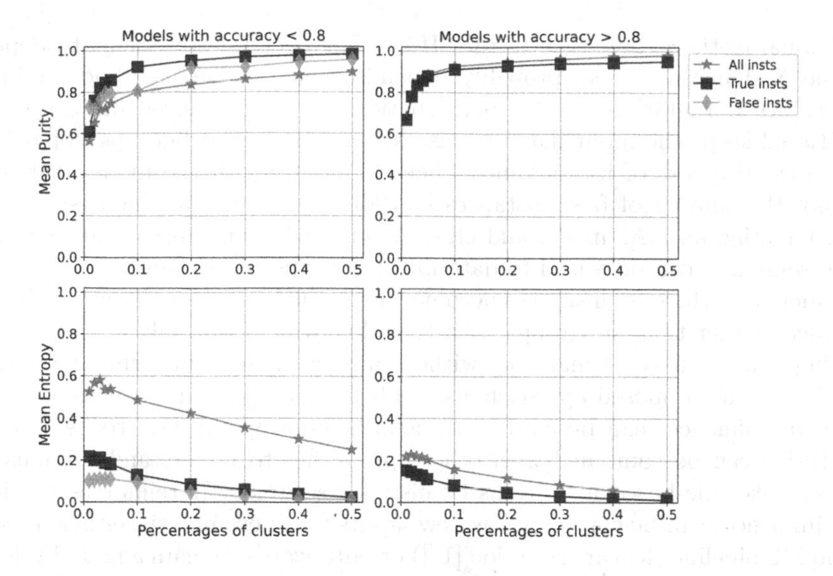

Fig. 5. Comparison of clustering of Spearman coalitional influences.

These two figures show that different XAI methods can lead to clusters with distinct qualities or behaviour based on the data subgroups selected. These methods can produce diverse and meaningful clusters to understand the modelling and dataset.

5 Discussion

Clustering on XAI influences showed better results than clustering on raw data, regardless of the percentage of clusters, the XAI method or the performance of the modelling. The influences seem to contain information allowing a better clustering, probably by highlighting the most significant attributes for each instance or removing noises from raw data. This finding seems consistent with the results of [3] while showing a more global approach, working with other XAI methods than KernelSHAP and a hundred of datasets.

Separating the instances correctly and incorrectly classified by the model also seems to give better results than keeping all the instances together. Since the information in the two subgroups is different, they each seem to create noise in the information of the other subgroup. Indeed, the misclassified instances are often outliers or critical instances in the dataset. Their behaviour is different from the general behaviour of the data, whereas correctly classified instances follow the behaviour that the model detects. However, as some misclassification may result from bias in a subgroup of the data or from the atypical behaviour of that subgroup compared to the whole dataset, it is of great interest to study them as a priority. When separating correctly and incorrectly classified instances, the differences in cluster quality seem to be more pronounced with the Spearman coalitional method than with KernelSHAP. The contribution seems to depend on the XAI method used, probably depending on the XAI method for influences since KernelSHAP creates perturbations on the instances and Spearman coalitional keeps the input data as it is. A limit to these subgroups' separation is also the decrease of its relevance when the accuracy of the model increases. Indeed, the number of false instances logically decreases with increasing accuracy. Creating an XAI model and clusters with a low instance count does not make sense and can only lead to data misunderstanding. However, as the accuracy increases, the false instances become mostly outliers of the dataset or biased instances rather than subgroups with their behaviours to analyse. Their small number can be analysed manually without any particular clustering method.

Finally, the proposed approach also adds another use of influences. Clusters based on influences can be used to focus on sub-groups of data to be studied. Clustering can be combined with other approaches to understand the clusters created, like rule-based algorithms or instance selection. It reinforces the idea that influences can be considered as new inputs for finer analysis, either directly in the ML pipeline (feature selection [15]) or, afterwards, to gain a more in-depth and concise understanding of the ML model and the underlying data. Examples of how explanation clusters can be used are available on the GitHub mentioned above.

6 Conclusion and Perspectives

We propose in this paper a general framework for clustering instances based on influences and predictions. We combine XAI local attribution methods with clustering to explore the space of influence data. We provide clusters of similar instances to assist in analysing modelling and dataset. Experiences validate the valuable contribution of influence-based clustering. The clusters from the influence-based framework are more homogeneous and of better quality, providing insight into the modelling. We also prove that the clusters formed are of good quality and pertinent, even for low-performance models. We also show the advantages of splitting the well- and misclassified instances by the model when studying a dataset as a whole, as it highlights the most important subgroups of data and the behaviour of outliers simultaneously.

Perspectives will first be focused on extending our approach for other supervised tasks. Clusters can also help select informative instances and provide a small number of instances to users. These instances can help to understand datasets and modelling using examples rather than statistical information. With users in the loop, the framework with instance selection added could be tested against example-based XAI methods. New information on the dataset and its subgroups may also provide feedback on the quality of the training data or the trained model to improve it. This idea of possible user feedback may be one way to improve data quality and modelling. Clustering based on influences may help to understand *why* the model is wrong and not just *where* the model is wrong, and allow for detecting bias in the data. Perspectives will then be focused on evaluating this feedback and how it can be implemented in the framework. A long-term perspective is to use the framework in a complete system where users can interact with the modelling and define typical instances to profile new data patterns for user testing.

Acknowledgment. We thank the French ANRT and Kaduceo company for providing us with PhD grants (no. 2020/0964).

References

1. Antoniadi, A.M., et al.: Current challenges and future opportunities for XAI in machine learning-based clinical decision support systems: a systematic review. Appl. Sci. **11** (2021)
2. Conrad, J.G., Al-Kofahi, K., Zhao, Y., Karypis, G.: Effective document clustering for large heterogeneous law firm collections. In: AIL Proceedings (2005)
3. Cooper, A., Doyle, O., Bourke, A.: Supervised clustering for subgroup discovery: an application to covid-19 symptomatology. In: ECML-PKDD Proceedings (2021)
4. Doumard, E., Aligon, J., Escriva, E., Excoffier, J., Monsarrat, P., Soulé-Dupuy, C.: A comparative study of additive local explanation methods based on feature influences. In: DOLAP Proceedings (2022)
5. Excoffier, J.B., Escriva, E., Aligon, J., Ortala, M.: Local Explanation-Based Method for Healthcare Risk Stratification. In: Medical Informatics Europe 2022. Studies in Health Technology and Informatics (2022)
6. Excoffier, J.B., Salaün-Penquer, N., Ortala, M., Raphaël-Rousseau, M., Chouaid, C., Jung, C.: Analysis of covid-19 in patients in France during first lockdown of 2020 using explainability methods. Med. Biol. Eng. Comput. **60** (2022)
7. Ferrettini, G., Aligon, J., Soulé-Dupuy, C.: Improving on coalitional prediction explanation. In: ADBIS Proceedings (2020)
8. Ferrettini, G., Escriva, E., Aligon, J., Excoffier, J.B., Soulé-Dupuy, C.: Coalitional strategies for efficient individual prediction explanation. Springer, ISF (2021)
9. Kaur, H., Nori, H., Jenkins, S., Caruana, R., Wallach, H., Wortman Vaughan, J.: Interpreting interpretability: understanding data scientists' use of interpretability tools for machine learning. In: CHI Proceedings (2020)
10. Lee, K., Ayyasamy, M.V., Ji, Y., Balachandran, P.V.: A comparison of explainable artificial intelligence methods in the phase classification of multi-principal element alloys. Sci. Rep. **12** (2022)
11. Lipton, Z.C.: The mythos of model interpretability. In: Machine Learning, the Concept of Interpretability is Both Important and Slippery. Queue, vol. 16 (2018)
12. Lundberg, S.M., Lee, S.I.: A unified approach to interpreting model predictions. In: NeurIPS Proceedings (2017)
13. Ribeiro, M.T., Singh, S., Guestrin, C.: "Why should I trust you?": explaining the predictions of any classifier. In: KDD Proceedings (2016)
14. Vanschoren, J., van Rijn, J.N., Bischl, B., Torgo, L.: Openml: networked science in machine learning. In: SIGKDD Explorations, vol. 15 (2013)
15. Wang, H., Doumard, E., Soulé-Dupuy, C., Kémoun, P., Aligon, J., Monsarrat, P.: Explanations as a new metric for feature selection: a systematic approach. IEEE J. Biomed. Health Inf. (2023)
16. Weerts, H.J., van Ipenburg, W., Pechenizkiy, M.: A human-grounded evaluation of shap for alert processing. arXiv preprint arXiv:1907.03324 (2019)

17. Zhang, Y., Liao, Q.V., Bellamy, R.K.: Effect of confidence and explanation on accuracy and trust calibration in AI-assisted decision making. In: FAccT Proceedings (2020)
18. Štrumbelj, E., Kononenko, I.: Explaining prediction models and individual predictions with feature contributions. Knowl. Inf. Syst. **41** (2014)

Data and Metadata Quality

Exploring Heterogeneous Data Graphs Through Their Entity Paths

Nelly Barret[1(✉)], Antoine Gauquier[2], Jia Jean Law[3], and Ioana Manolescu[1]

[1] Inria and Institut Polytechnique de Paris, Palaiseau, France
{nelly.barret,ioana.manolescu}@inria.fr
[2] Institut Mines Télécom, Paris, France
antoine.gauquier@etu.imt-nord-europe.fr
[3] Ecole Polytechnique, Palaiseau, France
jia-jean.law@polytechnique.edu

Abstract. Graphs, and notably RDF graphs, are a prominent way of sharing data. As data usage democratizes, users need help figuring out the useful content of a graph dataset. In particular, journalists with whom we collaborate [4] are interested in identifying, in a graph, the *connections between entities*, e.g., people, organizations, emails, etc. We present a novel, interactive method for exploring data graphs through *their data paths connecting Named Entities* (NEs, in short); each data path leads to a tabular-looking set of results. NEs are extracted from the data through dedicated Information Extraction modules. Our method builds upon the pre-existing ConnectionLens platform [4,5] and follow-up work on dataset abstraction [8,9]. The contribution of the present work is in the interactive and efficient approach to enumerate and compute NE paths, based on an algorithm which automatically recommends subpaths to materialize, and rewrites the path queries using these subpaths. Our experiments demonstrate the interest of NE paths and the efficiency of our method for computing them.

Keywords: Data graphs · Graph exploration · Information Extraction

1 Motivation and Problem Statement

Data graphs, including RDF knowledge graphs, as well as Property Graphs (PGs), are often used to represent data. More broadly, *any structured or semi-structured dataset can be viewed as a graph*, having: (*i*) an internal node for each structural element of the original dataset, e.g., relational tuple, XML element or attribute, JSON map or array, URI in an RDF graph; (*ii*) a leaf node for each value in the dataset, e.g., attribute value in a relational tuple, text node or attribute value in XML, atomic (leaf) value in a JSON document, or literal in RDF. The connections between the data items in the original dataset lead to edges in the graph, e.g. parent-child relationships between XML or JSON

A. Abelló et al. (Eds.): ADBIS 2023, LNCS 13985, pp. 163–179, 2023.
https://doi.org/10.1007/978-3-031-42914-9_12

nodes, edges connecting each relational tuple node with their attributes, etc. In a relational database, primary/foreign keys may lead to more edges, e.g., the node representing an Employee tuple "points to" the Company tuple representing their employer. This graph view of a dataset has been introduced to support unstructured (keyword-based) search on (semi)structured data, since [3,15] and through many follow-up works [20].

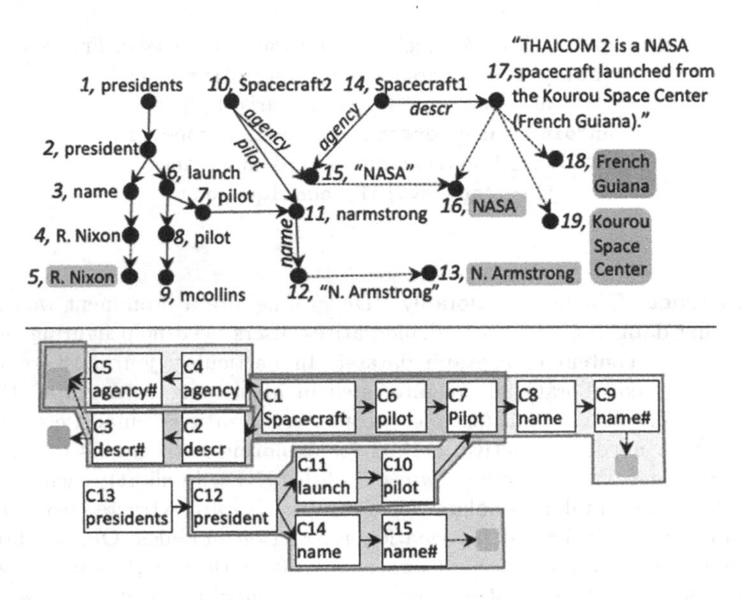

Fig. 1. Sample data graph (top), and corresponding collection graph (bottom) on which paths linking entities are explored (highlighted areas). (Color figure online)

Entity-Rich Graphs. Building on this idea, the ConnectionLens system [4,5] has been developed to facilitate, for non-IT users such as data journalists, the exploration of datasets of various models, including relational/CSV, XML, JSON, RDF, and PGs. ConnectionLens turns any (set of) datasets into a single graph as outlined above. For instance, the data graph at the top of Fig. 1 features RDF triples about NASA spacecrafts (labeled edges), and an XML document describing presidents who attended spacecraft launches (tree with labeled nodes and unlabeled edges). ConnectionLens also includes Information Extraction modules, which extract, from any leaf node in the data graph, Named Entities (People, Locations and Organizations) [5], as well as other types of entities that journalists find interesting: temporal moments (date, time), Website URIs, email addresses, and hashtags. We designate any of these pieces of information as *entities*, and we model them as extra nodes, e.g., in Fig. 1(top), organizations appear on a pink background, people on yellow and locations on green, respectively. Each entity is extracted from a leaf text node, to which it is connected by a dashed edge. When an entity is extracted from more than one text node, the edges connecting it to those nodes increase the graph connectivity, e.g., "NASA" extracted from the nodes 15 and 17.

Goal: Efficient, Interactive Exploration of Entity Connections. Journalists are interested in *data paths ending in entity pairs of certain kinds* in a given dataset, e.g. in Fig. 1, "how people are connected to places?". When shown the set of corresponding labeled paths, users may pick one to *further explore*: how many pairs of entities are connected by each path?, which entities are most frequent?, etc. Note that it is important to consider paths *irrespective of the edges directions in the data graph.* This is because, depending on how the data is modeled, we may encounter $x \xrightarrow{\text{boughtProperty}} y \xrightarrow{\text{locatedIn}} c$, or $x \xleftarrow{\text{hasOwner}} y \xrightarrow{\text{locatedIn}} c$; both paths are interesting.

Challenges and Contributions. The analysis outlined above raises two challenges. First, it requires *materializing the entity pairs connected by the paths*, which may be very costly, if (*i*) the graph is large, and/or (*ii*) there are many paths (the latter is almost always true, if the data is complex/heterogeneous, and/or if we allow paths to traverse edges in both directions). Second, the large number of paths may *overwhelm users.* Non-expert users, or users which are not familiar with the dataset structure, cannot be expected to state "only the paths that they would like to see", since they lack technical expertise and/or dataset knowledge. However, *if prompted by the system*, they can give valuable input on whether *certain links (or connections) are worth making*, or whether they are just spurious links that would generate uninteresting paths. An example of uninteresting path is: in a dataset of French national assembly members, all addresses are in France, thus the France named entity enables connecting any two people.

Our contributions towards addressing these challenges is as follows. (*i*) From the data, we *generate a small set of user questions*, which help us leverage the domain knowledge that users may have, to generate interesting paths (Sect. 2 and 3). (*ii*) To speed up the materialization of interesting paths, we *recommend a set of views (subpaths) to materialize*, and *rewrite each path query using these materialized views* (Sect. 4). This allows to identify subpaths that appear in more than one path, and compute the corresponding data paths only once, greatly improving performance (Sect. 5).

2 From Datasets to Data Graphs

In order to propose a general approach that works on any data format (e.g. RDF, JSON, XML, etc.), we start by building a *data graph* out of each input dataset (Sect. 2.1). These graphs may be large, thus enumerating paths on them would be inefficient. Therefore, we leverage a more compact structure, namely a *collection graph* (Sect. 2.2). The graph representation and the collection graph are based on prior work [8,9]. Finally, we explain how we produce a single collection graph out of *several datasets* (Sect. 2.3), which journalists may need to exploit together in a particular investigation.

2.1 From a Dataset to a Data Graph

We transform any dataset as a directed graph $G_0 = (N_0, E_0, \lambda_0)$ where N_0 is a set of nodes, E_0 is a set of vertices connecting N_0 nodes and λ_0 is a function assigning a label l to each node and edge (l may be empty). We map each data model on G_0 as follows.
RDF graphs are naturally mapped on G_0: each subject, respectively object, is turned into a node and an edge labelled with the property is connecting them. An XML document can be naturally viewed as a tree, with element nodes having element and attribute children. XML elements may carry #ID attributes whose values uniquely identify them; other XML elements may carry #IDREF attributes, whose values act as "foreign keys" referring to other elements by their #ID value. ID-IDREF information can be supplied in an optional Document Type Description (DTD) or XML Schema (XSD); when these are not available, ID-IDREF links can be found by profiling [1,16] techniques, which we also implemented. In the graph representation of an XML document, ID-IDREF links lead to more edges between elements (thus, the graph is no longer a tree). JSON documents are also modelled as trees, where each map, array, and leaf value is a node and parent-child edges are connecting them.

Named entities (NE) are extracted from each leaf of the data graph (RDF literal, XML text node or attribute value, or JSON value). For instance, in Fig. 1, the NASA organization (pink background) is extracted from both the RDF and XML datasets. Each extracted NE is materialized by a new node in the graph, connected via an extraction edge (dashed arrows in Fig. 1) to each leaf text node from which it has been extracted.

2.2 From a Data Graph to a Collection Graph

The collection graph is a compact representation of the data graph. It is based on an equivalence relation between the data graph nodes: the collection graph has exactly one node for each equivalence class of data graph nodes; further, whenever $n_1 \rightarrow n_2$ is an edge in the data graph, the collection graph comprises an edge $C_1 \rightarrow C_2$, where C_1, C_2 are the equivalence classes (also called collections) to which n_1, n_2 belong, respectively.

The most natural node equivalence relation differs across data models. Specifically, XML nodes we consider equivalent are elements with the same name; text nodes that are children of equivalent elements; and values of same-name attributes of equivalent elements. For instance, in Fig. 1, the pilot nodes 7 and 8 are equivalent. In JSON, we consider equivalent nodes found on the same labeled path, from the JSON document root, to the node. A path is a concatenation of node and edge labels, separated by . (dots), where we assign special labels: μ to each map node, α to each array node, and ϵ to each empty node or edge label. For instance, in the JSON snippet `[{"name": "Alice", "address": {"street": "Main", "city": "NY"}}]` the path to "NY" is: $\alpha.\epsilon.\text{address}.\mu.\text{city}$. In RDF, there are numerous ways to define node equivalence [10]. RDF collection graphs are built through the TypedStrong summarization method [12], working as follows. Whenever an RDF node has one or more types, all nodes with the same set of types are said equivalent (in RDF, a node can have several types, related or not, e.g. *Student* and *Employee*). For nodes without types, TypedStrong summarization relies on the properties (labels of incoming and outgoing edges) that the nodes have, by introducing a notion of *outgoing/incoming property cliques*: (*i*) two properties that a node have, are in the same outgoing clique, e.g., agency and pilot are in the same outgoing clique because they are both property of node 10 in Fig. 1, and also with descr because node 14 has agency and *descr*; incoming property cliques are symmetrically defined based on incoming properties; (*ii*) two nodes without types are equivalent if they have identical outgoing and incoming property cliques. For instance, nodes 10 and 14 (the two spacecrafts) are equivalent, since they have the same outgoing and incoming property cliques.

Figure 1 (bottom) shows the collection graph corresponding to the data graph. We named the collections C_1, C_2, etc. Note that some data models have *labeled* edges, e.g., RDF, while others have (mostly) *unlabeled* edges, e.g., XML. For uniformity, we transform any labeled edge into a node and extend our summarization also to such nodes. In Fig. 1, collection C_6 contains the nodes introduced instead of the pilot edges in the RDF dataset. Collection names in our figure are only for ease of explanation (they are not required in our method). Some collections, such as C_{12}, C_{13}, have nodes with identical names, in which case we use that name. For collections such as C_1, with RDF nodes each with a different URI, we use an intuitive name, e.g., "Spacecraft".

Entity Collection Profiles. Each leaf collection in the collection graph corresponds to a set of literals (strings), out of which various NEs may have been extracted. These collections' names end with a # to help distinguish them (e.g. C_5 agency#). For each such collection C, we compute an *entity profile* storing how many entities of each type were extracted from its string nodes. In Fig. 1, there are four such profiles, each shown as a box filled with the color of an entity type, e.g., the child of C_5 is pink reflecting the Organization entities extracted from agency values. In practice, long text nodes often lead to multiple NEs extracted, of several types. Knowledge of which leaf collections contain which kind of entities will be crucial to help users explore the graph (Sect. 3).

2.3 From Multiple Datasets to a Collection Graph

Journalists often need to work with several datasets, e.g. a JSON collection of political tweets, the list of French mayors in XML, and an RDF graph of public investment in companies across France. Such datasets often have *common values*, e.g., the cities that mayors represent are also the places where companies are situated. Interesting data paths can be found *across* data sources.

To obtain a single collection graph from a set of datasets, we proceed as follows. First, we build a separated collection graph from each dataset (as in Sect. 2.2). Then, whenever two collections C_1, C_2 from distinct datasets share values, we replace them by a new collection $C_{1,2}$, which contains the values of C_1 and C_2, and inherits all the incoming and outgoing edges of C_1 and C_2

in the collection graph they came from. Their original collection graphs are thus connected, and the entity profile of the new collection $C_{1,2}$ is built. In Fig. 1, the collection graph reflects this unification: the pink filled node reflects organizations from both RDF and XML.

3 Paths Between Entities

In this section, we discuss categories of paths that users might be interested in, and how to ask for their input.

An important first observation is: by the way we built the collection graph, *to each dataset path corresponds a path in the collection graph.* For instance, consider the data path $13 \leftarrow 12 \xleftarrow{\text{name}} 11 \xleftarrow{\text{pilot}} 10 \xrightarrow{\text{agency}} 15 \rightarrow 16$ in Fig. 1. The collection graph features the corresponding path ■ $\leftarrow C_9 \leftarrow C_8 \leftarrow C_7 \leftarrow C_6 \leftarrow C_1 \rightarrow C_4 \rightarrow C_5 \rightarrow$ ■.

Further, *some paths in the collection graph correspond to no paths in the data graph.* For instance, the path $C_6 \leftarrow C_1 \rightarrow C_2$ does not correspond to any path in the data, because no spacecraft (part of the collection C_1) has *both* the agency and descr properties. Such collection paths, with no support in the data, are introduced by summarization, which compresses the graph structure with some information loss. In our example, the fact that a spacecraft has agency and pilot, another has pilot and descr, and none has agency and descr is "simplified" into a collection C_1 that may have any combination of the three properties (represented by collections C_2, C_4, C_6). We consider this loss of information acceptable in exchange for the ability to work on a much smaller object (collection graph) instead of a potentially very large data graph.

Based on the above, our approach will be to (*i*) enumerate paths on the collection graph, then (*ii*) turn each path into a query, and (*iii*) evaluate this query on the data graph, and show users the resulting actual data connections (if any).

3.1 Characterizing Entity Paths in the Collection Graph

Each path between two entities is first, characterized by a **pair of entity types** of the form (τ_1, τ_2), where $\tau_1, \tau_2 \in \mathcal{E}$, with $\mathcal{E}$ being the set of supported entity types. $\mathcal{E}$ contains entity types such as Person, Location, Organization, Email, URI, Hashtag, Date, etc.

An entity path is also characterized by its **length**, i.e., the number of edges it contains. Depending on the application, interesting connections can be made by paths of different lengths; however, it appears likely that beyond a length such as 10 or 15, connections may become meaningless. Therefore, and also to control how many collection paths they want to inspect, users may specify a **maximum path length** L_{max}, whose default value we set to 10.

Path Directionality. By definition, each entity-to-entity collection path cp is of the form: $\square \leftarrow C_1 \rightsquigarrow C_2 \rightarrow \blacksquare$, where $\square, \blacksquare$ are two entity profiles, such that the first, respectively, the second, contains some entities of types τ_1, respectively,

τ_2, and C_1, C_2 are value collections such as C_5 and C_9 in the example at the beginning of Sect. 3. The directions of the leftmost and rightmost edges are by convention always towards □, ■, which represent entities. Let cp_0 denote the path $C_1 \leftrightsquigarrow C_2$. This path may be:

- *Unidirectional*, i.e., all cp_0 edges go from C_1 towards C_2, or the opposite;
- *Shared-sink*, i.e., cp_0 may contain a (collection) node C such that all edges between C_1 and C (if any) go from C_1 towards C, and all edges between C_2 and C (if any) go from C_2 towards C. A shared-sink path is $C_1 \rightarrow C_6 \rightarrow C_7 \leftarrow C_{10} \leftarrow C_{11}$.
- *Shared-root*, i.e., cp_0 may contain a (collection) node C such that all edges between C and C_1 (if any) go from C towards C_1, and all edges between C and C_2 (if any) go from C towards C_2. A shared-root path is $C_9 \leftarrow C_8 \leftarrow C_7 \leftarrow C_6 \leftarrow C_1 \rightarrow C_4 \rightarrow C_5$.
- *General*, i.e., the edges may be in any direction.

Unidirectional paths are quite rare. This is because entity-connecting paths must have at each end a node from which an entity is extracted. Most of the time, these are *two literal (string) nodes* (as opposed to internal nodes structuring the dataset). Literals have incoming edges, but not outgoing ones (other than those towards extracted entities); thus, *there is no unidirectional path from a literal to another.* However, in some RDF datasets, *NEs are extracted from URIs*, e.g., the triple https://dbpedia.org/Facebook $\xrightarrow{\text{locatedIn}}$ http://dbpedia.org/California is a unidirectional data path from an Organization to a Location. Similarly, shared-sink paths only occur when nodes in C_1 and C_2 have outgoing edges, and NEs appear in their labels; this only happens in RDF URIs.

Low-Specificity Connections. Some edges in the data graph reflect connections that can be seen as weak, or *non-specific.* In details, let's first consider data edges with non-empty labels, e.g., RDF triples. Let e be the edge $n_1 \xrightarrow{a} n_2$ for some URIs n_1, a, n_2. The *specificity* of e, denoted e_s, is computed as $2/(N_{1,a} + N_{2,a})$, where $N_{1,a}$, $N_{2,a}$ are the numbers edges labeled a outgoing, resp. incoming n_1, resp. n_2 [5]. The highest $N_{1,a}$ and/or $N_{2,a}$, the lowest e_s. For instance, the specificity of each agency edge in Fig. 1 is $2/(1+2) = 2/3$. For our purposes, we extend specificity to unlabeled edges as follows: the specificity of an edge $n_1 \xrightarrow{\epsilon} n_2$ is $2/(1+n_{1,2})$ where $n_{1,2}$ is the number of ϵ (empty-label) edges outgoing n_1, towards nodes having the same label as n_2. For instance, the specificity of the edge between nodes 6 and 7 is also 2/3.

In the collection graph, the edges with a non-empty label, connecting nodes from two equivalence classes lead to a collection, e.g., agency triples lead to C_4. We attach to this *collection, the average specificity of all the data edges it comes from*, e.g., to C_4 we attach 2/3. Empty-label edges connecting graph nodes from two equivalence classes lead to an edge in the collection graph, e.g., $C_{11} \rightarrow C_{10}$. We attach the average specificity of the original edges to this *edge* between two collections.

3.2 Eliciting User Input and Collection Path Enumeration

First, we ask users which two entity types they want to connect (thus selecting τ_1, τ_2). Next, we ask the maximum path length L_{max} (with a default value, currently 10). Then, we ask *how many types of data edges they are willing to review*, in order to possibly decide to disallow paths to traverse such connections; the default value is 5. By *type of data edge*, we mean either an edge label, such as agency, or a pair of (parent name, child name), in case the edge is unlabeled, such as launch-pilot in Fig. 1. Once the user chooses this value, say n_{spec}, we show her successively the n_{spec} collections (or collection graph edges) with the lowest average specificity, and ask if such a collection (or edge) should be excluded from the paths we search for, or not.

Finally, based on the collection graph, we *enumerate all the paths connecting entities* of types τ_1, τ_2, of maximum length L_{max}, and respecting the exclusion constraints which users may have specified. We use a simple dynamic programming algorithm, running in memory, as the collection graph is typically much smaller than the data. We then inform the user: "There are N_{uni} unidirectional paths, N_{sink} shared-sink paths, N_{root} shared-root paths, and N_{gen} general paths between entities of type τ_1 and τ_2." The user can then either (*i*) chose to materialize one of these path sets, or (*ii*) change settings, e.g., L_{max}, n_{spec}, exclude more or less edge types, etc., until the user is satisfied with the predicted number of paths and decides to trigger materialization of a path set (Sect. 4).

4 Materializing Data Paths

At this point, we have a set of *collection paths*, which must be transformed into queries and evaluated *on the data graph*. Each such query matches similar-structure data paths, thus its results are shown to users as a table: the first and last attribute of such a table comprise entities of type τ_1, τ_2, while the intermediary attributes are the nodes and edges connecting these entities in the data graph. For instance, let τ_1 be Person, τ_2 be Organization: the light-yellow, respectively, light-pink background shapes in Fig. 1 materialize the two paths which, in this graph, connect the pink child of C_5 (■) with the yellow children (■) of C_9, respectively, C_{15}.

4.1 From a Collection Path to a Query over the Data Graph

Each collection path translates into a chain-shaped conjunctive query. For instance, the path on yellow background in Fig. 1, going through C_5 and C_9, becomes:

$$q_1(\bar{x}) \text{ :- } n(x_1, \tau_{\text{Org}}, ■), e(x_2, x_1, \_), n(x_2, \_, C_5), e(x_3, x_2, \text{agency}), n(x_3, \_, C_1),$$
$$e(x_3, x_4, \text{pilot}), n(x_4, \_, C_7), e(x_4, x_5, \text{name}), n(x_5, \_, C_9), e(x_5, x_6, \_),$$
$$n(x_6, \tau_{\text{Person}}, ■)$$

This query refers to two relations: $n(ID, type, equiv)$, describing nodes, with the last attribute denoting their equivalence class, and $e(s, d, label)$, describing edges between nodes s and d and carrying a certain label. Each x_i is a variable; $\bar{x}$ in the query head denotes all the x_i variables, $1 \leq i \leq 6$. We use _ to denote a variable which only appears once, in a single query body atom. Finally, τ_{Org} and τ_{Person} denote the node types of extracted Organization, respectively, extracted Person entity. Similarly, the pink-background collection path translates into:

$$q_2(\bar{x}) \text{ :- } n(x_1, \tau_{\mathrm{Org}}, \blacksquare), e(x_2, x_1, \_), n(x_2, \_, C_3), e(x_3, x_2, \mathsf{descr}), n(x_3, \_, C_1), e(x_3, x_4, \mathsf{pilot}), n(x_4, \_, C_7), e(x_5, x_4, \_), n(x_5, \_, C_{11}), e(x_6, x_5, \_), n(x_6, \_, C_{12}), e(x_6, x_7, \_), n(x_7, \_, C_{14}), e(x_7, x_8, \_), n(x_8, \_, C_{15}), e(x_8, x_9, \_), n(x_9, \tau_{\mathrm{Person}}, \blacksquare)$$

Each of these queries can be evaluated through any standard graph database. However, evaluating dozens or hundreds of path queries on large graphs can get very costly. Further, since we do not know which paths may result from the user choices, we cannot establish path indexes beforehand.

View-Based Optimization. To address this problem, we propose an optimization, based on the observation that *queries resulting from collection paths may share some subpaths.* For instance, the subquery $s(x_3, x_4)$:- $n(x_3, \_, C_1), e(x_3, x_4, \mathsf{pilot}), n(x_4, \_, C_7)$ is shared by $q_1(\bar{x})$ and $q_2(\bar{x})$. Therefore, we decide to (*i*) evaluate s and store its results; (*ii*) rewrite $q_1(\bar{x})$ and $q_2(\bar{x})$ by replacing these atoms in each query, by a single occurrence of the atom $s(x_3, x_4)$. The next sections formalize this for larger query sets, also showing how to handle different *alternatives* that may arise as to which shared subpaths to materialize.

4.2 Enumerating Candidate Views

A first question we need to solve is enumerating, based on a set $\mathcal{Q}$ of path queries, the possible subqueries that we could materialize, and based on which we could rewrite some workload queries.

Let $q \in \mathcal{Q}$ be a path query: it is an alternating sequence of node (n) and edge (e) atoms. We denote by n_q the number of edge atoms, then the number of node atoms is $n_q + 1$. We denote by $n_{\mathcal{Q}}$ the highest n_q over all $q \in \mathcal{Q}$.

Without loss of generality, our first heuristic **(H1)** is: we only consider **connected subpaths** of q as candidate subqueries. If q is of the form $q(\bar{x})$:- $n_1, e_1, \ldots, e_{n_q}, n_{n_q+1}$, each connected subpath of q, denoted sq, is determined by two integers $1 \leq i \leq n_q$, $i < j \leq n_q + 1$, such that $sq(x_i, x_j)$:- $n_i, e_i, \ldots, n_j, e_j$, and x_i, x_j are the IDs of the nodes in the atoms n_i, n_j, respectively. We denote by $q|^{i,j}$ the subquery of q determined by the positions i, j. For instance, when q_1 is the sample query in Sect. 4.1, $q_1|^{3,4}$ is the subquery $s(x_3, x_4)$ introduced there. Considering connected (cartesian-product free) candidate views is common in the literature, too (see Sect. 6).

Each query $q \in \mathcal{Q}$ has $O(n_q^2)$ connected subpaths, that can be easily enumerated from q's syntax. A second heuristic **(H2)** we adopt is: we only consider **shared subpaths**, that is, those subpaths s for which there exist $q', q'' \in \mathcal{Q}$, $q' \neq q''$, and integers i', j', i'', j''' such that $s = q'|^{i',j'} = q''|^{i'',j''}$, possibly after some variable renaming. For the queries q_1, q_2 in refsec:collectionspspathspsquery, the subquery $s_{3,4}$ is $q_1|^{3,4}$ and also $q_2|^{3,4}$. (H2) restricts the number of candidate

Algorithm 1: Selecting views to materialize and the respective view-based rewritings

Input : Queries $\mathcal{Q}$, candidate materialized views $\mathcal{V}$
Output: Materialized views $\mathcal{M}$ and rewritings $\mathcal{R}$ for some $\mathcal{Q}$ queries

```
1  M ← ∅; R ← ∅
2  while V ≠ ∅ do
3      for v ∈ V do
4          ben(v) ← 0; cost(v) ← cost to compute and store the view v
5          for q ∈ Q, q can be rewritten using v do
6              (ben(v, q), r_{q,v}) ← the cost of evaluating q based on v, through the
               rewriting r_{q,v}, minus the cost of evaluating q directly on the graph
7              ben(v) ← ben(v) + ben(v, q)
8      (v_max, b_max) ← a view v_max maximizing ben(v) − cost(v), and its benefit
9      if b_max − cost(v_max) < 0 then
10         exit
11     Add v_max to M
12     for q ∈ Q, q can be rewritten using v_max do
13         if ben(v_max, q) > 0 then
14             Add r_{q,v_max} to R
15             Remove q from Q
16     Remove v_max from V
```

views from $|\mathcal{Q}| \times n_{\mathcal{Q}}^2$ to a number that depends on the actual workload $\mathcal{Q}$, and which decreases when $\mathcal{Q}$ paths look more like each other. Another interest of (H2) is: the benefit of using a view v to rewrite one query q is likely offset by the cost of materializing v; actual performance improvements start when v is *used twice (or more)*, which is exactly the case for subqueries shared by several $\mathcal{Q}$ queries.

Our third heuristic **(H3)** is: among the possible subqueries shared by two queries q', q'', **consider only the longest ones**. That is, if s_1, s_2 are two shared subqueries of q' and q'' such that $n_{s_1} > n_{s_2}$, do not consider the subquery s_2.

Our heuristics (H1), (H2), (H3) lead to **building the candidate view set** $\mathcal{V}$ as follows. For each pair of distinct queries (q', q'') where $q', q'' \in \mathcal{Q}$, add to $\mathcal{V}$ the longest, shared, connected subqueries of q' and q''. The complexity of this algorithm is $O(|\mathcal{Q}|^2 \times n_{\mathcal{Q}}^2)$, while $|\mathcal{V}|$ is in $O(|\mathcal{Q}|^2)$.

4.3 Selecting Materialized Views and Rewriting Path Queries

Knowing the path queries $\mathcal{Q}$ and the candidate view set $\mathcal{V}$, we need to determine: a set $\mathcal{M} \subseteq \mathcal{V}$ of views which we actually materialize, in order to rewrite some $\mathcal{Q}$ queries. We collect the rewriting of each such queries in $\mathcal{R}$. The decision to materialize a view incurs a *cost*, since the view data must be computed and stored. We denote $cost(\cdot)$ the cost of evaluating a view (or query), and assume it can be computed without actually computing it. Materializing a view is more attractive if (*i*) rewritings using it reduce significantly query evaluation costs, and (*ii*) its own materialization cost is small.

In the most general case, a query could be rewritten based on any number of views, and also involving the base graph. For instance, query q_1 from Sect. 4.1 could be rewritten as: $q_1|^{1,3} \bowtie q_1|^{3,4} \bowtie q_1|^{4,6}$, where each $\bowtie$ denotes a natural join, on the variables x_3, respectively, x_4. However, enumerating all such alternatives makes the query rewriting problem NP-hard [14]. Instead, we adopt another pragmatic heuristic **(H4)**: **rewrite each query using not more than one view**. This simple choice keeps the view selection complexity under control, all the while providing good performance.

Algorithm 1 depicts our **greedy** method for finding $\mathcal{M}$ and $\mathcal{V}$. It computes the *benefit* of each view v for each query that may be rewritten using v, as well as the cost of v. In a greedy fashion, it decides to materialize the view v_{max} maximizing the *overall* benefit (for all $\mathcal{Q}$ queries), and uses it to rewrite all queries whose evaluation cost can be reduced thanks to v_{max}, via the rewriting $r_{q,v_{max}}$. These queries are then removed from $\mathcal{Q}$, the benefits of the remaining views are recomputed over the diminished $\mathcal{Q}$, and the process repeats until no profitable view to materialize can be found.

Estimating Costs. Algorithm 1 needs to compute: (*i*) $cost(\cdot)$, the cost to evaluate a query q or materialize a view v; (*ii*) $r_{q,v}$, the rewriting of q using a view v; and (*iii*) $cost(r_{q,v})$, the cost of such a rewriting. All these costs must be estimated *before* any query or view results are computed. We do this as follows. For (*i*), we *use the cost estimation of the graph data management system* (GDBM, in short) *storing the graph.* Our implementation relies on PostgreSQL, whose `explain` command returns both the estimated number of results of certain query (or view), denoted $size(q)$, and the cost of computing those results. For (*ii*), recall (Sect. 4.2) that when v is used to rewrite q, v is a subpath of q, thus there exist i, j such that $v = q|^{i,j}$. The rewriting $r_{q,v}$ is easily obtained by replacing, in the body of q, the atoms from the ith to the jth, with the head of v. Handling (*iii*) is more complex than (*i*). This is because the cost of a query (or view) is estimated based on *statistics* the GDBM has about the stored graph. In contrast, the GDBM cannot estimate cost of $r_{q,v}$, because v *has not been materialized* yet, thus the GDBM cannot reason about v like it does about the graph. To compensate, we proceed as follows: we compute the cost of reading the hypothetical view v_{max} from the database, by multiplying $size(v_{max})$, the estimation of the view size, with a constant (we used Postgres' own CPU_TUPLE_COST); then, we estimate the cost of $r_{q,v_{max}}$ as this reading cost plus the cost of estimating the parts of q not in v_{max} plus the cost of joining v_{max} with these (one or two) remaining query parts. We estimate the cost of each such join by adding their input sizes, which we then multiply with another (GDBM) constant. This reflects the fact that modern databases feature efficient join algorithms, such as memory-based hash joins, whose complexity is linear in the size of their inputs.

Complexity. Algorithm 1 makes at most $O(|\mathcal{V}| \times |\mathcal{Q}|)$ iterations, which can be simplified into $O(|\mathcal{Q}|^3)$. Forming a rewriting takes $O(n_\mathcal{Q})$, bringing the total to $O(|\mathcal{Q}|^3 \times n_\mathcal{Q})$.
Impact of Heuristics. As previously discussed, (H1) is universally adopted in the literature: no candidate view features cartesian products. (H2), imposing that views benefit at least two queries, preserves result quality, i.e., cost savings, under every *monotone cost model*, ensuring that the cost of evaluating a query q is at least that of evaluating s, when s is a subquery of q. In contrast, (H3) and (H4) may each divert from the globally optimal solution. However, as our experiments show, our chosen rewritings perform well in practice, and the algorithm itself is very efficient.

5 Experimental Evaluation

Our approach is fully implemented in Java 11, on top of ConnectionLens [4,5] which builds the data graph (Sect. 2.1) and Abstra [8,9] which builds the collection graph (Sect. 2.2); these are stored in PostgreSQL. We experimented on a Linux server with an Intel Xeon Gold 5218 CPU @ 2.30 GHz and 196 GB of RAM. We used PostgreSQL v9.6. Our system is available at: https://team.inria.fr/cedar/projects/pathways/. Our evaluation seeks to answer the two following questions: (*i*) *how are entities connected in each dataset?* (Sect. 5.2) and (*ii*) *does Algorithm 1 help to evaluate paths queries over the data graph?* (Sect. 5.3).

Table 1. Dataset overview.

Dataset name	$\|N\|$	$\|E\|$	$\|\tau_P\|$	$\|\tau_L\|$	$\|\tau_O\|$	$min(e_s)$
PubMed	63,052	89,710	5,993	2,151	5,096	0.001
Nasa	59,408	128,068	634	690	4,530	0.0002
YelpBusiness	57,963	61,627	322	427	1,437	0.001
YelpBusiness4	229,949	247,074	1,099	1,230	4,199	0.0002

5.1 Datasets and Settings

We present experiments on an XML, an RDF and a JSON dataset. They all come from real-life applications (as opposed to synthetic) to stay close to application needs, and to ensure realistic Named Entities (NEs). Indeed, synthetic datasets are often generated with an interest on structure, while the leaf (text) values lack interesting information.

We used an **XML** PubMed subset describing scientific articles from PubMed, a database of biomedical publications. Each article is described by its title, journal, link, year, DOI, keywords list and authors list. Authors are identified by their name and their affiliation; they may declare their conflicts of interest in the <COIStatement> tag. We used the **RDF** Nasa dataset, describing NASA flights, spacecrafts involved in launches, related space missions and the participating agencies. Finally, we used the **JSON** YelpBusiness dataset where each business has an id, a name, an address, a city, a state, a postal code and coordinates (latitude and longitude). It also has a set of categories (e.g. bakery, shoe store, etc.) and a set of attributes (e.g. `acceptCreditCards`), the latter may be deeply nested. They also received reviews from customers modeled as a number of stars (from 0 to 5) and the number of reviews. **YelpBusiness4**, 4 times larger than YelpBusiness, allows studying the scalability of our algorithm. Table 1 shows for each dataset: its number of nodes $|N|$, edges $|E|$, numbers of extracted NEs $|\tau_P|$, $|\tau_L|$, $|\tau_O|$ and the minimum edge specificity $min(e_s)$. Without loss of generality, we experiment with the NE types Person, Location, Organization, whose types are denoted τ_P, τ_L, τ_O, respectively. We set L_{max} to 10, and avoided connections whose assigned specificity (Sect. 3) was less than 5% of the average specificity over all node/edge collections.

5.2 Path Enumeration

For each dataset and pair of entity types, Fig. 2 and 3 report the number of paths of each directionality (Sect. 3.1), the minimum and maximum length L_p of each path, and the minimum and maximum data path support (number of results when evaluated on the data), this is denoted S_p. For the PubMed (XML) and YelpBusiness (JSON) datasets, we obtained only shared-root paths: this is because of the tree structure of these datasets, where text values (leaves) are only connected by going through a common ancestor node. In the RDF Nasa dataset, we also found general-directionality paths. The JSON datasets are more irregular, leading to more paths. In almost every case, a few collection paths had 0 support, due to dataset summarization (Sect. 3). The maximum support may be high, e.g. 15,181 in the PubMed dataset.

(τ_1, τ_2)	N_{root}	$\min(S_p)$	$\max(S_p)$
(τ_P, τ_O)	21	0	13,988
(τ_P, τ_L)	21	0	15,181
(τ_L, τ_O)	21	0	5,054
(τ_P, τ_P)	21	0	389
(τ_L, τ_L)	21	0	1,214
(τ_O, τ_O)	21	3	3,090

(τ_1, τ_2)	N_{root}	N_{gen}	$\min(S_p)$	$\max(S_p)$
(τ_P, τ_O)	99	1	0	629
(τ_P, τ_L)	95	5	0	137
(τ_L, τ_O)	97	3	0	603
(τ_P, τ_P)	97	3	0	89
(τ_L, τ_L)	97	3	0	3,050
(τ_O, τ_O)	97	3	0	8,960

Fig. 2. Entity paths in PubMed (left) and Nasa (right). PubMed: $min(L_p) = 5$; $max(L_p) = 8$. Nasa: $min(L_p) = 5$; $max(L_p) = 9$.

(τ_1, τ_2)	N_{root}	max(L_p)	min(S_p)	max(S_p)
(τ_P, τ_O)	41	7	0	651
(τ_P, τ_L)	33	7	0	193
(τ_L, τ_O)	21	5	0	1,412
(τ_P, τ_P)	28	7	0	35
(τ_L, τ_L)	15	5	2	158
(τ_O, τ_O)	21	5	0	1,232

(τ_1, τ_2)	N_{root}	max(L_p)	min(S_p)	max(S_p)
(τ_P, τ_O)	48	7	0	2,593
(τ_P, τ_L)	39	7	0	760
(τ_L, τ_O)	39	5	0	258
(τ_P, τ_P)	36	7	0	207
(τ_L, τ_L)	15	5	0	674
(τ_O, τ_O)	21	5	0	4,889

Fig. 3. Entity paths in the YelpBusiness (left) and YelpBusiness4 (right) datasets. YelpBusiness and YelpBusiness4: $min(L_p) = 5$.

Table 2. Data paths evaluation on the PubMed and Nasa datasets.

(τ_1, τ_2)	T_O	$\|\mathcal{Q}_{TO}\|$	$\|\mathcal{Q}_{NV}\|$	$\|\mathcal{V}\|$	$\|\mathcal{M}\|$	T_R	$T_{\mathcal{Q}_{NV}}$	$T = T_R + T_{\mathcal{Q}_{NV}}$	$s = T_O/T$
PubMed									
(τ_P, τ_O)	250.36	5	1	16	5	3.78	0.32	4.10	61×
(τ_P, τ_L)	37.29	0	1	16	5	19.06	0.32	19.38	2×
(τ_L, τ_O)	151.29	2	2	16	5	11.88	8.59	20.47	7×
(τ_P, τ_P)	152.59	3	1	16	5	44.19	0.08	44.27	3×
(τ_L, τ_L)	169.64	2	1	16	5	71.32	0.31	71.63	2×
(τ_O, τ_O)	317.92	5	1	16	5	22.99	0.25	23.24	13×
Nasa									
(τ_P, τ_O)	195.47	1	0	80	10	54.14	N/A	54.14	3×
(τ_P, τ_L)	254.26	3	0	68	10	44.57	N/A	44.57	5×
(τ_L, τ_O)	1073.55	32	0	77	9	131.58	N/A	131.58	8×
(τ_P, τ_P)	278.95	4	0	76	10	92.01	N/A	92.01	3×
(τ_L, τ_L)	1103.48	30	0	77	9	101.35	N/A	101.35	10×
(τ_O, τ_O)	1318.78	37	0	77	9	247.43	N/A	247.43	5×

These results show that numerous interesting entity paths exist in our datasets, of significant length (up to 9), and some with high support, bringing the need for an efficient evaluation method.

5.3 Efficiency of Path Evaluation

We now study the efficiency of data path computations over the graph. Table 2 and Table 3 show, for each dataset and entity type pair, T_O is the time to evaluate the corresponding path queries without the view-based optimization of Sect. 4.2 and 4.3, referred to as **VBO** from now on. $|\mathcal{Q}_{TO}|$ is the number of queries whose execution we stopped (time-out of 30s) without VBO. $|\mathcal{Q}_{NV}|$ is the number of queries for which Algorithm 1 did not recommend a view. T_R is the time to evaluate the rewritten queries on the data graph, while $T_{\mathcal{Q}_{NV}}$ is the time to evaluate the non-rewritten queries $\mathcal{Q}_{NV}$; $T = T_R + T_{\mathcal{Q}_{NV}}$ is the (total) execution time to evaluate queries using VBO. Finally, $s = T_O/T$ is the speed-up due to

Table 3. Data path evaluation on the YelpBusiness datasets.

(τ_1, τ_2)	T_0	$\|\mathcal{Q}_{TO}\|$	$\|\mathcal{Q}_{NV}\|$	$\|\mathcal{V}\|$	$\|\mathcal{M}\|$	T_R	$T_{\mathcal{Q}_{NV}}$	$T = T_R + T_{\mathcal{Q}_{NV}}$	$s = T_0/T$
YelpBusiness									
(τ_P, τ_O)	205.95	2	0	22	6	4.20	N/A	4.20	49×
(τ_P, τ_L)	410.87	7	1	19	5	40.87	1.27	42.12	9×
(τ_L, τ_O)	239.90	0	1	20	10	1.15	0.6	1.75	137×
(τ_P, τ_P)	466.58	9	2	23	5	15.33	12.02	27.35	17×
(τ_L, τ_L)	450.00	15	1	8	4	9.89	<0.01	9.89	45×
(τ_O, τ_O)	334.22	4	1	10	5	2.83	<0.01	2.83	118×
YelpBusiness4									
(τ_P, τ_O)	804.70	26	0	23	6	62.52	N/A	62.52	12×
(τ_P, τ_L)	454.19	10	1	20	5	92.50	<0.01	92.50	5×
(τ_L, τ_O)	242.57	5	1	10	5	62.74	6.61	69.35	3×
(τ_P, τ_P)	317.00	7	1	27	7	14.35	1.08	15.43	20×
(τ_L, τ_L)	395.49	10	1	8	4	2.62	18.15	20.77	19×
(τ_O, τ_O)	347.23	8	1	10	5	42.93	2.34	45.27	7×

VBO. We do not report times to materialize views because they were all very short (less than 0.01s). All times are in seconds.

The evaluation time T_0 without VBO ranges from 100 s to 2000 s; these path queries require 5 to 9 joins, on graphs of up to more than 200,000 edges (Table 1). $|\mathcal{Q}_{NV}|$, the number of queries that could not make use of any views, is rather small, which is good. The number of candidate views, respectively, materialized views depend on the complexity of the dataset, and thus on the complexity of the paths. The total path evaluation time T is reasonable. Finally, the VBO speed-up is at least 2× and at most 137×, showing that our view-based algorithm allows to evaluate path queries much more efficiently.

5.4 Experiment Conclusion

On graph leading to entity paths of various lengths and support, the view-based rewriting significantly reduces the path query evaluation time over the data graph.

6 Related Work and Conclusion

Many graph exploration methods exist, see, e.g., [18]. Modern graph query languages such as GPML [11] (no implementation so far) or the JEDI [2] SPARQL extension allow *checking* for paths between query variables, if users can specify a regular expression that the path labels match. Other systems interact with users to incrementally build SPARQL queries, In keyword-based search (KBS, in short) [3,5,20], one asks for connections between two or more nodes matching specific keywords. KBS is handy when users *know keywords (entities) to search for.* In [6], graph queries are extended with a KBS primitive. The algorithms proposed there work directly on the graph; finding such trees in general is NP-hard, since it is related to the Group Steiner Tree problem. Complementary to these, we focus on *identifying, and efficiently computing, all paths between certain extracted entities*, to give a first global look at the dataset content, for graphs obtained from multiple data models.

Our view selection problem is a restriction (to path-only queries) of those considered, e.g., in [13,14,17,19]. This allows for its low complexity, while being very effective. Our algorithms rely on collection graphs built by Abstra [8,9], the interactive path enumeration approach, including VBO, is novel; it has been demonstrated in [7].

Acknowledgments. This work has been funded by the DIM RFSI PHD 2020-01 project and the AI Chair SourcesSay (ANR-20-CHIA-0015-01) chair.

References

1. Abedjan, Z., Golab, L., Naumann, F., Papenbrock, T.: Data Profiling. Synthesis Lectures on Data Management. Morgan & Claypool Publishers (2018)
2. Aebeloe, C., Setty, V., Montoya, G., Hose, K.: Top-K diversification for path queries in knowledge graphs. In: ISWC Workshops (2018)
3. Agrawal, S., Chaudhuri, S., Das, G.: DBXplorer: a system for keyword-based search over relational databases. In: ICDE (2002)
4. Anadiotis, A., Balalau, O., Bouganim, T., et al.: Empowering investigative journalism with graph-based heterogeneous data management. IEEE Bull. Tech. Committee Data Eng. **45**, 12–26 (2021)
5. Anadiotis, A., Balalau, O., Conceicao, C., et al.: Graph integration of structured, semistructured and unstructured data for data journalism. Inf. Syst. **104**, 101846 (2022)
6. Anadiotis, A.C., Manolescu, I., Mohanty, M.: Integrating connection search in graph queries. In: ICDE (2023)
7. Barret, N., Gauquier, A., Law, J.J., Manolescu, I.: Pathways: entity-focused exploration of heterogeneous data graphs (demonstration). In: ESWC (2023)
8. Barret, N., Manolescu, I., Upadhyay, P.: Abstra: toward generic abstractions for data of any model (demonstration). In: CIKM (2022)
9. Barret, N., Manolescu, I., Upadhyay, P.: Computing generic abstractions from application datasets. In: EDBT (2024)
10. Cebiric, S., et al.: Summarizing semantic graphs: a survey. VLDB J. **28**(3), 295–327 (2019). https://doi.org/10.1007/s00778-018-0528-3
11. Deutsch, A., Francis, N., Green, A., Hare, K., Li, B., Libkin, L., et al.: Graph pattern matching in GQL and SQL/PGQ. In: SIGMOD (2022)
12. Goasdoué, F., Guzewicz, P., Manolescu, I.: RDF graph summarization for first-sight structure discovery. VLDB J. **29**(5), 1191–1218 (2020). https://doi.org/10.1007/s00778-020-00611-y
13. Goasdoué, F., Karanasos, K., Leblay, J., Manolescu, I.: View selection in semantic web databases. PVLDB **5**(2) (2011)

14. Halevy, A.Y.: Answering queries using views: a survey. VLDB J. **10**(4), 270–294 (2001). https://doi.org/10.1007/s007780100054
15. Hristidis, V., Papakonstantinou, Y., Balmin, A.: Keyword proximity search on XML graphs. In: ICDE (2003)
16. Jiang, L., Naumann, F.: Holistic primary key and foreign key detection. J. Intell. Inf. Syst. **54**(3), 439–461 (2020). https://doi.org/10.1007/s10844-019-00562-z
17. Le, W., Kementsietsidis, A., Duan, S., et al.: Scalable multi-query optimization for SPARQL. In: ICDE (2012)
18. Lissandrini, M., Mottin, D., Hose, K., Pedersen, T.B.: Knowledge graph exploration systems: are we lost? In: CIDR (2022). http://www.cidrdb.org/
19. Mistry, H., Roy, P., Sudarshan, S., Ramamritham, K.: Materialized view selection and maintenance using multi-query optimization. SIGMOD Rec. **30**(2), 307–318 (2001)
20. Yang, J., Yao, W., Zhang, W.: Keyword search on large graphs: A survey. Data Sci. Eng. **6**(2), 142–162 (2021). https://doi.org/10.1007/s41019-021-00154-4

Managing Linked Nulls in Property Graphs: Tools to Ensure Consistency and Reduce Redundancy

Jacques Chabin[1], Mirian Halfeld-Ferrari[1], Nicolas Hiot[1,2](✉), and Dominique Laurent[3]

[1] LIFO – Université d'Orléans, INSA CVL, Orléans, France
[2] EnnovLabs – Ennov, Orléans, France
nhiot@ennov.com
[3] ETIS – CNRS, ENSEA – CY Université, Cergy-Pontoise, France

Abstract. Ensuring the provision of consistent and irredundant data sets remains essential to minimize bugs, promote maintainable application code and obtain dependable results in data analytics. A major challenge in achieving consistency is handling incomplete data, *i.e.*, missing information that may be provided later, comes from the fact that, the use of marked (or linked) nulls is required in many applications to express unknown but connected information. In this context, it is well known that maintaining the data consistent and irredundant is not an easy task. This paper proposes a query-driven incremental maintenance approach for consistent and irredundant incomplete databases. Can graph databases improve the efficiency of this operation? How can graph databases manipulate linked nulls? What is the impact of using graph databases on other essential maintenance operations? This paper presents an innovative approach to answering these questions, highlighting the proposal's strengths and weaknesses and offering avenues for further research.

Keywords: graph databases · incremental maintenance · incomplete data · constraints · tuple generating dependencies · updates

1 Introduction

Consistent data stores lead to fewer bugs, easier maintenance, and reliable analytics. Developers can focus on their tasks without being bogged down by data consistency issues. Our current research projects focus on data extracted from clinical cases. We deal with databases which are incomplete but consistent with respect to some given integrity constraints expressed as tuple-generating dependencies (tgd). Consistency is our first concern. For example, let $c1 : Pat(x), SOSY(x, y) \rightarrow PrescExam(x, z)$ be a constraint which implies that if a patient x exhibits a symptom y, then they should pass an exam z. If we consider the fact that Lea is a patient with pain in her hands, *i.e.*, $Pat(Lea)$ and $SOSY(Lea, pain\ on\ hands)$, the database must also include the

A. Abelló et al. (Eds.): ADBIS 2023, LNCS 13985, pp. 180–194, 2023.
https://doi.org/10.1007/978-3-031-42914-9_13

atom $PrescExam(Lea, N_1)$ for consistency. This atom indicates that an exam has been prescribed to Lea, but we do not yet know which one.

It is also crucial to address incompleteness, which can take various forms. Our focus in this work is on a database perspective, where incompleteness arises when values are missing. We adopt Reiter's approach [20] that provides First-Order Logic (FOL) semantics to null values of type 'value exists but is currently unknown'. If null values are linked, then the missing data is linked as well. For instance, the set $\mathfrak{D}_1 = \{PrescExam(Lea, N_1), ExamResult(Lea, N_1, N_2)\}$ seen as a database instance, indicates that an examination is prescribed for Lea, with unknown type denoted by N_1, and with unknown result represented by N_2.

Several sources can provide cleaned and well-formatted data to our database. Integrating this new data while maintaining consistency and avoiding redundancy is a challenge - this also involves matching the new data with the missing data to avoid redundancy. For example, if we add the new information $PrescExam(Lea, x\text{-}ray)$ to $\mathfrak{D}_1$ above, we cannot replace N_1 by *x-ray* in $PrescExam(Lea, N_1)$, because N_1 also appears in $ExamResult(Lea, N_1, N_2)$, which indicates that N_1 is a linked null. If later we receive the new information $ExamResult(Lea, x\text{-}ray, join\ inflammation)$, then we replace N_1 by *x-ray* since the instantiation of N_1 is the same in all atoms where it appears. The resulting database instance in this case is $\mathfrak{D}_1'' = \{ExamResult(Lea, x\text{-}ray, join\ inflammation), PrescExam(Lea, x\text{-}ray)\}$.

Work in [6] is well-suited to tackle the aforementioned challenge. However, in this paper, we take it a step further by introducing an *incremental* version of the approach and applying it to a *graph database* system. Our proposal leverages the advantages of data access, manipulation, and management capabilities provided by database systems, contrary to the in-memory version of [6]. Moreover, it aims to improve the process of creating sets of atoms that are linked by their null values, which was an expensive step in the earlier approach.

To achieve our objective, we undertook a study to investigate the potential of using a graph database. Our research aims to overcome the limitations of the original approach by designing a graph database model that facilitates the exploration of relationships between null values, along with proposing incremental update routines. One type of graph model that we have been using in our projects is the labelled property graph (LPG), which includes named relationships and properties. We have been working with LPG-based graph databases like Neo4J, and Cypher is the most widely used query language for LPGs. It is also the basis for the development of GQL, an ISO standard in progress.

In summary, our paper introduces an innovative approach that leverages the advantages of graph databases to enhance the creation of linked atoms with null values. We also examine the impact of this approach on other essential maintenance operations, providing insights into its overall effectiveness. Our paper also presents a critical analysis of the use of graphs in this context, highlighting the strengths and weaknesses of this approach and offering suggestions for future research directions. Additionally, we provide an overview of the advantages of our incremental version. The rest of this section showcases the contribution of our work and discusses the representation of null information in LPG-graphs.

Incremental Database Evolution. Database evolution is initiated by update requests. Its maintenance involves two main actions: *chasing* and *simplification.* The *chase* procedure is used to ensure consistency with a set of integrity constraints, expressed as rules, and which may generate new null values. The simplification process eliminates redundancies by removing null values that can be instantiated without breaking their links. This action corresponds to the computation of the *core* [9]. We adopt the policy in [6] to define the evolution of our database, but the proposal in this paper differs from the in-memory version in [6] in the following aspects:

1. *Incrementality is the kernel of our approach:* (a) Chasing is performed only on constraints concerned by the update, whereas in [6], all constraints are checked. (b) Simplification is guided by the null values potentially simplifiable due to the update, in contrast to [6], where simplification considers all null values.
2. *Our approach is query-driven as it deals with data stored in database systems, unlike the in-memory version in* [6]: (a) A constraint is a rule with a conjunction of atoms in its body. It is triggered, to produce the atom in its head, only when its body is fully instantiated based on the update atoms and the database instance. A *chase query*, partially instantiated with update constants, searches the database to fully instantiate the body of that constraint. (b) Queries, denoted as q_{bucket}, retrieve the null values appearing in the database instance and connected to the update being performed. (c) The process relies on the results of query q_{bucket} to identify atoms with nulls linked to those in q_{bucket}'s results. Then, a conjunction of these atoms is used to construct a new query, denoted by q_{core}, which guides the simplification decisions.
3. *When working with database systems, we assess which database model is better suited for our approach.* (a) Relational database model is used as our baseline. Each atom $R(A)$ is represented as a tuple A in table R. In this context, it is observed that the most costly task is the extraction of sets of atoms that are linked through their null values. Indeed, this operation requires scanning all tables, since null values can be associated to any attribute in a table, and thus cannot be indexed. (b) Query engines typically assume that graphs in graph databases are complete, but this assumption is not valid in practice due to missing data. Neo4J's approach of treating nulls as 'value does not exist' is however not adequate in case of linked nulls. To address this challenge, we propose a novel database design that treats nulls as first-class citizens. (c) Our graph database model enables null values to be treated as first-class citizens and indexed. This model simplifies operations in which we can identify all atoms that are directly or indirectly linked to a null value by simply selecting that null value.

Our proposal follows the database evolution semantics from [6], which has been shown to be effective[1], deterministic, and adhere to a minimal change property. Let $\mathfrak{D}$ be an incomplete but consistent database instance and U be a set of updates. The notation $\mathfrak{D} \Diamond U$ represents the insertion or deletion of the required updates in or from $\mathfrak{D}$. In [6], a from-scratch approach generates a new database instance denoted as $\mathfrak{D}' = core(upd(\mathfrak{D} \Diamond U))$, where *upd* is the update process described in [6]. In contrast, our paper describes an *incremental* version of the update process denoted as $upd_{|U}$. Therefore, the new database instance is represented as $\mathfrak{D}' = core_{|NullBucket}(upd_{|U}(\mathfrak{D} \Diamond U))$, where $NullBucket$ refers to the set of nulls affected by the update ($upd_{|U}$) applied to $\mathfrak{D} \Diamond U$.

Paper Organization. Section 2 presents the background necessary to understand our approach described in Sects. 3 and 4. Here, we recall the logic formalism used in our algorithms, while more practical considerations about the design of our graph database aim to highlight the implementation of the queries on Cypher. Section 5 presents our experimental study, and Sect. 6 overviews related work and presents future work.

2 Preliminary Considerations: Theory and Application

Theoretical Background. Assuming familiarity with FOL, we consider atoms as $P(t_1, \ldots, t_n)$ where P is a predicate of arity n and $t_1, \ldots, t_n$ are terms (constants, nulls, or variables). A fact is an atom with only constants and an instantiated atom has no variables. $null(A)$ denotes the set of nulls in an instantiated atom A. Homomorphisms between sets of atoms A_1 and A_2 map terms in A_1 to A_2, such that: (i) $h(t) = t$ if t is a constant, (ii) if $P(t_1, \ldots, t_n)$ is in A_1, then $P(h(t_1), \ldots, h(t_n))$ is in A_2. If h_1 is a homomorphism from A_1 to A_2 with an inverse homomorphism, then A_1 is isomorphic to A_2.

Φ denotes the set of existentially quantified formulas ϕ, which are conjunctions of atomic formulas. The set of atomic formulas in ϕ is denoted by $atoms(\phi)$. A *model* M of a formula ϕ in Φ is a set of facts that has a homomorphism from $atoms(\phi)$ to M. If each model of ϕ_1 is a model of ϕ_2, then we write $\phi_1 \Rightarrow \phi_2$, and ϕ_1 and ϕ_2 are said to be *equivalent*, denoted by $\phi_1 \Leftrightarrow \phi_2$, if $\phi_1 \Rightarrow \phi_2$ and $\phi_2 \Rightarrow \phi_1$ hold. ϕ_1 is said to be *simpler than* ϕ_2, denoted by $\phi_1 \sqsubseteq \phi_2$, if $\phi_1 \Leftrightarrow \phi_2$ and $atoms(\phi_1) \subseteq atoms(\phi_2)$. A simplification ϕ_1 of ϕ_2 is *minimal* if $\phi_1 \sqsubseteq \phi_2$ and there is no ϕ_1' such that $\phi_1' \sqsubset \phi_1$. For instance, if ϕ is $(\exists x, y)(P(a, x) \wedge P(a, y))$, then $(\exists x)(P(a, x))$ and $(\exists y)(P(a, y))$ are distinct but *equivalent* simplifications of ϕ. It is proven in [6] that if ϕ is in Φ and ϕ_1 and ϕ_2 are *minimal* simplifications of ϕ, then $atoms(\phi_1)$ and $atoms(\phi_2)$ are isomorphic. Minimal simplifications are called *cores* and the core of a given formula ϕ is denoted by $core(\phi)$.

A *database instance* is a set of instantiated atoms written as $atoms(Sk(\phi))$, where $Sk(\phi)$ is the Skolem version of a formula ϕ in Φ such that $core(\phi) = \phi$. A *constraint* (or rule) is a tuple-generating dependency (tgd) of the form

[1] If an update is not rejected, the updated database contains the inserted data and does not contain the deleted ones.

$(\forall X,Y)(body(X,Y) \rightarrow (\exists Z)head(X,Z))$, more simply written $body(X,Y) \rightarrow head(X,Z)$, where X, Y, and Z are vectors of variables, $body(X,Y)$ is a conjunction of atoms and $head(X,Z)$ is an atom. Constraint satisfaction is defined as usual: given a set I of instantiated atoms, $I \models c$ if for every homomorphism h such that $h(body(c)) \subseteq I$, there is a homomorphism h' such that $h(body(c)) = h'(body(c))$ and $h'(head(c))$ belongs to I. Here, by *homomorphism* we mean any mapping from the variables in c to constants or nulls. If $\mathbb{C}$ is a set of constraints, $I \models \mathbb{C}$ if for every c in $\mathbb{C}$, $I \models c$.

Graph Database Design in Neo4J. Our incremental approach is implemented through Cypher queries on a database, with the goal of optimizing the costly process of creating sets of atoms linked by null values. To make such retrieval efficient, given an atom $P(t_1,\ldots,t_n)$ our graph model represents P as a node, linked to other nodes representing the terms $t_1,\ldots,t_n$. All nodes have a *symbol* property and are classified into three types distinguished by their labels: *Atom* has the label **:Atom** and the value of their *symbol* property corresponds to the predicate symbol P *Constant* represent constant values and have two labels, **:Element** and **:Constant**. The value of their *symbol* property is the constant itself *Null* represent nulls and have two labels, **:Element** and **:Null**. The *symbol* property of such nodes is the null name, which is prefixed with '_'.

Atom are connected by an edge to *Constant* and *Null* having a *rank* property that identifies the position of the term within the atom. The model of the atom $P(t_1,\ldots,t_n)$ is illustrated in Fig. 1a, where t_i represents constants and t_j represents nulls. The notation on the right of the edges indicates the relationship cardinality between an atom and its terms: each element is connected to at least one atom, while atoms may have no terms. Figure 1b illustrates part of our database instance (rectangular nodes are atoms and circular nodes are elements) where optimization labels and attributes are omitted.

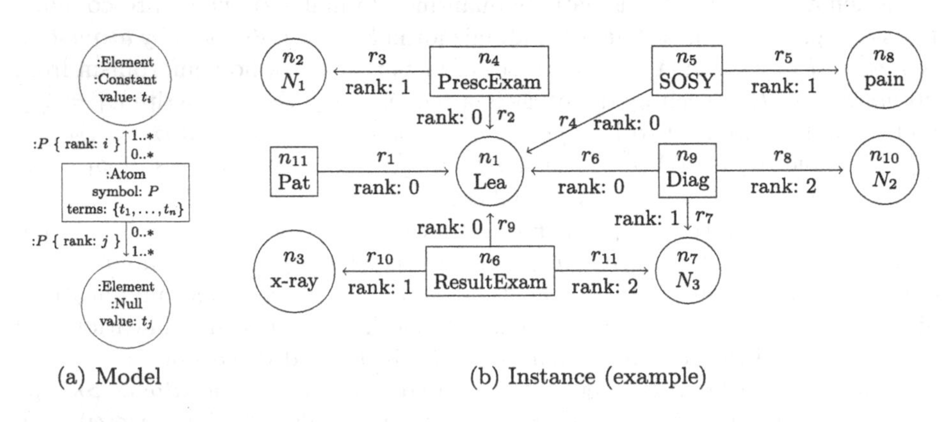

Fig. 1. Graph database model and an instance

Our model offers advantages for certain operations, but it increases the cost of converting between graph and logic formats for atoms. These conversions are essential for communication between the database and local computation procedures. To optimize conversions and graph traversals, we introduce the following design redundancies that significantly improve performance.

• *To avoid edge traversal:*

(a) Each *Atom* stores, as a property called *terms*, an ordered list containing the terms of the atom (rectangular node in Fig. 1a); *e.g.*, to obtain atom $SOSY(Lea, pain)$ in Fig. 1b starting with the node n_5, instead of traversing edges r_4 and r_5, we retrieve the attributes *terms* of node n_5.

(b) Each edge is assigned a label P named as the attribute *symbol* of its source *Atom* node (edges with label P in Fig. 1a); *e.g.*, to obtain all atoms of the form $SOSY(Lea, \_)$, starting from node n_1 in Fig. 1b, we just have to traverse r_4. Edges r_1, r_2, r_9 and r_6 have not to be visited.

• *To allow efficient access to nodes:*

(a) A uniqueness constraint is added on the property *symbol* of nodes with label **:Element** (implying that, *e.g.*, there is a unique node representing *Lea*). (b) An index is built on the property *symbol* of each node with label **:Atom**, and a uniqueness constraint is defined on the pair of properties *symbol/terms* (implying that, *e.g.*, there is a unique node representing atom $SOSY(Lea, pain)$).

3 Incremental Redundancy Reduction

In our approach, a database $\mathfrak{D}$ is expected to be equal to its core to avoid data redundancy. Formally, given a set of atoms I and a set of nulls ν occurring in I, we look for a homomorphism h such that $h(N) = N$ if $N \notin \nu$ and $h(I)$ is minimal so as $h(I) \subseteq I$. For instance, let $I_1 = \{PrescExam(Lea, N_1),\ ExamResult(Lea, N_1, N_2),\ PrescExam(Lea, x\text{-}ray),\ ExamResult(Lea, x\text{-}ray, N_3),\ ExamResult(Lea, scanner, N_4)\}$, and $\nu = \{N_1, N_2\}$. For h_1 such that $h_1(N_1) = x\text{-}ray$ and $h_1(N_2) = N_3$, we have $I_1' = h(I_1) = \{PrescExam(Lea, x\text{-}ray),\ ExamResult(Lea, x\text{-}ray, N_3),\ ExamResult(Lea, scanner, N_4)\}$. Notice that N_4 is *not* involved in the simplification.

Given I and ν_0, nulls *linked* in I to nulls in ν_0 have to be identified. We do so by computing for every N in ν_0, the set $\mathsf{LinkedNull}_{I,N}$ obtained as the limit of the sequence $\left(\mathsf{LinkedNull}^k_{I,N}\right)_{k\geq 0}$ defined by: (i) $\mathsf{LinkedNull}^0_{I,N} = \{A_i \in I \mid N \in null(A_i)\}$ and (ii) $\mathsf{LinkedNull}^k_{I,N} = \{A_i \in I \mid (\exists A_j \in \mathsf{LinkedNull}^{k-1}_{I,N})(null(A_i) \cap null(A_j) \neq \emptyset)\}$. It is indeed easy to see that for every $k \geq 0$, $\mathsf{LinkedNull}^k_{I,N} \subseteq \mathsf{LinkedNull}^{k+1}_{I,N}$ and $\mathsf{LinkedNull}^k_{I,N} \subseteq I$. Thus the sequence $\left(\mathsf{LinkedNull}^k_{I,N}\right)_{k\geq 0}$ is monotonic and bounded by I. As I is finite, this sequence has a unique limit, which is precisely the sub-set of I denoted by $\mathsf{LinkedNull}_{I,N}$.

It therefore turns out that redundancy has only to be checked with respect to the atoms in $\bigcup_{N\in\nu_0} \mathsf{LinkedNull}_{I,N}$ and the set ν of nulls occurring in this set.

Algorithm 1: $Simplify(I, \nu_0)$

1: $PSet := \{\mathsf{LinkedNull}_{I,N} \mid N \in \nu_0\}$
2: **for all** $P \in PSet$ **do**
3: Build the query q_{core} and compute its answer $q_{core}(I)$
4: **if** $\mid (q_{core}(I)) \mid > 1$ **then**
5: $h_m :=$ $ChooseMostSpec(q_{core}(I))$
6: $I := (I \setminus P) \cup h_m(P)$
7: **return** I

Algorithm 1 shows how redundancies are dealt with, given input I and ν_0: $\mathsf{LinkedNull}_{I,N}$ is computed for each N in ν_0 (line 1), and so, the nulls in $PSet$ constitute the set ν with respect to which I is simplified. On line 3, for each P in $PSet$, a query $q_{core} : ans(X) \leftarrow A_1(X_1), \ldots, A_n(X_n)$ is built by replacing each occurrence of N_i in P by a variable x_i. Thus, assuming that p nulls occur in P, when evaluating the answer $q_{core}(I)$ of q_{core}, the tuple $(N_1, \ldots, N_p)$ is obviously returned. However, it may happen that the answer contains other tuples, each of which defining a possible instantiation of the nulls in P, meaning that P is redundant. To implement these remarks, when the evaluation of q_{core} over I returns more than one tuple (line 4), one most specific tuple is chosen (line 5), and denoting by h_m the associated homomorphism, I is simplified (line 6) by replacing all atoms A in P by $h_m(A)$. Let us consider the set I_1 as above and $\nu_0 = \{N_1\}$. Then, $\mathsf{LinkedNull}_{I,N_1} = \{\ PrescExam(Lea, N_1),\ ExamResult(Lea, N_1, N_2)\ \}$. Thus: $q_{core}\ :\ ans(x_1, x_2)\ \leftarrow\ PrescExam(Lea, x_1), ExamResult(Lea, x_1, x_2)\}$ and $q_{core}(I_1) = \{(N_1, N_2), (x\text{-}ray, N_3)\}$. Hence, h_m such that $h_m(N_1) = x\text{-}ray$ and $h_m(N_2) = N_3$ is returned (line 5), and I_1 is simplified into I_1' as above.

To compute the most specific homomorphism h_m, we construct a table H_P with p columns and q rows, where p is the number of nulls in $P = \mathsf{LinkedNull}_{I,N}$, q is the number of answers returned by $q_{core}(I)$, and each h_i is an answer of $q_{core}(I)$ with h_1 being the identity. A cell $H_P[i, j]$ in H_P is set to $h_i(N_j)$. Given h_1 and h_2 over the same set of symbols Σ, h_1 is said to be *less specific than* h_2, denoted by $h_1 \preceq h_2$, if there exists a homomorphism h over Σ such that $h \circ h_1 = h_2$. We use H_P for identifying among the answers $h_1, \ldots h_q$, one most specific homomorphism h_m (see [5] for more details). Our approach is comparable to query optimization techniques [2,7] because we use tableau optimization. However, our approach differs from [2] in two fundamental ways: (1) our tableau is based on query answers rather than on the query body, and (2) we generate one most specific homomorphism, whereas in [2] non-most specific homomorphisms are discarded.

We compute LinkedNull with a Cypher query whose template is shown in Fig. 2. The `UNWIND` clause is used to convert a list into individual rows. A `MATCH` clause is used to identify patterns through homomorphisms in the LPG graph and returns a table of variable instantiations. Here, the `MATCH` clause looks for paths starting with the null of the `nullValueNode` to any other node representing an atom which is not `nullValueNode` itself (condition imposed by the

WHERE clause). On line 6, the WITH clause performs a 'group by' to structure the table with tuples where each null nullValueNode is associated to a list of endNodes (the nodes reached by paths pathP). On line 7 a new organisation is built: linkedNodes is divided into two lists, one containing nodes that represent predicate symbols (linkedAtoms) and one for those representing nulls (linkedNulls). Notice that we place the initial node nullValueNode in the first position of the latter list. In the resulting table, each atom is associated to a list of nulls.

4 Incremental Chase and Update Procedures

```
1  UNWIND $nulls AS nullPredName
2  MATCH (nullValueNode:Element:Null {value: nullPredName}),
3      pathP = (nullValueNode)-[*1..maxPathLength]-(endNode)
4  WHERE endNode <> nullValueNode AND
5      ALL(n IN nodes(pathP) WHERE NOT (n:Constant))
6  WITH COLLECT(DISTINCT endNode) AS linkedNodes, nullValueNode
7  WITH [n IN linkedNodes WHERE (n:Atom)] AS linkedAtoms,
8     [nullValueNode] + [n IN linkedNodes WHERE (n:Null)] AS linkedNulls
9  UNWIND linkedAtoms AS a RETURN a.symbol as a, a.terms as e, linkedNulls
```

Fig. 2. Cypher template to find LinkedNull sets

Insertions. Algorithm 2 inserts atoms from set iRequest into $\mathfrak{D}$. The side-effects are computed using a chase procedure (line 1), and the core of $\mathfrak{D} \cup ToIns$ is computed by considering only the nulls in *NullBucket* (retrieved through q_{bucket} line 2) and their linked nulls. $q_{bucket}(I)_{[S]}$ searches for null values that appear in atoms that are less specific than one in S. Instead of imposing restriction on the constraint format, we use a maximal degree δ_{max} to control null value depth and avoid infinite chasing. At each insertion, null degrees are set to 0. When a constraint is applied, generated nulls are assigned a degree of $\delta + 1$, where δ is the maximal degree of nulls in the constraint body, or 0 if no null occurs. If $\delta(N) \geq \delta_{\max}$, insertion is stopped, and $\mathfrak{D}$ is not changed. In Algorithm 2, if all nulls in the simplified instance have degree less than $\delta_{\max}$ (check through q_{degree}

Algorithm 2: Insert($\mathfrak{D}, \mathbb{C}, \delta_{max}$, iRequest)

1: $ToIns := Chase4Insert(\mathfrak{D}, \mathbb{C}, \delta_{max}, \text{iRequest})$
2: $NullBucket := \{N_j \mid N_j \text{ is a null obtained by } q_{bucket}(\mathfrak{D} \cup ToIns)\}$
3: $\mathfrak{D}' := Simplify(\mathfrak{D} \cup ToIns, NullBucket)$
4: **if** $q_{degree}(\mathfrak{D}')_{[NullBucket, \delta_{max}]}$ **then**
5: $\quad q_\delta(\mathfrak{D}')_{[NullBucket, 0]}$ {for each N in *NullBucket*, if in $\mathfrak{D}'$, sets $\delta(N)$ to 0}
6: $\quad$ **return** $\mathfrak{D}'$
7: **else**
8: $\quad$ **return** $\mathfrak{D}$

---

line 4), null degrees are set to 0 (through q_δ line 5), and $\mathfrak{D}'$ is returned since it is always consistent (as proven in [6]); otherwise, the database is not modified.

Deletions. Our incremental algorithm for the deletion from $\mathfrak{D}$ of atoms in dRequest is displayed in Algorithm 3. On line 1, all atoms in $\mathfrak{D}$ isomorphic to one in the set dRequest are retrieved through the query q_{Iso}. For instance, if dRequest $= \{P(a, N_1)\}$ and $\mathfrak{D}_1 = \{P(a, N_5)\}$, then query q_{Iso} returns $\{P(a, N_5)\}$. On line 2, the *Chase4Delete* function incrementally computes the side-effects by generating two sets of atoms, $ToDel$ and $ToIns$, which represent atoms that should be deleted and inserted as side-effects, respectively. Once these side-effects have been incorporated in $\mathfrak{D}$ to produce $\mathfrak{D}'$ (line 3), this new instance is simplified as in the case of insertion: impacted nulls are generated on line 4 and the simplified instance is computed on line 5. We notice that, contrary to insertions, deletions are *never* rejected.

Algorithm 3: $Delete(\mathfrak{D}, \mathbb{C}, \delta_{max}, \text{dRequest})$

1: isoDel $:= q_{Iso}(\mathfrak{D})_{[\text{dRequest}]}$
2: $ToDel, ToIns := Chase4Delete(\mathfrak{D}, \mathbb{C}, \delta_{max}, \text{isoDel})$
3: $\mathfrak{D}' := (\mathfrak{D} \cup ToIns) \setminus ToDel$
4: $NullBucket := \{N_j \mid N_j \text{ is a null obtained by } q_{bucket}(\mathfrak{D}')_{[ToIns \cup ToDel]}\}$
5: $\mathfrak{D}' := Simplify(\mathfrak{D}', NullBucket)$
6: **return** $\mathfrak{D}'$

Chasing. Different chasing versions have been suggested in literature [19]. Our method uses the parameter $\delta_{\max}$ to deal with constraints without any limitations. Our approach is an incremental version of the chasing method presented in [6], which is closely related to standard and core chase procedures. Two similar routines implement the chasing reasoning by using a query denoted by q_{ch}.

Chase4Insert is called on line 1 of Algorithm 2. It avoids the generation of unnecessary side effect atoms by activating a constraint c only if: (i) $body(c)$ contains at least one atom being inserted, (ii) atoms in $body(c)$ that don't meet (i) map to atoms in the database instance $\mathfrak{D}$, and (iii) the instantiation of $head(c)$ is inserted only if no equivalent atom already exists in $\mathfrak{D}$. In logical terms, the query is $q_{ch} : Q(\alpha) \leftarrow L_1(\alpha_1), \ldots, L_m(\alpha_m), not\ L_0(\alpha_0)$, where α is the list of variables corresponding to variables in $body(c)$. If $h_t(body(c)) \subseteq \mathfrak{D}$, then q_{ch} has a non-empty answer only if $h'_t(L_0(\alpha_0)) \notin \mathfrak{D}$ for any extension h'_t of h_t. Furthermore, in *Chase4Insert*, during a chase step, the degree of newly created nulls is computed and the condition $\delta(h'(head(c))) \leq \delta_{max}$ is verified *i.e.*, we do not trigger rules having nulls in $body(c)$ that do not meet this condition.

$c_1 : Pat(x), SOSY(x, y)^- \rightarrow PrescExam(x, z)$
$c_2 : PrescExam(x, z)^-, PlaceOfExam(z, w) \rightarrow ExamResult(x, z, y)$
$c_3 : ExamResult(x, y, z)^- \rightarrow Diag(x, y, w)$

Fig. 3. Set of (general) constraints

Chase4Delete is called at line 2 of Algorithm 3. To determine the constraints affected by the deletion of atom A, a *backward chase* is performed. This chase identifies an instantiation h such that $h(head(c)) = A$. Then the homomorphism restriction $h|body(c)$ is applied to c to generate an extended instantiation h'. We check if $h'(head(c))$ is isomorphic to the atom $h(head(c))$ being deleted, using the following reasoning: (1) If an isomorphic atom is generated, we insert the atom marked as '−' from $h(body(c))$ into $ToDel$, as at least one atom in $h(body(c))$ must be deleted to prevent c from being triggered. Notice that, to avoid non-determinism, one atom in the body of the constraints is marked as '−' during constraint design (to indicate deletion priority). (2) If no isomorphic atom is generated, we add $h'(head(c))$ to $ToIns$ along with its side effects. However, we also check whether δ_{max} is respected when computing the side effects of $h'(head(c))$. If not, we insert the marked atom from $h(body(c))$ into $ToDel$.

Example 1. In the context of clinical cases, let $\mathbb{C}$ be the abridged version of constraints depicted in Fig. 3. Constraint c2 links prescribed exams, that have taken place in a medical center, to results, while constraint c3 connects exam results to diagnoses.

Let $\mathfrak{D}_1 = \{Pat(Lea), SOSY(Lea, \textit{pain on hands}), PrescExam(Lea, N_1)\}$ be a database instance consistent with respect to $\mathbb{C}$ and $\delta_{max} = 3$. Given iRequest = $\{PrescExam(N_2, testCovid), PlaceOfExam(testCovid, LabA), PrescExam(Lea, \textit{x-ray})\}$, only constraints c2 and c3 are triggered. Algorithm 2 (line 1) returns the set $ToIns = \{PrescExam(Lea, \textit{x-ray}), PrescExam(N_2^0, testCovid), Diag(N_2^0, covidTest, N_4^2), PlaceOfExam(testCovid, LabA), ExamResult(N_2^0, covidTest, N_3^1)\}$ where exponents represent null degree. Let $I = \mathfrak{D}_1 \cup ToIns$.

For core computation, the query q_{Bucket} find null values in I that concerns the predicates in $ToIns$. Thus, $NullBucket = \{N_1, N_2, N_3, N_4\}$ and by Algorithm 1, we obtain that $\mathsf{LinkedNulls}_{I,N_1} = \{N_1\}$ and $\mathsf{LinkedNulls}_{I,N_2} = \{N_2, N_3, N_4\}$. The simplification of I (line 3 of Algorithm 2) results in: $\mathfrak{D}_2 = \{Pat(Lea), PrescExam(Lea, \textit{x-ray}), SOSY(Lea, \textit{pain on hands}), PlaceOfExam(covidTest, LabA), Diag(N_2, covidTest, N_4), PrescExam(N_2, covidTest), ExamResult(N_2, covidTest, N_3)\}$.

Consider now Algorithm 3 applied to $\mathfrak{D}_3 = \{SOSY(Lea, \textit{pain on hands}), Pat(Lea), PrescExam(Lea, \textit{x-ray})\}$. The deletion of $PrescExam(Lea, \textit{x-ray})$ implies $ToDel = \{PrescExam(Lea, \textit{x-ray})\}$ and $ToIns = \{PrescExam(Lea, N_1)\}$, since re-applying c1 on $\mathfrak{D}_3 \setminus \{PrescExam(Lea, \textit{x-ray})\}$ generates an atom with a null value. The new updated instance is $\mathfrak{D}_4 = \{PrescExam(Lea, N_1), Pat(Lea), SOSY(Lea, \textit{pain on hands})\}$. If we now require the deletion of $PrescExam(Lea, N_1)$ from $\mathfrak{D}_4$, re-applying c1 on $\mathfrak{D}_4 \setminus \{PrescExam(Lea, N_1)\}$ generates an atom isomorphic to the one being deleted. Regarding side-effect deletions, $ToDel = \{PrescExam(Lea, N_1), SOSY(Lea, \textit{pain on hands})\}$ and $ToIns = \emptyset$. The updated instance is $\mathfrak{D}_5 = \{Pat(Lea)\}$. □

5 Experimental Study

We tested using three data sets: *Movie*[2] and *GOT*[3] are Neo4J instances with 7 and 19 predicate symbols, respectively, and *Social*[4] is a data set from the Linked Data Benchmark Council with 23 predicate symbols. We refer to [5] and our gitlab repository [17] for a detailed explanation of how we generated additional data from the original sources, ensured consistency in our database, and controlled the generation of nulls. We have 9 instances as illustrated in Table 1, where 8 of them contains nulls and 1 instance without nulls.

Runs are built from instances in Table 1 by varying the update type (insertion or deletion), adjusting the size of the update (1, 5, 10, 20 atoms), and artificially increasing the number of facts by duplicating data n-times (scales 1, 2, 5). During each run, 10 iterations were performed, with 3 warm-up iterations to preload the system and database cache. The database is restored between each iteration and the Java garbage collector is triggered to ensure consistent timing. The benchmarks were implemented in Java 16 with MySQL 8 and Neo4J 4.4, running on a Rocky Linux 8.7 with 4 vCPU and 16 GB of memory, achieving an average of 1 GBs^{-1} read/write on disk, using docker 20.10.21.

Table 1. Datasets used in experiments

Database	Nb of facts	Nb of nulls	Nb of rules	Null/Facts (τ)
Movie	604	340	12	0.56
GOT	24818	17232	32	0.69
$Social_{1K}$	2248	190	39	0.08
$Social_{10K}$	16559	1183	39	0.07
$Social^{0N}_{10K}$	16559	0	39	0.00
$Social^{50N}_{10K}$	16559	50	39	0.00
$Social^{100N}_{10K}$	16559	100	39	0.01
$Social^{500N}_{10K}$	16559	500	39	0.03
$Social^{1000N}_{10K}$	16559	1000	39	0.06

Impact of Our Incremental Maintenance Approach. The results demonstrate the effectiveness of using a DBMS that implements incremental update processing for efficiently updating large databases. Due to the high memory requirements of the from-scratch method, the comparison is limited to the *Movie* database only. When tested on *Movie*-scale1, these updating approaches are comparable. We obtain, on average: 9017 ms for the in-memory, from-scratch version and, for the query-based, incremental ones, 151 ms for MySQL and 2380 ms for Neo4J. When considering an instance five times larger, we get an average of 888966 ms for the in-memory version, 595 ms for MySQL and 2706 ms for Neo4J.

Impact of Using a Graph Database. The primary goal of our experiments is to leverage the benefits of graph databases in enhancing the linked null search operation. An implementation of our incremental approach over a relational database (with nulls represented by special (Skolem) constants) is settled as

[2] https://github.com/neo4j-graph-examples/movies.
[3] https://github.com/neo4j-graph-examples/graph-data-science.
[4] https://ldbcouncil.org/benchmarks/graphalytics/.

our baseline. The results demonstrate that our graph data model significantly enhances retrieving LinkedNull sets, but its impact on other essential maintenance operations was more negative than expected. The retrieval of LinkedNull sets was found to be the most expensive operation of our updating policy in the relational model, as shown in Fig. 4 where outsiders with more than 30 s differences are removed. Our graph model, discussed in Sect. 2, focuses on optimizing this aspect. The results are amazing with the retrieval of LinkedNull sets being 25 times less expensive in the graph model than in the relational model (a reduction of 96%). We expected our proposed model to have a bad performance for chasing, as pattern matching is known to be time-consuming, and our model generates queries with more complex patterns. However, the actual results were even worse than anticipated, with the chasing operation being 170 times more expensive in the graph model than in the relational model. As a result, the overall performance of the relational model outperforms the graph model.

Detailed Performance Analysis. We analyze incremental updating performance with respect to database size, the number of nulls and the number of queries generated to interact with the DBMS. In Fig. 5, each plot's right axis represents the total number of facts in the instance. The curves indicate the average resulting values for all runs corresponding to the displayed abscissa. To enhance readability, the plots disregard outcomes for *GOT* occurrences with over 17000 nulls. Figures 5a and b show that for MySQL databases, the updating runtime increases with the number of nulls, but it is slightly impacted by the database instance size. For Neo4J, the situation is the completely inverse of what was observed in MySQL. The type of the update (insert or delete) does not affect the performance of our approach, regardless of the database system. The number of generated queries increases with the increase of the number of nulls. The increase in nulls implies an augmentation in the number and size of LinkedNull sets. As a consequence, the MySQL version generates a large amount of queries that impacts its performance while the impact on the Neo4J version is negligible. Our approach is better suited for data sets with predicates of high arity, rather than those composed of binary atoms, such as the *Social* dataset.

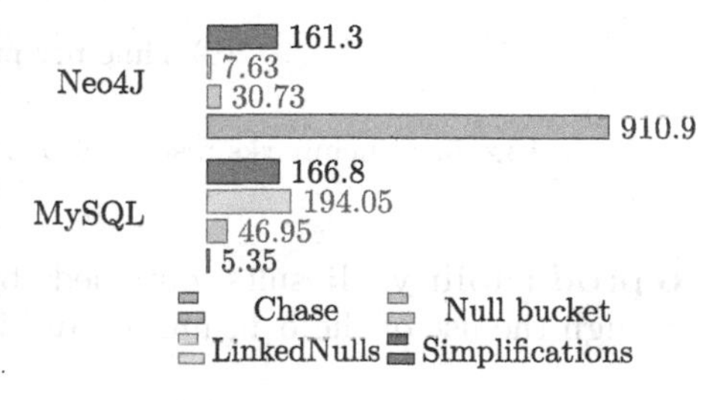

Fig. 4. Average time (ms) per DBMS

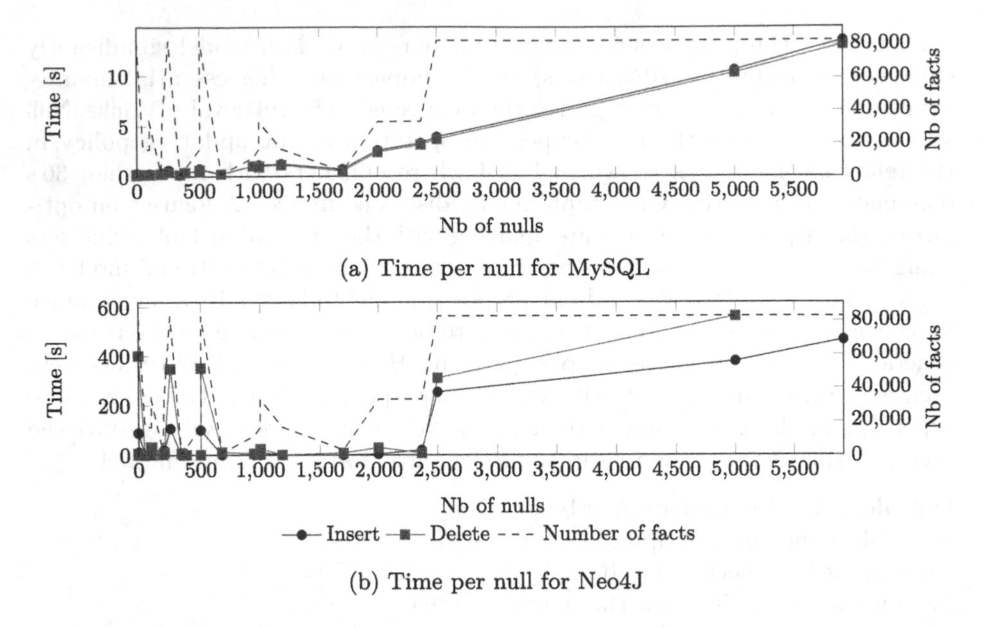

(a) Time per null for MySQL

(b) Time per null for Neo4J

Fig. 5. Benchmarks results of 540 scenarios, average over 10 runs

Reproducibility. Results obtained by our experiments are reproducible through the use of the benchmarks and implementation available in [17].

6 Related and Future Work: Conclusions

Given our experimental results, how can we address the questions posed in our abstract? Yes, utilizing a graph database and our data model can significantly improve the efficiency of retrieving linked null partitions. However, this improvement comes at a high cost, as the chasing operation is considerably less efficient on a graph compared to the relational model. A potential solution is to add design redundancy by overlaying different graph designs, such as connecting *Atoms* (e.g., connecting *patient* and *exam* nodes with an edge *prescribeTo*). Yet, the graph's complexity and update difficulty should be noted.

Our study confirms the pros and cons of using graph databases. They perform poorly when it comes to handling intricate pattern matching. The design of a graph database heavily relies on queries, unlike a relational database that presents a uniform data structure to work with. Consequently, queries that are not optimized for the database model may result in poor performance. However, they excel at path traversal queries, as nodes store information concerning their neighbourhood. As a result, they are an attractive option for exploring data relationships and for considering data analytics techniques such as predicting node connections. But how accurate are the results of data analytics on possibly inconsistent data? Constraints in graph databases are not widely used, but Neo4J has proposed some options. Starting with key for graphs [12], the graph entity dependencies (GEDs) have been proposed and their static analysis properties have been studied [13]. We refer to [3,4] for excellent overviews on the subject.

Updates are often less prioritized than queries in scientific research, despite evidence that maintaining database coherence in a dynamic environment can be complicated (see *e.g.*, [15,21]). Handling incomplete data in databases is also a difficult problem. Today, it deserves attention, especially in light of the increasing interest in *certain* answers [8]. While there is a solid foundation for addressing incompleteness in relational databases [1,10,11,14,18,20,23,24], incompleteness beyond the relational data model has received less attention [22]. As a result, updating with respect to constraints is rarely considered in this context.

Our approach is a step in this direction and raises the question of how representing linked nulls on graphs. It employs Reiter's semantics for unknown data to address the consistency maintenance problem from a logical standpoint. Our exclusively positive constraints allow for proven correction and completion of our updating policy (in [6]). It is possible to encounter null values in the query answers, which implies that, for the moment, nothing in our database allows us to provide their instantiation. They may also indicate a connection with other data. This aligns with the needs of our projects, but we must also consider modern applications that require data analytics. Which model should we adopt to meet these needs? A hybrid database model [16] could potentially be a solution, but it needs to be flexible enough to handle multiple representations of the data in each data model. Our incremental approach is a valuable tool in this context because it is designed to be independent of the data model.

Acknowledgements. Work partially supported by projet SENDUP (ANR-18-CE23-0010) and developed in the context of the DOING action (MADICS and DIAMS). We express our gratitude to the interns who contributed to this project, in particular Lucas Moret-Bailly for his valuable suggestions.

References

1. Abiteboul, S., Grahne, G.: Mise-à-jour des bases de données contenant de l'information incomplète. In: Journées Bases de Données Avancés, 6–8 Mars 1985, St. Pierre de Chartreuse (Informal Proceedings) (1985)
2. Aho, A.V., Sagiv, Y., Ullman, J.D.: Efficient optimization of a class of relational expressions. ACM Trans. Database Syst. **4**(4), 435–454 (1979)
3. Angles, R., et al.: PG-Keys: keys for property graphs. In: SIGMOD Conference, pp. 2423–2436. ACM (2021)
4. Bonifati, A., Fletcher, G., Voigt, H., Yakovets, N.: Querying Graphs. Synthesis Lectures on Data Management, Morgan & Claypool Publishers (2018)
5. Chabin, J., Halfeld Ferrari, M., Hiot, N., Laurent, D.: Incremental consistent updating of incomplete databases (extended version - technical report). Technical report, LIFO- Université d'Orléans, (2023). www.hal.science/hal-03982841
6. Chabin, J., Halfeld Ferrari, M., Laurent, D.: Consistent updating of databases with marked nulls. Knowl. Inf. Syst. **62**(4), 1571–1609 (2020)
7. Chandra, A.K., Merlin, P.M.: Optimal implementation of conjunctive queries in relational data bases. In: Symposium on the Theory of Computing (1977)

8. Console, M., Guagliardo, P., Libkin, L., Toussaint, E.: Coping with incomplete data: recent advances. In: PODS, pp. 33–47. ACM (2020)
9. Fagin, R., Kolaitis, P.G., Popa, L.: Data exchange: getting to the core. ACM Trans. Database Syst. **30**(1), 174–210 (2005)
10. Fagin, R., Kuper, G.M., Ullman, J.D., Vardi, M.Y.: Updating Logical Databases. Advances in Computing Research, vol. 3, pp. 1–18 (1986)
11. Fagin, R., Ullman, J.D., Vardi, M.Y.: On the semantics of updates in databases. In: Proceedings of the 2nd ACM SIGACT-SIGMOD Symposium on Principles of Database Systems, Atlanta, Georgia, USA, pp. 352–365 (1983)
12. Fan, W., Fan, Z., Tian, C., Dong, X.L.: Keys for graphs. Proc. VLDB Endow. **8**(12), 1590–1601 (2015)
13. Fan, W., Lu, P.: Dependencies for graphs. In: Proceedings of the 36th ACM SIGMOD-SIGACT-SIGAI Symposium on Principles of Database Systems, PODS, Chicago, USA, pp. 403–416 (2017)
14. Grahne, G.: The Problem of Incomplete Information in Relational Databases. Lecture Notes in Computer Science, vol. 554. Springer, Heidelberg (1991). https://doi.org/10.1007/3-540-54919-6
15. Alves, M.H.F., Laurent, D., Spyratos, N.: Update rules in datalog programs. J. Log. Comput. **8**(6), 745–775 (1998)
16. Hassan, M.S., Kuznetsova, T., Jeong, H.C., Aref, W.G., Sadoghi, M.: GRFusion: graphs as first-class citizens in main-memory relational database systems. In: SIGMOD Conference, pp. 1789–1792. ACM (2018)
17. Hiot, N., Moret-Bailly, L., Chabin, J.: www.gitlab.com/jacques-chabin/UpdateChase (2023)
18. Imielinski, T., Lipski, W., Jr.: Incomplete information in relational databases. J. ACM **31**(4), 761–791 (1984)
19. Onet, A.: The chase procedure and its applications in data exchange. In: Data Exchange, Integration, and Streams, pp. 1–37 (2013)
20. Reiter, R.: A sound and sometimes complete query evaluation algorithm for relational databases with null values. J. ACM **33**(2), 349–370 (1986)
21. Schewe, K., Thalheim, B.: Limitations of rule triggering systems for integrity maintenance in the context of transition specifications. Acta Cybern. **13**(3), 277–304 (1998)
22. Sirangelo, C.: Representing and querying incomplete information: a data interoperability perspective (2014). www.tel.archives-ouvertes.fr/tel-01092547
23. Winslett, M.: Updating Logical Databases. Cambridge University Press, New York (1990)
24. Zaniolo, C.: Database relations with null values. J. Comput. Syst. Sci. **28**(1), 142–166 (1984)

Assessment of Data Quality Through Multi-granularity Data Profiling

Claudia Diamantini, Alessandro Mele, Domenico Potena, and Emanuele Storti(✉)

DII, Polytechnic University of Marche, Ancona, Italy
{c.diamantini,d.potena,e.storti}@univpm.it

Abstract. The management of modern solutions for Big Data management and analytics, most notably Data Lakes and Data Lakehouses, is faced with new challenges stemming from the versatility offered by such technologies, as well as the continuously evolving variety and volume of data sources, necessitating the tracking of data quality concerns. In this scenario, this paper proposes a metadata management framework for summary data sources with the capability to generate data profiles at various levels of detail. The approach leverages a Knowledge Graph, which defines dimensions and measures according to the multidimensional model. Profiles are then exploited to efficiently assess a set of quality properties of sources in a Big Data framework, including completeness, coverage and consistency that are formally defined and evaluated.

Keywords: Big Data management · Metadata · Knowledge graph · Multidimensional model · Data Profiling · Data Quality · Data Lake · Data Lakehouse

1 Introduction

Modern solutions for Big Data management and analytics are characterized by great versatility and flexibility, which also pose significant challenges, particularly for data quality. Without an efficient metadata management system and data quality assessment, data stored in Big Data repositories may be corrupted, incomplete, or inconsistent. Poor information and inaccurate analysis can have detrimental effects on business decisions and consequently, business performance, innovation, and competitiveness. In fact, according to Gartner, 40% of all business initiatives fail to achieve their targeted benefits as a result of poor data quality [11].

This particularly holds for BI applications operating on top of Data Lakes storing both structured and unstructured data, without any predefined structure. In this architecture, often a subset of data is ETLed to a downstream Data Warehouse for analytical purposes. More recently, the notion of Data Lakehouse is emerging to combine the key benefits of Data Lakes and Data Warehouses: a

A. Abelló et al. (Eds.): ADBIS 2023, LNCS 13985, pp. 195–209, 2023.
https://doi.org/10.1007/978-3-031-42914-9_14

scalable, flexible, and cost-effective solution for managing large amounts of data while providing powerful management and optimization features for advanced analytics, including the management of integrated ETL processes and Data Warehouses into the Data Lakehouse. As such, while traditional databases primarily focus on raw data, Data Lake(houses) often require to manage summary data, i.e. statistical measures or indicators derived from raw data, either as the result of ingestion of external sources[1], or as the result of ETL processes. An example is the COVID 19 Data Lake maintained by Microsoft Azure[2], which includes statistical data sources on the effects of the pandemic. These kinds of data have a specific structure, suitably described by the multidimensional model typical of Data Warehouses, and rise specific issues on metadata management and data quality that have not been taken into account by the literature.

In this work, we propose a framework for data quality assessment of summary data sources. The framework includes a model of source metadata that is semantically enriched by mappings to concepts of a Knowledge Graph. Collectively, we call it the Semantic Data Lake model. We introduced the semantic model in a previous work to deal with data integration and answering in Data Lakes [9], using the COVID Data Lake as a real-world validation workbench. In the present work, the model is enriched with data statistics and leveraged to define and calculate quality measures. In particular, we focus on completeness, coverage and consistency. The former is calculated by comparing data to the domain knowledge provided by the Knowledge Graph, which by definition is business-focused. Depending on what the organization defines as relevant concepts, it is then possible to determine how much available data covers such concepts. In other terms, it is possible to get an automated measurement of coverage by defining, at least partially, the "universe of discourse". Finally, consistency aims to validate data w.r.t. the domain knowledge.

Specifically, the contribution of the present work is as follows:

- we introduce the concept of multi-granularity profile, based on a hierarchical dimension representation within the Metadata layer of the framework;
- we define mechanisms for computation of the profile across different granularities;
- we formally define a number of measures to assess properties for data quality evaluation, namely completeness, coverage and consistency, based on multi-granularity profiles, providing also formal demonstration of upper and lower bounds.

The rest of the paper is structured as follows: in Sect. 2 we survey relevant research in the literature. Section 3 is devoted to introduce the Semantic Data Lake model, focusing on the definition of multi-granularity profile. Measures for the assessment of Data Lake (house) quality properties are discussed in Sect. 4.

[1] For example open data by public bodies reporting economic trends or the results of governmental initiatives.

[2] https://docs.microsoft.com/en-us/azure/open-datasets/dataset-covid-19-data-lake.

Section 5 provides an evaluation of the approach, while Sect. 6 concludes the work and draws future research lines.

2 Related Work

Unlike data warehouses or DBMS, Data Lake (house)s may not be accompanied with descriptive and complete data catalogs, which are essential to on-demand discovery and integration. A critical requirement is hence extracting metadata from sources and enriching data with meaningful metadata (such as detailed data description and integrity constraints), to avoid the risk of data swamps. Among the several metadata management systems proposed in the literature, Google Dataset Search (GOODS) [4] collects and extracts metadata on owners, timestamps, schema, and relationships among multiple datasets, exposing them in dataset profiles, while Constance also enriches data sources by annotating data and metadata with semantic information [13]. Indeed, semantic models and technologies can play a valuable role by providing standardized ways to represent data, offering a comprehensive view of underlying data sources, modeling relationships and dependencies.

A particular type of metadata regards the content of data sources, in terms of statistics, or sketches providing a summary representation. Several profiling techniques can be used [2], for both single-columns (typically for supporting dataset discovery) or multiple-columns (for data exploration), with a comprehensive or approximate approach (e.g., sampling, sketches, wavelets). Such techniques can be applied on parallel architectures in order to speed up the computation in case of Big Data e.g. sampling-based approaches for incremental maintenance of approximate histograms [12].

In the Data Lake (house) domain, only a few work addressed the issue of defining a data quality model. However, much effort has been put on modeling and measuring data quality in databases [3] and data warehouses [1]. Although no general agreement exists on the set of dimensions capable to best define the quality of data, the majority of the authors focus on a basic set, i.e. accuracy, completeness, consistency, and timeliness. Only some of them, however, can be managed automatically in an open scenario such as the Data Lake (house), e.g. accuracy would require a comparison with real-world entities, which may not be known apriori. Data quality measures can be used for a variety of tasks, with a passive or pro-active approach aimed to enhance global data quality. In the DL field, the majority of work rely, as we do, on data quality measures to support discovery and integration, although some work investigated approaches enforcing quality constraints on the data right after ingestion and extraction. CLAMS [10] is an early approach to discover, design expressive integrity constraints over large unstructured and semi-structured datasets with very limited schema information, and enforce them through scalable, distributed algorithms to detect data inconsistencies.

Validation of data can be performed following multiple approaches, e.g. declarative data constraints describing how valid data should look like in the data pipelines[3] or by detecting dependencies, either through a (partially) supervised [14] or unsupervised approach (e.g., [18]). Different types of validation logics can be adopted when types are complex and data refer to a finite set of values from a given domain, as in the case of statistical data sources. Here, dictionary-based or semantic-based approaches can be adopted, e.g. based on shared vocabularies or through methods for automatic type identification, e.g., through deep learning as in Sherlock [15], possibly using contextual information [19].

3 Semantic Data Lake: Data Model

In this Section, we introduce the model for a Semantic Data Lake. We define a Semantic Data Lake as a tuple $SDL = \langle \mathcal{S}, \mathcal{G}, \mathcal{K}, \mathcal{O}, f \rangle$, where $\mathcal{S} = \{S_1, \ldots, S_n\}$ is a set of multidimensional data sources, $\mathcal{G} = \{G_1, \ldots, G_n\}$ is a set of metadata graph, one for each source, $\mathcal{K}$ is the Knowledge Graph, providing definitions of measures and dimensions according to a reference ontology $\mathcal{O}$, and $f \subseteq \mathcal{G} \times \mathcal{K}$ is a function mapping source metadata to Knowledge Graph concepts.

In Subsect. 3.1 we discuss the Knowledge Layer, which includes the reference ontology, and the Knowledge Graph. Then, in Subsect. 3.2 we introduce the Metadata Layer, which includes the graph representing the source metadata. Finally, in Subsect. 3.3 we discuss the mapping function linking the two layers.

Our approach is agnostic w.r.t. the degree of structuredness of the sources, ranging from relational datasets to semi-structured (e.g., XML, JSON) documents. Its application can be envisaged as part of data curation activities after data ingestion as well as after the execution of ETL processes.

3.1 Knowledge Layer

The knowledge layer of the Semantic Data Lake comprises the Knowledge Graph and the KPIOnto ontology. The *Knowledge Graph* is a directed edge-labelled graph, defined as a set of nodes representing entities and a set of directed labelled edges representing binary relations between nodes. We refer to Resource Description Framework (RDF) as a standard data model, thus enabling standard graph access and query mechanisms. The Graph provides a representation of the domain knowledge in terms of definitions of indicators, dimension hierarchies and dimension members.

On top of the graph we rely on the *KPIOnto* ontology, an OWL2-RL ontology aimed to provide the terminology to model indicators, dimensions and their properties. To the purpose of the present work, relevant classes and properties are those representing multidimensional hierarchies:

- `Dimension`: the class includes the definition of dimensions along which indicators are measured. It is aligned to the class `qb:MeasureProperty` in the RDF Data Cube Vocabulary [6].

[3] An example is Tensor Flow Data Validation: https://www.tensorflow.org/tfx/guide/tfdv.

- `Level`: the class represents a specific level, which is related to the corresponding dimension through the property `inDimension`. A level is also linked to the corresponding upper level through property `rollup`, which enables to encode the hierarchical relations among levels.
- `Member`: the class of members. Each instance belongs to a certain level through the property `inLevel`. A member can `rollup` to other members.

The full ontology specification can be reached at http://w3id.org/kpionto, while for a general description we refer the interested reader to [8]. In the following, we will refer to KPIOnto elements by using the prefix `kpi`.

The right part of Fig. 1 shows the fragment of a Knowledge Graph representing the *Geo* dimension, with some corresponding levels and members.

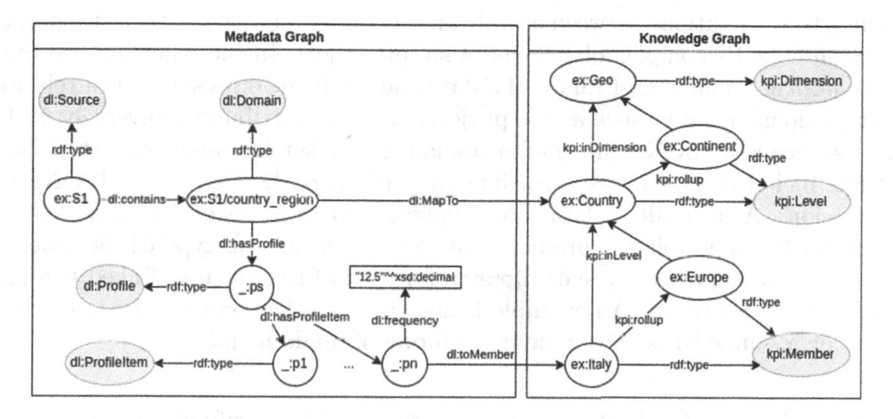

Fig. 1. A fragment of the Metadata and Knowledge graphs for the case study. Nodes representing classes are in yellow. (Color figure online)

3.2 Metadata Layer

The objective of the metadata layer is to represent and store a set of metadata enriching the data sources. Various metadata can be collected for a source, that can be classified under a number of categories [7,17].

In the following, we first discuss how a source is represented along with its schema in the metadata layer. Then, we focus on the representation of the data source profile. The graph is implemented as an RDF graph, where nodes are represented through a URI which is local to the Data Lake (house) with prefix `ex`. On the other hand, prefix `dl` is used for terms from a custom Data Lake (house) vocabulary we developed for this work.

Structural Metadata. Since the representation of schemas is highly model-dependent (e.g., relational databases vs. XML documents), a preliminary step is required for the management of a Data Lake (house), which uniformly represents this structural metadata into a common model. For this, we rely on a network-based approach proposed in [7]. The procedure for generating the structural metadata depends on the typology of data source, e.g., a relational database has tables with attributes, while XML/JSON documents include complex/simple elements and their attributes. For each source S_k, structural metadata are represented as a directed graph $G_k = \langle N_k, A_k, \Omega_k \rangle$, where N_k are nodes, A_k are edges and $\Omega_k : A_k \rightarrow \Lambda_k$ is a mapping function s.t. $\Omega_k(a) = l \in \Lambda_k$ is the label of the edge $a \in A_k$. The graph is built incrementally starting from the definition of a node s_k representing the data source S_k. Then, for each structural element of the source (e.g. a column in a relational database or an object in a JSON document), a new node is added. An edge $(n_x, n_y) \in A_k$ is defined to represent the *structural* relation existing between the elements x and y, e.g., this corresponds to the relations between a table and a column of a relational database, or between a JSON complex object and a simple object. In the following, we refer to *domain* as a multiset of values of a data source. If the data source is a relation table, a domain can be seen as the projection of one attribute. Conversely, if the data source is a JSON collection, a domain is the set of values extracted from all the included documents according to a given path (e.g., using JSONPath expression). As a result, a data source corresponds to a set of domains.

In RDF, the graph is represented by defining a node of type `dl:Source` for the source and nodes of type `dl:Domain` for each of its domains, linked through property `dl:contains`. An example is shown in the left part of Fig. 1, where a node for a source S1 is shown along with one of its domains.

General Metadata. Further metadata are added to specify information useful for exploration and analysis of the Data Lake (house). Each metadata element is represented as an additional node that is then linked to the node representing the source. They include DCMI metadata[4] (with prefix `dcterms`) such as the `title` of the source, its `description`, the file `format`, a number of `subjects` (typically linking to DBPedia resources), the `creator`, the `publisher` and possible `contributors`, the date of creation, the `licence`, along with several others properties. Access metadata are useful to define the complete path to the data source, and is expressed through a property `dl:location`. Furthermore, basic statistics such as the number of dimensions are represented through the property `dl:domains`, while the number of items through the property `dl:items`.

Multi-granularity Profile. The metadata graph of a data source is enriched with further information for domains that represent dimensions. The identification of such domains is detailed in the next subsection. We define *profile set* the metadata providing summary information on the value distribution of a specific domain of a data source. Each summary element is defined as a *profile item.*

[4] https://www.dublincore.org/specifications/dublin-core/dcmi-terms/.

Definition 1 *(Profile item). Let S be a source, d be a domain mapped to level L_j of the Knowledge Graph, let m_i be a member in L_j, then $p_i = \langle m_i, o_i \rangle$ is a profile item, where $o_i = |\{v \in d : v = m_i\}|/|d|$ is the number of occurrences of m_i in the domain d divided by the cardinality of d.*

For example, the domain "country_region" of source S1 includes a profile item $\langle Italy, 0.235 \rangle$, as the value "Italy" represents the 23.5% of the domain values.

Domain profiles are represented in the metadata graph of a data source by adding a node of class `dl:ProfileItem` for each profile item, which is linked (1) to the corresponding member node in the Knowledge Graph through property `dl:toMember` and (2) to a literal value (integer) specifying the relative frequency, through property `dl:frequency`. A domain is then associated to a node of type `dl:Profile` representing the profile, through property `dl:hasProfile`. In turn, the profile node is connected to each profile item through property `dl:hasProfileItem`, as shown in the left part of Fig. 1. If errors or noise occur in the data, there may be values that do not match any member of the level. In such a case, the profile will include also the percentage of non-matched values as a literal through property `dl:unrecognized`.

Given the mapping between each profile item and a member of the mapped level in the Knowledge Graph, the profile for a domain can be easily computed for upper levels of the dimension hierarchy.

Definition 2 *(α-level granularity profile). Given a source S, a domain d mapped to a level L_j of a dimension D, a profile set $ps^j = \{p_1^j, \ldots, p_n^j\}$ with $p_i^j = \langle m_i^j, o_i^j \rangle$, a level $L_k > L_j$ in the dimensional hierarchy, the L_k-granularity profile for d is given by $ps^k = \{\langle m^k, o^k \rangle : inLevel(m^k, L_k), o^k = \sum_{i=1}^{n} o_i^j\}$.*

Practically, this can be done through aggregation, by exploiting `rollup` relations in the Knowledge Graph, as shown in Listing 1.1. To make an example, the domain "country_region" in source S1 is mapped to level *Geo.Country*. As such, it is possible to calculate its profile at the *Geo.Continent* granularity level, by aggregating the profile items by continent.

Listing 1.1. SPARQL query for generation of upper profile for a given domain.

```
SELECT ?upper_m (SUM(?f) as ?o)
WHERE {<domain> dl:hasProfileElement ?b.
        ?b dl:toMember ?mroll.
        ?mroll kpi:mRollup ?upper_m.
        ?b dl:frequency ?f}
GROUP BY ?mroll
```

Finally, we refer to *source profile* as the set of all the domain profiles of a data source.

The complexity of calculating profiles is linear with respect to the number of values of the domain. The cardinality of each profile set is always less than or at most equal to the number of members in the related mapped level. As an example, if the level *Geo.Country* includes 1000 members, the profile set of domain "country_region" of source S1 will have a cardinality ≤ 1000.

Although profiles are computed only for domains representing dimensional levels (see next for dimension identification), calculation may be too costly in a Data Lake (house) context. Since the relative occurrence of values is what is needed, one way to reduce computation time is to estimate profiles only on a subset of the data. Simple random sampling without replacement is a flexible, general method commonly adopted in most practical cases. It can be easily applied in distributed computation frameworks, e.g. Spark, through data partitioning [16]. Clearly, the accuracy of the estimate will increase as more samples are analyzed, defining a balance between speed and accuracy of the result.

3.3 Mapping Function

The data model is completed by defining the mapping function $f \subseteq \mathcal{G} \times \mathcal{K}$, which links the metadata layer to the knowledge layer. We consider three types of mappings, to semantically enrich source elements representing indicators, dimension levels, and their members, respectively.

Specifically, let $G_i \in \mathcal{G}$ be the metadata graph of source S_i:

- if node $n_d \in G_i$ represents a dimension, it is mapped to $L \in \mathcal{K}$ representing the dimension level semantically equivalent to n_d. This relation is discovered following the approach summarized in the next subsection and is represented through the `dl:mapTo` property.
- if node $n_m \in G_i$ represents a measure, it is mapped to $ind \in \mathcal{K}$ representing the indicator semantically equivalent to n_m.
- For each node $n_{p_i} \in G_i$ representing the profile $p_i = \langle v_i, o_i \rangle \in ps^{n_d}$, the node is mapped to $m \in \mathcal{K}$ representing the member of the level L semantically equivalent to the value v_i. This relation is represented through the `dl:toMember` property.

The process to identify domains representing dimensions, given a new data source, and to properly map them to the Knowledge Graph, was discussed in [9]. It relies on the evaluation of the similarity between a domain and a dimensional level, based on an efficient estimation of set containment among them. This is an asymmetric variant of the Jaccard coefficient: formally, given two sets X,Y, the set containment is given by $c(X, Y) = |X \cap Y|/|X|$. Set containment is particularly suited to deal with skewed sets (i.e., with very different cardinality), as the measure is independent on the dimension of the second set. This is exactly the case of comparing a dimensional level, that may include a very large number of values (e.g., cities or regions), to a domain, which on the other hand typically has a much smaller number of distinct values. Instead of a complete, expansive computation through a complete comparison, we rely on an approximate evaluation based on MinHashes [5] and LSH Ensemble [20], a sublinear approximate algorithm.

For what concerns measures, since they are numerical domains and are not constrained to a finite number of values, solutions based on comparison cannot be applied. Several approaches can be considered ranging from string-based ones to those based on dictionary, semantic similarity (e.g., [7]) or frequency distribution and will be discussed in future work.

4 Data Quality Assessment

On top of the model of the Data Lake, in this Section we discuss a number of measures aimed to assess the quality of a data source, which exploits data profiles and other metadata. In particular, hereby we focus on measures for completeness, coverage and consistency, by providing definitions and methods for their computation. In our scenario, the evaluation can be performed dynamically either at source loading, discovery or query time, since it is based on profile metadata that are stored in the Metadata graph during the loading phase.

4.1 Data Completeness

Data completeness is recognized as one of the most relevant data quality dimensions as incomplete and/or inaccurate data may deeply impact data analytic tasks [21]. This measure can provide useful insights also to evaluate how much a source is general with respect to the Knowledge Graph, i.e. to what extent actual data focus on a particular subset of members (e.g. data referring only to a particular country). The completeness of a domain is defined as the ratio between the number of distinct values that can be mapped to members of the level (which is equivalent to the cardinality of the profile set), and the cardinality of the mapped level.

Definition 3 *(Completeness of a domain). Given a data source S, a domain $d \in S$ mapped to a level L having a number of members $|L|$, with a profile set $ps(d)$, the completeness of d is given by $com(d) = \frac{|ps(d)|}{|L|}$.*

This corresponds to calculating the set containment of L against d. To give an example, if *Geo.Country* includes 1000 different members and the distinct values for a given domain d that matches elements in L is 250, then $com(d, Geo.Country) = \frac{250}{1000} = 0.25$.

It is possible to extend the definition to evaluate the completeness for a whole data source. By considering that a source has a dimensional schema given by the set of domains that are mapped to corresponding levels in the Knowledge Graph, its theoretical maximum cardinality is given by the cardinality of the cartesian product of such levels. Indeed, for each combination of the members for all level, at most one item can be available in the data source.

Definition 4 *(Maximum theoretical cardinality for a source). Given a data source S, a set of domains $\{d_1, \ldots, d_n\}$, a set of mapped levels $\{L_1, \ldots, L_n\}$ such that d_i is mapped to L_i, its maximum cardinality is defined as $MaxCard(S)=\prod_{i=1}^{n} |L_i|$.*

To give an example, if a data source has a dimensional schema given by *Geo.Country* and *Time.Day*, it will contain at most a number of records equal to all possible combinations of countries and days in the Knowledge Graph. The following definition of completeness of a source stems from the previous one: the completeness is computed as the relative cardinality w.r.t. the maximum theoretical cardinality.

Definition 5 *(Completeness of a source). Given a data source S, a set of domains $\{d_1, \ldots, d_n\}$, a corresponding set of mapped levels $\{L_1, \ldots, L_n\}$ such that d_i is mapped to L_i, a maximum theoretical cardinality $MaxCard$, the completeness of S is given by $com(S) = \frac{|S|}{MaxCard(S)}$.*

4.2 Data Coverage

By only considering completeness as a quality measure does not enable to completely assess how much the Data Lake (house) is rich in terms of data source content. It may happen that, for instance, sources cover only a small percentage of the levels/members defined at Knowledge Graph level. We propose here a data coverage measure which focuses on the comparison of the Knowledge Graph to the data sources. The coverage of a dimensional level w.r.t. the set of all domains mapped to it is calculated as the percentage of the whole set of members that are present at least in the profile of one of such domains.

Definition 6 *(Coverage of a dimensional level). Given a level L having a number of members $|L|$, the set $\{d_1, \ldots, d_n\}$ of domains in the sources of the Data Lake such that d_i is mapped to L, the coverage of L is $cov(L) = \frac{|\bigcap_{i=1}^{n} ps(d_i)|}{|L|}$.*

By generalizing the previous measure, it is possible to evaluate the extended coverage of a level L by exploiting the notion of multi-granular profiles. In this case, also domains mapped to a sub-level of L are considered and L-granularity profiles are calculated through Definition 2. This measure is motivated by the fact that indicators not explicitly stored at a given level L can, in some cases, be calculated by aggregating indicators calculated at lower levels.

Definition 7 *(Extended coverage of a dimensional level). Given a level L_k having a number of members $|L|$, a set $D_L = \{d_1, \ldots, d_n\}$ of domains such that $\forall d_i \in D_L$, d_i is mapped to L_k, a set $X = \{x_1, \ldots, x_m\}$ of domains such that $\forall x_i \in X$, $\exists L_j$ such that x_i is mapped to $L_j < L_k$, let $U_L = \{\bigcup_{i=1}^{n} ps(d_i)\}$ and $U_X = \{\bigcup_{i=1}^{m} ps^k(x_i)\}$ respectively be the union of the set of profiles for D_L and the union of the set profiles for X calculated at granularity L. The extended coverage of L_k is $cov(L_k) = \frac{|U_D \bigcup U_X|}{|L|}$.*

4.3 Consistency

We refer to the term consistency to indicate the property of a data source to be correctly aligned with the Knowledge Graph. In particular, we focus on the set of domains representing the dimensional schema of a data source, and define a consistency index which measures the ratio of the data source items (e.g., records) whose values are validly mapped to members in the Graph.

Definition 8 *(Consistency of a level and a data source).*
Given a data source S including $|S|$ items, with a dimensional schema $D_S = \{d_1, \ldots, d_n\}$, where each d_i is mapped to a level L_j, with $ps_i = \langle m_i, o_i \rangle$. The consistency index of a domain d_i is $\mathit{Con}(d_i) = \sum_{i=1}^{n} o_i$. An item $t \in S$ is consistent if $\forall d_i \in D_S, d_i(t) = m$, where $m \in \mathcal{K}$ is such that `inLevel(m,L_j)`. Let N_t be the number of such consistent items in S. The consistency index of S is $\mathit{Con}(S) = \frac{N_t}{|S|}$.

The actual value of the consistency index can be determined only by a complete lookup of the data source. However, bounds can be determined analytically as follows, by considering that the index depends on the consistency level of each domain, which can be directly obtained from the domain profiles.

Theorem 1 (Lower and upper bound to consistency). *Given a source S and its dimensional schema represented by a set of domains $\{d_1, \ldots, d_n\}$ with $n > 1$, such that each d_i is mapped to a level L_j, with consistency $\mathit{Con}(d_i)$, for the consistency index $\mathit{Con}(S)$ the following property holds:*

$$max(0, 1 + \sum_{i=1}^{n}(\mathit{Con}(d_i) - 1)) \leq \mathit{Con}(S) \leq min(\mathit{Con}(d_i), \ldots, \mathit{Con}(d_n)). \quad (1)$$

Proof Let $X_i = \{t \in S : d_i(t) = m, inLevel(m, L_j\}$ be the set of items in S such that the restriction to d_i includes a valid member in the Knowledge Graph. By definition of consistency of a domain, $\mathrm{Con}(d_i) = \frac{|X_i|}{|S|}$. Let $X_1, \ldots, X_n$ be the sets corresponding to the domains $d_1, \ldots, d_n$. The intersection of such sets includes the items having valid members for all domains, i.e. $\mathrm{Con}(S) = \frac{|X_1 \cap X_2 \cap \ldots X_n|}{|S|}$.

An upper bound for $\mathrm{Con}(S)$ is obtained as follows. Each item in the intersection belongs to each of the sets, i.e. $\forall t \in S, t \in X_1 \cap X_2 \cap \ldots X_n \implies t \in X_1, t \in X_2, \ldots t \in X_n$, from which $|X_1 \cap X_2 \cap \ldots X_n| \leq |X_1|, \ldots, |X_1 \cap X_2 \cap \ldots X_n| \leq |X_n|$. As a consequence, $|X_1 \cap X_2 \cap \ldots X_n| \leq min(|X_1|, \ldots, |X_n|)$. As such, $\mathrm{Con}(S) = \frac{|X_1 \cap X_2 \cap \ldots X_n|}{|S|} \leq min(\frac{|X_1|}{|S|}, \ldots, \frac{|X_n|}{|S|}) = min(\mathrm{Con}(d_i), \ldots, \mathrm{Con}(d_n))$.

A lower bound for $\mathrm{Con}(S)$ is obtained as follows. From the inclusion-exclusion principle, given a set S and a number $k = 2$ of subsets $X_1, X_2 \subseteq S$, $|X_1 \cap X_2| = |X_1| + |X_2| - |X_1 \cup X_2|$. Given that $|X_1 \cup X_2|$ is bound to $|S|$, we have that $|X_1 \cap X_2| = |X_1| + |X_2| - |X_1 \cup X_2| \geq |X_1| + |X_2| - |S|$ if the right-hand expression of the inequality is greater than 0, or 0 otherwise. Let us now define $J = X_1 \cap X_2$. Let us consider a case with $k + 1$ subsets, i.e. 3, with X_3. Then $|J \cap X_3| = |J| + |X_3| - |J \cup X_3| \geq |J| + |X_3| - |S|$. By replacing J, $|X_1 \cap X_2 \cap X_3| \geq (|X_1| + |X_2| - |S|) + |X_3| - |S| = |X_1| + |X_2| + |X_3| - 2|S|$ if the right-hand expression is greater than 0, or 0 otherwise. By generalizing to n subsets $X_1, \ldots, X_n$, $|\bigcap_{i=1}^{n} X_i| \geq \sum_{i=1}^{n} |X_i| - (n-1)|S|$ if the right-hand expression is greater than 0, or 0 otherwise. By dividing by $|S|$, $\frac{|\bigcap_{i=1}^{n} X_i|}{|S|} \geq \sum_{i=1}^{n} \frac{|X_i|}{|S|} - (n-1)$. From this, we obtain that $\frac{|\bigcap_{i=1}^{n} X_i|}{|S|} = \mathrm{Con}(S) \geq max(0, 1 + \sum_{i=1}^{n}(\mathrm{Con}(d_i) - 1))$.

5 Evaluation

In this section we first report on the cost assessment for profile generation, in order to evaluate its applicability in real Data Lake scenarios. As a preliminary step, a Knowledge Graph has been synthetically generated with 10 dimensions, 5 levels per dimension, 10 members for the first level, 100 for the second and so forth (i.e., with drill-down factor of 10). Datasets have been generated from the Knowledge Graph with cardinalities ranging from 10^3 to 10^7 and a number of domains equal to 10, 30 and 50 (20% of which are dimensions). LSH Ensemble was configured by using the following values: 128 permutations, 32 parts, with a threshold of 0.8. In Fig. 2(a) we report results on execution time by dataset size, focusing on the time for mapping discovery (i.e. execution of LSH Ensemble and graph creation) and profile calculation[5]. As it is clear from the results, the profile generation time requires a marginal share of the overall time for the complete loading of a source, which can be shortened through the application of sampling techniques, possibly on parallel architectures.

As for the calculation of quality measures, we setup an experimental Data Lake including 10 data sources. Each of them includes 100k items and 20 domains, 4 of which are mapped to a randomly chosen dimensional level, with an increasing percentage of noise per domain ranging from 0% to 90% to simulate non-perfectly matching data[6]. We report results for the evaluation of the consistency index in Fig. 2(b), where the actual consistency index is shown along with analytically computed minimum and maximum bounds. As expected, in this case since for each data source, domains have been generated from randomly picking the same number of members with the given percentage of noise, the maximum bound is equal to the noise degree. On the other hand, the coverage of each level varies according to the number of domains mapped to it. As an example, level $L4$ in dimension $D2$ contains 100k members and only 1 domain is mapped to it in our experimental set, namely for source with noise equal to 20%. Since the number of profile items is 55084, the resulting coverage is 0.55084, while the completeness of the source is equal to 10^{-8}. Datasets, Knowledge Graph and the code used for dataset generation are available at the project repository[7].

[5] Tests have been carried out on a machine with the following configuration: QEMU Virtual CPU version 1.5.3 - 2.26 GHz (8 processors), 32 GB RAM, 64-bit Windows 10 Pro.

[6] As it does not affect the evaluation, we assume a random distribution of noise.

[7] https://github.com/KDMG/DataLakes_material.

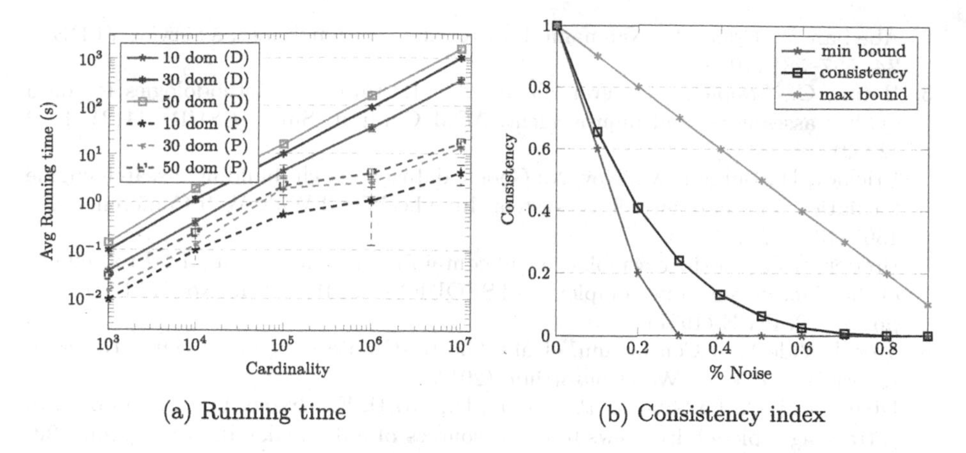

(a) Running time

(b) Consistency index

Fig. 2. (a) Execution time (log scale) for mapping discovery (D) and profiling of dimensional domains (P), for datasets with increasing cardinality (log scale) and domains (20% of which are dimensions). Average values over 10 repetitions. (b) Consistency index for sources with 100000 items and 20 domains, with an increasing % of noise in domains.

6 Conclusions

In this paper we presented a metadata management framework which extends previous work with the capability to annotate multidimensional data sources with profiles at various levels of granularity. The approach builds on top of a Knowledge Graph including the definition of dimension hierarchies, and an efficient mechanism for automatically discoverying mappings between domains of a data source, e.g., a column, and dimensional levels. The paper also proposes a number of data quality measures based on the defined metadata framework, which aims to evaluate completeness of a data source, the coverage of a dimensional level and the consistency of a domain with respect to the Knowledge Graph, also defining theoretical bounds.

Future work will aim to validate the proposal on real-world data sources. Furthermore, we are planning to exploit the notion of data profile for an efficient evaluation of the similarity among sources. This can be useful to support dataset discovery, data exploration and query answering. Finally, the compilation of profiles and evaluation of quality measures can in principle be defined as a by-product of ETL processes. Future work will be devoted to explore the possibility to endow ETL with such a capability.

References

1. Jarke, M., Jeusfeld, M.A., Quix, C., Vassiliadis, P.: Architecture and quality in data warehouses: an extended repository approach. Inf. Syst. **24**(3), 229–253 (1999). 10th International Conference on Advanced Information Systems Engineering

2. Abedjan, Z., Golab, L., Naumann, F.: Profiling relational data: a survey. VLDB J. **24**, 557–581 (2015)
3. Batini, C., Cappiello, C., Francalanci, C., Maurino, A.: Methodologies for data quality assessment and improvement. ACM Comput. Surv. (CSUR) **41**(3), 1–52 (2009)
4. Brickley, D., Burgess, M., Noy, N.: Google dataset search: building a search engine for datasets in an open web ecosystem. In: The World Wide Web Conference, pp. 1365–1375 (2019)
5. Broder, A.Z.: On the resemblance and containment of documents. In: Proceedings of the Compression and Complexity of SEQUENCES 1997 (Cat. No. 97TB100171), pp. 21–29. IEEE (1997)
6. World Wide Web Consortium, et al.: The RDF data cube vocabulary. Technical report, World Wide Web Consortium (2014)
7. Diamantini, C., Lo Giudice, P., Potena, D., Storti, E., Ursino, D.: An approach to extracting topic-guided views from the sources of a data lake. Inf. Syst. Front. **23**, 243–262 (2021)
8. Diamantini, C., Potena, D., Storti, E.: SemPI: a semantic framework for the collaborative construction and maintenance of a shared dictionary of performance indicators. Fut. Gen. Comput. Syst. **54**, 352–365 (2016)
9. Diamantini, C., Potena, D., Storti, E.: A knowledge-based approach to support analytic query answering in semantic data lakes. In: Chiusano, S., Cerquitelli, T., Wrembel, R. (eds.) Advances in Databases and Information Systems, ADBIS 2022. LNCS, vol. 13389. Springer, Cham (2022). https://doi.org/10.1007/978-3-031-15740-0_14
10. Farid, M., Roatis, A., Ilyas, I.F., Hoffmann, H., Chu, X.: CLAMS: bringing quality to Data Lakes. In: Proceedings of the International Conference on Management of Data (SIGMOD/PODS 2016), San Francisco, CA, USA, pp. 2089–2092. ACM (2016)
11. Friedman, T., Smith, M.: Measuring the business value of data quality. Technical report, Gartner (2011)
12. Gibbons, P.B., Matias, Y., Poosala, V.: Fast incremental maintenance of approximate histograms. ACM Trans. Database Syst. (TODS) **27**(3), 261–298 (2002)
13. Hai, R., Geisler, S., Quix, C.: Constance: an intelligent data lake system. In: Proceedings of the International Conference on Management of Data, SIGMOD 2016, San Francisco, CA, USA, pp. 2097–2100. ACM (2016)
14. Hai, R., Quix, C., Wang, D.: Relaxed functional dependency discovery in heterogeneous data lakes. In: Laender, A.H.F., Pernici, B., Lim, E.-P., de Oliveira, J.P.M. (eds.) ER 2019. LNCS, vol. 11788, pp. 225–239. Springer, Cham (2019). https://doi.org/10.1007/978-3-030-33223-5_19
15. Hulsebos, M., et al.: Sherlock: a deep learning approach to semantic data type detection. In: Proceedings of the 25th ACM SIGKDD International Conference on Knowledge Discovery & Data Mining, pp. 1500–1508 (2019)
16. Mahmud, M.S., Huang, J.Z., Salloum, S., Emara, T.Z., Sadatdiynov, K.: A survey of data partitioning and sampling methods to support big data analysis. Big Data Min. Anal. **3**(2), 85–101 (2020)
17. Oram, A.: Managing the Data Lake. O'Reilly, Sebastopol (2015)
18. Song, J., He, Y.: Auto-validate: unsupervised data validation using data-domain patterns inferred from data lakes. In: Proceedings of the 2021 International Conference on Management of Data, pp. 1678–1691 (2021)
19. Zhang, D., Suhara, Y., Li, J., Hulsebos, M., Demiralp, Ç., Tan, W.-C.: Sato: contextual semantic type detection in tables. arXiv preprint arXiv:1911.06311 (2019)

20. Zhu, E., Nargesian, F., Pu, K.Q., Miller, R.J.: LSH ensemble: Internet-scale domain search. Proc. VLDB Endow. **9**(12), 1185–1196 (2016)
21. Zouari, F., Kabachi, N., Boukadi, K., Guegan, C.G.: Data management in the data lake: a systematic mapping. In: Proceedings of the 25th International Database Engineering & Applications Symposium, pp. 280–284 (2021)

Data Ingestion Validation through Stable Conditional Metrics with Ranking and Filtering

Niels Bylois(✉), Frank Neven, and Stijn Vansummeren

Hasselt University and transnational University of Limburg, Data Science Institute, Diepenbeek, Belgium
niels.bylois@uhasselt.be

Abstract. We present a data ingestion quality validation approach using conditional metrics, a novel form of metrics that compute data quality metrics over specific parts of the ingestion data. We propose a method that automatically derives conditional metrics from historical ingestion sequences, using stability as a selection criterion for implementing these metrics as data unit tests. If an ingestion batch fails any unit tests, we show how conditional metrics can be utilized to identify potential errors. We show the effectiveness of our approach through an evaluation on a real world data set under various error scenarios.

Keywords: data cleaning · data profiling · dynamic data

1 Introduction

Modern data analytics pipelines continuously collect and ingest new data. Validating the quality of collected data at ingestion time is crucial in such pipelines, for a number of reasons. First, and foremost, the quality of the derived insights, and the decisions driven by them, depend directly on the quality of the collected data [13]. Second, as more and more of the data analysis process is automated, small errors in source data risk propagating to later data consumers (such as machine learning models), which may themselves act as data sources in other processing pipelines—thereby potentially magnifying the error [4]. Finally, data errors may even cause data processing pipelines to crash (e.g., because of null pointer exceptions due to missing data). Machine learning platforms are therefore including explicit data validation components into their pipelines [2,3,6].

In recognition of the importance of data quality validation in modern analytics pipelines, several tools have been proposed to aid in automatic validation [4,11,13]. Broadly speaking, these tools allow specification, either manually or automatically, of so-called *data unit tests*. When a new batch of data is to be ingested, the registered tests are executed to gauge the batch's quality where failing tests highlight data quality problems. The tests themselves entail computing certain metrics on the data batch (e.g., the minimum or maximum value

A. Abelló et al. (Eds.): ADBIS 2023, LNCS 13985, pp. 210–223, 2023.
https://doi.org/10.1007/978-3-031-42914-9_15

appearing in a numerical column or the number of distinct elements appearing in a column) and checking that these fall within an expected range.

Unfortunately, these tools suffer from two limitations. First, the data unit tests that they support are based on *global* metrics: metrics that are computed on the entire data batch, or on an entire column in the batch. As the following example shows, they hence only provide *coarse-grained* signals of data quality and are unable to detect *fine-grained* errors, i.e. errors that occur only in a specific (potentially small) part of the batch. Batches with fine-grained errors hence go unnoticed. Second, even when a data unit test signals that a batch has a data quality problem, it does not provide a principled method to identify the part of the batch that is responsible for a test's failure. As such, either the entire batch must be discarded or a human expert must manually identify and subsequently remove or correct the erroneous tuples.

Example 1. A public railway company has equipped all of its trains with measurement sensors that record the train's arrival and departure time at each train station. By comparing these times with the time schedule, the train's software computes the corresponding delays (if any). At the end of each day, the measurements of all trains are collected, and ingested in the railway company's data lake. The delays are used to identify hotspot routes, as well as computation of service quality indicators to the government. Train 5437runs daily from Hasselt (Belgium) to Blankenberge (Belgium). This route is notorious for the delays that it incurs when it passes through the busy Brussels railway stations. As such, train 5437 normally reports non-zero delay. Due to a hardware malfunction on March 15, however, it consistently reports zero delay for this train.

The metrics used by state-of-the-art tools are unable to detect this error. Indeed: because zero delay is *not* an uncommon value when considering the entire ingestion batch (some trains run on time), metrics such as MIN(*Delay*), MAX(*Delay*), and AVG(*Delay*) will not consider zero delay as an anomaly. We have verified this experimentally with the method from Redyuk et al. [11] using the same experimental setup as described in Sect. 5. These metrics are unable to detect these fine-grained errors. It is important to observe that, even if one of the global metrics signals a data quality problem, it is not clear which train or set of trains in the batch cause the problem. For instance, if the unit tests based on MIN(*Delay*) or MAX(*Delay*) signal an anomaly then of course we can easily identify the erroneous trains: simply compute the trains whose delay value is below (or above) the expected minimum (resp. maximum) value. However, if a unit test based on AVG(*Delay*) signals a problem, then it is unclear how to identify the trains that caused the delay to deviate. □

The method presented in this paper is an approach for data quality valudation that is not only capable of detecting the *fine-grained errors* illustrated above, but also helps in *identifying* responsible erroneous tuples. We focus on the setting where a data pipeline regularly ingests batches of external data and introduce *conditional metrics* (CM for short) that, in contrast to a global metric, only computes its value on a specified subset of tuples in the ingestion batch. In the railway example above, the conditional metric AVG(*Delay* | *Train* = 5437) computes the average delay of train 5437 in the batch, as opposed to the global metric AVG(*Delay*) which computes the average delay of all trains.

Our approach consists of two phases: (i) a unit test discovery phase and (ii) a monitoring and error identification phase, as schematically illustrated in Fig. 1. In the *unit test discovery phase*, we are given a sequence $\overline{R}$ of previously ingested batches and our objective is to automatically derive a set Θ of CM-based data unit tests from $\overline{R}$ such that a yet-to-be-ingested batch B can be considered to be of acceptable quality if it passes all tests in Θ. We propose a simple notion of *stability* as a means to decide on the set of CMs to promote to unit tests: a CM is *stable* for $\overline{R}$ w.r.t. a chosen anomaly detection method A when the number of anomalies detected by the classifier produced by A in $\overline{R}$ does not exceed a predetermined threshold. Only stable CMs are promoted to unit tests.

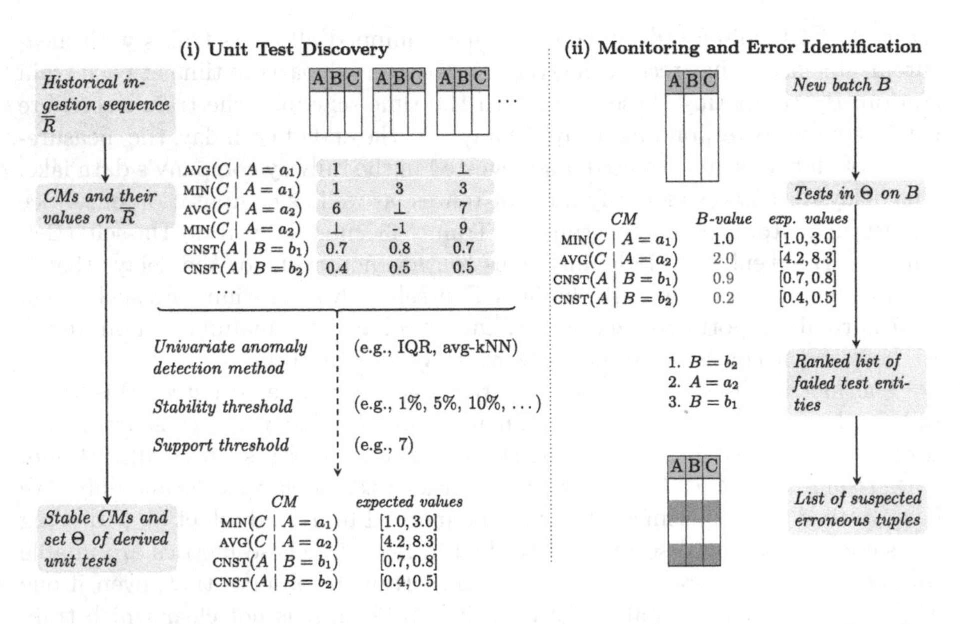

Fig. 1. Overview of approach: (i) *discovery phase* to derive data unit tests based on stable conditional metrics; (ii) *monitoring and error identification phase* where each new batch is validated and erroneous tuples are reported.

In the *monitoring and error identification phase*, we take the set Θ of data unit tests derived in the discovery phase, and use this to validate each new ingestion batch B. If all tests in Θ succeed on B then B is deemed to be of acceptable quality. When at least one test in Θ rejects B then our objective is to identify the tuples in B with suspected errors. Every conditional unit test specifies a set of erroneous tuples in a natural way: the subrelation of B that its CM refers to. Simply flagging *all* the tuples of violated unit tests selects too much, however (i.e., it results in high recall but very low precision). The reason is that violating tests are often correlated: a single error in B may cause multiple tests (each selecting different subrelations) to fail. The key challenge in the monitoring phase, therefore, lies in *ranking* the violated unit tests according to relevance, and from this ranked list of tests *filter* a list of suspected erroneous tuples for further inspection.

Our Contributions are as Follows. (1) We propose *conditional metrics* as a means for providing fine-grained data quality signals (Sect. 2), and the two-phase framework summarized in Fig. 1 as a general framework for data ingestion validation as well as identification of erroneous tuples. We refer to this framework as SCMRF (*Stable Conditional Metrics with Ranking and Filtering*). (2) We study the design space for identifying erroneous tuples in the monitoring phase and formulate the problem of ranking unit tests according to relevance as a problem of ranking nodes in a special kind of bipartite graph. We establish that likely candidates for ranking nodes, based on popular methods for identifying authorative or central nodes, work poorly in this respect. In response, we develop a new, tailored, ranking measures *correlation density* (*cd*)) together with an associated threshold-determination algorithm that can successfully identify erroneous tuples (Sect. 4). (3) We present an extensive experimental validation of our approach on a real world dataset under various error scenarios (Sect. 5) and compare with global metrics.

2 Conditional Metrics

We assume that the reader is familiar with the relational data model and the standard notation of relational algebra (cf., e.g., [1]). Because metrics may yield different results when computing on sets as opposed to bags, we will work with the bag-based version of relational algebra.

Metrics. A *global metric* (GM) *on attribute* A is a function that maps non-empty relations over the single attribute A to a real number. If μ is a GM, we will use the notation $\mu(A)$ to stress that μ is a GMon attribute A. Global metrics are extended to apply to non-empty relations with arbitrary schema (containing attribute A) by first projecting on A, i.e., $\mu(R) := \mu(\pi_A(R))$.

We assume given a fixed finite set $\mathcal{M}$ of GMs on a fixed set of attributes. Concrete GMs that we will consider in our experiments are, for example, $\text{COUNT}(A)$ which counts the number of rows in a relation R, $\text{COUNT-DIST}(A)$ which counts the number of distinct A-values, $\text{AVG}(A)$ that computes the average, and so on. Note that the availability of a global metric on A crucially depends on the type of A (i.e., $\text{AVG}(\cdot)$ only makes sense on numerical attributes, and $\text{MIN}(\cdot)$ only on attributes that are equipped with an order).

Conditional Metrics. A *conditional metric* (CM) m is an expression of the form $\mu(Y \mid X = x)$ where X, Y are attributes, μ is a GMon Y, and x is a value in the domain of X. If R is a relation whose schema includes at least the attributes X and Y, then the result of evaluating m on R is defined as

$$m(R) := \begin{cases} \mu(\pi_Y \sigma_{X=x}(R)) & \text{if } x \in \pi_X(R), \\ \bot & \text{otherwise.} \end{cases}$$

That is, $m(R)$ applies μ on the part of R where $X = x$ if x occurs in R, and is $\bot$ otherwise.

We call X the *conditioning attribute* (also known as the *group by* attribute), and Y the *metric attribute* (also known as the *aggregation* attribute) of m. We also refer to x as the *key* of m.

Example 2. We further develop the scenario mentioned in Example 1. Consider the conditional metrics $m_1 = \text{AVG}(\textit{Delay} \mid \textit{Train} = 5437)$, $m_2 = \text{AVG}(\textit{Delay} \mid \textit{Train} = 2891)$ and $m_3 = \text{AVG}(\textit{Delay} \mid \textit{Train} = 6061)$ that compute the average delay for three different trains. Here, *Train* is the conditioning attribute, *Delay* is the metric attribute, AVG is the global metric and the keys are 5437, 2891, and 6061, respectively. □

Data Unit Tests. A CM-*based data unit test* (hence-forth simply called unit test) ϕ is a pair (m, c) where m is a CMand $c\colon \mathbb{R} \to \{\mathsf{true}, \mathsf{false}\}$ is a function that classifies the range of values produced by m into expected values ($c(m(R)) = \mathsf{true}$) and anomalies ($c(m(R) = \mathsf{false}$). A relation R *satisfies* ϕ, denoted $R \models \phi$, if $c(m(R)) = \mathsf{true}$ or $m(R) = \bot$. When, R *fails* ϕ; we also say that ϕ *rejects* R.

Note in particular that R vacuously satisfies ϕ if the key x of m is not present in the conditioning attribute in R (since $m(R) = \bot$ in that case). This makes sense because m (and, hence ϕ) is meant to only investigate the part of R where $X = x$. If such part does not occur, there is nothing to test.

It is important to stress that current data ingestion validation tools [4,11–13] do not use CM-based data unit tests, but GM-*based* unit tests: these are based on global metrics. Formally, they are pairs (μ, c) with μ a global metric and $c\colon \mathbb{R} \to \{\mathsf{true}, \mathsf{false}\}$ a function that classifies the range of μ into expected values and anomalies.

Entities. The amount of possible CMs (and therefore associated data unit tests) can be huge as it depends on the number of values occurring in the input data that can be used as keys in conditional metrics. To limit the number of possible CMs, we will only consider CMs in which the conditioning attributes refer to real-world entities. This restriction to entities helps improve the relevance of data unit tests: it makes little sense for instance to monitor CMs like $\text{COUNT-DIST}(\textit{Train} \mid \textit{Delay} = 8)$ that measure deflections of a specific delay value. Furthermore, such nonsensical CMs can correlate in unexpected ways with entity CMs, thereby unnecessarily complicating the identification of erroneous tuples.

3 Discovering Conditional Metrics

Recall from the Introduction that our approach to data ingestion validation consists of two phases. A *discovery phase* where the objective is to automatically derive a set of CM-based data unit tests from a historical ingestion sequence, and a *monitoring and error identification phase* where these tests are used to determine whether a newly arrived ingestion batch is acceptable as well as identify erroneous tuples in the batch when that is not the case. We focus on the discovery phase in this section, and describe the monitoring phase in Sect. 4.

Ingestion Sequences. We are interested in validating batches of data that are ingested over time into a single relation.[1] Formally, an *ingestion sequence* is a sequence $\overline{R} = R_1, \ldots, R_n$ of relations, all over the same schema. Every R_i is a batch of tuples to be ingested. In what follows, we write $\overline{R}[i]$ for R_i, Ri for the prefix $R_1, \ldots, R_{i-1}$, and $\overline{R}_{\leq i}$ for $Ri + 1$.

CMs on Ingestion Sequences. Given an ingestion sequence $\overline{R}$ and a conditional metric $m = \mu(Y \mid X = x)$, we write $m(\overline{R})$ for m applied point-wise to those elements of $\overline{R}$ that contain key x, i.e., $m(\overline{R}) := m(\overline{R}[i_1]), \ldots, m(\overline{R}[i_k])$ where $i_1 < \ldots < i_k$ and $\{i_1, \ldots, i_k\} = \{i \in [1, |\overline{R}|] \mid x \in \pi_X(\overline{R}[i])\}$. This is hence the sequence of real numbers that records how the values computed by m evolve over the batches containing $X = x$.

Support. The *support* of a CMm on $\overline{R}$ is defined as the number of batches in $\overline{R}$ containing the key of m. Formally, $\text{supp}(m, \overline{R}) := |m(\overline{R})|$. To avoid edge cases where keys occur too infrequently, we are only interested in deriving classifiers for CMs whose support is higher than some predefined *support threshold* t_{supp}.

Anomaly Detection and Stability. Anomaly detection refers to "the problem of finding patterns in data that do not conform to expected behavior" [7]. Such unexpected patterns are also referred to as anomalies, exceptions, or outliers. In our setting, the data is a sequence $m(\overline{R})$ of real values, and we ask that an anomaly detection method A uses $m(\overline{R})$ to deliver a classifier $c = A(m(\overline{R}))$ that can distinguish between expected values of m and unexpected ones (anomalies). Because each point in the data $m(\overline{R})$ is a single real value, we are hence interested in *univariate* anomaly detection methods, which are significantly less complex than multivariate methods [7].

In principle, any univariate anomaly detection method A can be used. Because we assume, however, that $\overline{R}$ (and therefore, $m(\overline{R})$) is mostly error-free, we require that the classifier c produced by A is *consistent* with this assumption: when we classify the elements of $m(\overline{R})$ according to c, we expect that most (if not all) of them are expected values. When c does not satisfy this condition, it makes little sense to use $\phi = (m, c)$ as a unit test, a such a test would already mark most batches in $\overline{R}$ itself as invalid.

Formally, we say that CM m is *stable* for $\overline{R}$ w.r.t. anomaly detection method A when the percentage of anomalies detected by the classifier $c = A(m(\overline{R}))$ in $m(\overline{R})$ is below a predefined *stability threshold* t_{stable}, i.e., when:

$$\frac{|\{i \in [1, |\overline{R}|] \mid m(\overline{R}[i]) \neq \bot, \overline{R}[i] \not\models (m, c)\}|}{|m(\overline{R})|} \leq t_{\text{stable}}.$$

For a fixed anomaly detection method A, only CMs that are stable for $\overline{R}$ w.r.t. A will be promoted to unit tests.

[1] The techniques in this paper can easily be generalized to multiple relations. We focus on one relation, however, to keep the presentation simple.

In this paper, we consider the inter quartile range (IQR) [2] univariate anomaly detection method. Our experiments in Sect. 5, show that, despite the existence of more complex outlier detection methods, this simple method is already effective. We also considered *average k-nearest neighbour method* (avg-kNN) (as, e.g., used in [11]) but found that overall it performs worse than IQR.

4 Monitoring and Error Identification

In the monitoring phase, we assume given a fixed set Θ of unit tests. Whenever a new ingestion batch B arrives, we apply all tests in Θ to it to check if B is of acceptable quality. If B satisfies all tests then it is judged of good quality, and ingestion can proceed. By contrast, as soon as B fails at least one test in Θ there is a potential data quality problem that needs to be resolved. In that case, we want to use the failed unit tests in Θ to help identify the tuples of B with potential errors. Every failed unit test naturally identifies an error region of B, namely the subrelation of B that it conditions on. Unfortunately, the same erroneous tuples can be picked up by several unit tests, each selecting a distinct region of possibly widely varying size. Therefore, returning the regions of all failed unit tests is too coarse in the sense that it has high recall, but low precision. We illustrate this claim by means of the following example.

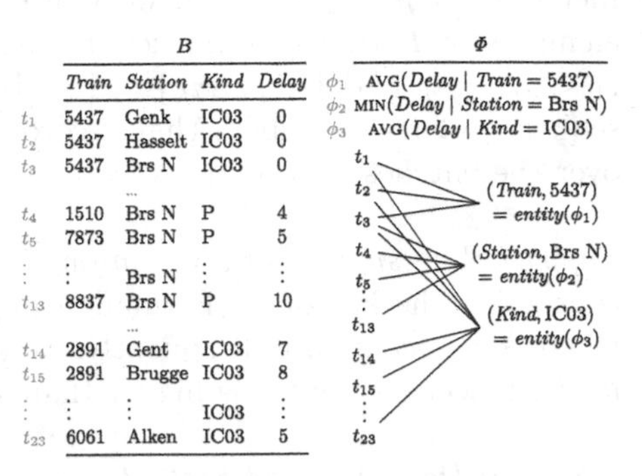

	Train	Station	Kind	Delay
t_1	5437	Genk	IC03	0
t_2	5437	Hasselt	IC03	0
t_3	5437	Brs N	IC03	0
	...			
t_4	1510	Brs N	P	4
t_5	7873	Brs N	P	5
⋮	⋮	Brs N	⋮	⋮
t_{13}	8837	Brs N	P	10
	...			
t_{14}	2891	Gent	IC03	7
t_{15}	2891	Brugge	IC03	8
⋮	⋮	⋮	IC03	⋮
t_{23}	6061	Alken	IC03	5

Fig. 2. Illustration of a to-be-ingested batch B, failed unit tests Φ, and the corresponding entity-tuple graph.

Example 3. Recall the scenario outlined in Example 1, where train 5437consistently reports zero delay. Consider the to-be-ingested batch B shown in Fig. 2 as well as the failed unit tests ϕ_1, ϕ_2, and ϕ_3 listed there. We note that for the sake of this example, we have added an extra attribute *Kind* that lists the kind of service that the train offers. Unit test ϕ_1 fails because the average delay of train 5437is now zero, which is unexpected given the historical sequence. However, in this example, the zero delay of train 5437also causes the minimum delay in station Brussels North (Brs N) to become zero, which is also unexpected, causing ϕ_2 to fail. Similarly, it also causes the average delay of route kind IC03 to become unexpected, failing ϕ_3. The root cause here is the zero delay of train 5437, and ideally we would hence like to return only the tuples t_1–t_3 of train 5437as suspected erroneous tuples. By contrast, the region of all failed unit tests comprises the larger set of tuples t_1–t_{23}. □

The previous example shows that failing unit tests may be correlated: they jointly reject B due to the same underlying set of erroneous tuples. To help

[2] https://en.wikipedia.org/wiki/Interquartile_range

identify this set of tuples, we hence desire a method to *filter* the error regions down to a subset of "representative" error regions. The tuples to report to the user for inspection are then the tuples in this subset of representative regions. We next develop such a method, based on *ranking* the regions in failing tests such that higher-ranked regions are expected to provide a better explanation of the possible erroneous tuples than lower-ranked ones. Based on this ranking, we subsequently select the subset of representative regions.

We formalize and approach this as follows. Throughout the section, let Θ be the set of all unit tests to monitor, let B be the batch to be ingested, and let $\Phi \subseteq \Theta$ be the set of all failed unit tests.

Entities. An *entity* is a pair (X, x) with X an entity attribute and x a particular value in the domain of X. If ϕ is a unit test with CM m then we write *entity*(ϕ) for the entity (X, x) where X is the conditioning attribute of m and x is the key. Let E be the set of all entities occurring in failed unit tests, $E := \{\mathit{entity}(\phi) \mid \phi \in \Phi\}$. We will refer to these entities as the *failed entities* of B.

Every failed entity $e = (X, x)$ selects a non-empty set of tuples in B, which we denote by *tuples*(e) where $\mathit{tuples}(e) := \{t \in B \mid t[X] = x\}$. Conversely, for a tuple t, let *entities*(t) be the (possibly empty) set of all failed entities in which t occurs, $\mathit{entities}(t) := \{(X, x) \in E \mid t[X] = x\}$. For a set $C \subseteq E$, we denote by $\mathit{tuples}(C) := \bigcup_{e \in C} \mathit{tuples}(e)$. In what follows, we let $T = \mathit{tuples}(E)$ be the set of all tuples in B that are selected by a failed entity; we will also refer to these tuples as the *failed tuples* of B.

Entity-tuple Graph. The failed entities define error regions in B. Rather than simply returning the set of all failed tuples as being potentially erroneous, we wish to return a more fine-grained set of tuples based on the structure of correlations between the failed entities. To model these correlations, we introduce the following graph structure. The *entity-tuple* graph is the bipartite graph G that has the failed tuples and failed entities as nodes, and where there is an undirected edge between tuple t and entity e if $t \in \mathit{tuples}(e)$ (or, equivalently, $e \in \mathit{entities}(t)$). See Fig. 2 for an illustration.

Two entities e_1 and e_2 are *directly correlated* if there is a tuple t that is selected by both, causing the existence of the path $e_1 - t - e_2$ in the entity-tuple graph. Entities that are not directly correlated may still be *indirectly* correlated because of paths of longer length. For example, in a path of the form $e_1 - t - e_2 - u - e_3$, the entities e_1 and e_3 are indirectly correlated through e_2. In this sense, every connected component of the entity-tuple graph hence represents a subset of failed entities that are (directly or indirectly) correlated. Entities in different connected components are uncorrelated.

Ranking. Intuitively, we want to exploit the correlation topology exhibited within a connected component C of G to rank C's entities according to their "relevance" such that highly-ranked entities provide a better representation of why the unit tests associated to C failed on B.

But, how can we measure this "relevance"? Intuitively, this seems similar to that of existing measures for quantifying the authority of a node in a graph, such as, e.g., node degree, closeness centrality [9], betweenness centrality [8], PageRank [5], and HITS [10]. As we will see experimentally in Sect. 5, however, these existing measures fail to satisfactorily quantify "relevance". The reason is that they are generic methods that do not take into account the specific semantics of entity-tuple graphs, where there are two kinds of nodes (entities and tuples), and direct correlation between entities is encoded by paths of length 2.

We therefore next propose a tailored ranking measures on entity-tuple graphs. This measure takes into account whether different tuples cause different sets of entities to become directly correlated. Intuitively, when all tuples $t \in \mathit{tuples}(e)$ have distinct entity sets $\mathit{entities}(t)$, this indicates a richer nucleus of correlations than when each tuple has the same set $\mathit{entities}(t)$. Formally, for a given entity e, we define its *correlation density* by: $\mathit{cd}(e) := \frac{|\{\mathit{entities}(t) | t \in \mathit{tuples}(e)\}|}{|\mathit{tuples}(e)|}$. This hence measures how many different entity sets $\mathit{entities}(t)$ are exhibited by e's neighbors in G, normalized by $|\mathit{tuples}(e)|$ which is the maximum number of such sets. High values of $\mathit{cd}(e)$ indicate that a larger fraction of the neighboring tuples t of e have distinct sets $\mathit{entities}(t)$—which we take as a proxy for richer correlation structure and therefore more relevance—while lower values of $\mathit{cd}(e)$ indicate that more tuples share the same sets $\mathit{entities}(t)$, and hence poorer correlation structure.

Example 4. To illustrate cd, reconsider Example 3 and Fig. 2. Denote $e_T = (\text{Train}, 5437)$, $e_S = (\text{Station}, \text{Brs N})$ $e_K = (\text{Kind}, \text{IC03})$. Then, we have

$$\mathit{cd}(e_T) = \frac{2}{3} > \mathit{cd}(e_S) = \frac{2}{11} > \mathit{cd}(e_K) = \frac{3}{13}$$

The correlation structure can be interpreted as follows: $\mathit{tuples}(e_T) = \{t_1, t_2, t_3\}$ with $\mathit{entities}(t_1) = \mathit{entities}(t_2) = \{e_T, e_S\}$ and $\mathit{entities}(t_3) = \{e_T, e_K\}$. This means that $\mathit{tuples}(e_T)$ is partitioned in two sets $\{t_1, t_2\}$ and $\{t_3\}$, each 'explained' by $\{e_T, e_S\}$ and $\{e_T, e_K\}$, respectively. Similarly, $\mathit{tuples}(e_S) = \{t_3, \ldots, t_{13}\}$ is partitioned into the sets $\{t_3\}$ and $\{t_4, \ldots, t_{13}\}$, and $\mathit{tuples}(e_K) = \{t_{14}, \ldots, t_{23}\}$ is partitioned into the sets $\{t_1, t_2\}$, $\{t_3\}$ and $\{t_{14}, \ldots, t_{23}\}$. The measure cd favors partitions into a larger number of subsets (relative to the total number of tuples selected by the entity). The underlying expectation is that when an entity e is the root cause (like e_T in this example), more entities will interfere in $\mathit{tuples}(e)$. That is, $\mathit{tuples}(e)$ will be 'explained' by a larger number of (unique) entity sets, relative to $|\mathit{tuples}(e)|$. □

Filtering. The measures introduced above allow us to rank entities on relevance, where higher values indicate more relevance. In practice, it does not suffice to only rank entities: we must also determine, based on this ranking, which entities we consider to select potentially erroneous tuples: it is these tuples that should be further inspected by a human expert. We use the following simple method for this purpose. Fix a ranking measure f. Let C be a connected component of the entity-tuple graph G. Rank the entities of C according to the chosen measure from high to low. Let $e_1, e_2, \ldots, e_n$ be this ranking (where entities with high measure values appear first). For every $1 \leq i < n$, let $gap(i) = f(e_i) - f(e_{i+1})$. Then return only the entities $e_1, \ldots, e_i$ where i is the smallest index that maximizes $gap(i)$. We denote the latter set by $gap(f, C)$.

This procedure is repeated for every connected component of G. Note that we hence select at least one entity for every connected component. The set SET(B) of *suspected erroneous tuples* of batch B is then defined as

$$\mathrm{SET}(B) := \bigcup\{tuples(e) \mid e \in gap(f, C), C \text{ a connected component of } G\}.$$

5 Evaluation

In this section we perform an extensive evaluation of the overall effectiveness of our proposed methodology. As our methodology supports both fine-grained error signals and error identification, which are not considered in the state of the art, there is no prior work that we can compare to directly. As such, letting SCMRF(A,f) refer to our method with entity ranking measure f, we fix A as IQR and focus on assessing the different choices for f. We first describe our testing methodology and detail the used dataset and testing scenarios. We then report the overall effectiveness of SCMRF(IQR,*cd*) on this dataset and scenarios and assess the different choices for f. Finally, we compare our approach based on conditional metrics to the state of the art based on global metrics.

Evaluation methodology. We employ the following experimental setup. We evaluate our framework on a real world dataset and three synthetic error scenarios that are detailed in Sect. 5. We consider the prefix of the first ℓ ingestion batches to comprise the historical ingestion sequence $\overline{R}_{\leq \ell}$, from which data unit tests are derived using the method of Sect. 3. The batch $\overline{R}[\ell + 1]$ is then used as the *test* batch, i.e., the batch that needs to be validated and where potential errors need to be identified. Because the dataset is not labeled with the ground truth of errors, we create a *modified version* of the test batch, referred to as B_{mod}. That is, we manipulate $\overline{R}[\ell + 1]$ by synthetically generating errors based on the different error scenarios, which capture various kinds of errors that may occur. For each scenario, we also specify the set of modified tuples, which we denote by MT(B_{mod}).

Recall from Sect.4 that SET(B_{mod}) denotes the set of suspected erroneous tuples. We employ precision, recall and F1-score as performance metrics to gauge how well SET(B_{mod}) corresponds to MT(B_{mod}). These are defined as follows (S refers to the selected set, R to the relevant set):

$$\text{precision} = \frac{|R \cap S|}{|S|}, \quad \text{recall} = \frac{|R \cap S|}{|R|}, \quad \text{F1} = 2 \times \frac{\text{precision} \times \text{recall}}{\text{precision} + \text{recall}}$$

with $S = \mathrm{SET}(B_{mod})$ and $R = \mathrm{MT}(B_{mod})$.

Golden Standard. It is well-known that the F1-score always lies in the interval [0,1] with values closer to 1 being better and F1= 1.0 serving as the golden to-achieve standard. In our context, however, such a perfect F1-score cannot always be obtained. Indeed, by design and definition, $\text{SET}(B_{mod})$ is always a union of the form $tuples(e_1) \cup \cdots \cup tuples(e_n)$ for different entities e_i, whereas $\text{MT}(B_{mod})$ does not need to be such a union of entities. For instance, in scenario N1 (see Sect. 5), $p\%$ of the tuples of a *single* entity of B are modified to obtain B_{mod}. As such, when $p < 100\%$ there is no set of entities (and hence, no set of conditional metrics or unit tests) that can select *exactly* those modified tuples. Towards a fair interpretation of F1-scores, the golden standard to be used here is therefore not F1=1.0, but the maximal F1-score that can be achieved by any union of entities. Formally, we define $\text{F1}^{\max}$as the maximal F1-score over all subsets $C \subseteq E$ of failed entities where the selected set is $S := tuples(C)$. When the F1-score of $\text{SET}(B_{mod})$ is equal or close to $\text{F1}^{\max}$it means that no other choice of ranking measure (or filtering) can improve SCMRF. Unfortunately, it is intractable to compute $\text{F1}^{\max}$. Therefore, we approximate $\text{F1}^{\max}$for each experiment via simulating annealing over the search space of subsets of failed entities. In particular, we generated for each experiment 5000 random entity sets as starting points on which we performed 10000 iterations that modify entity sets to improve the F1-score.

Datasets and Scenarios. The NMBS dataset is a real-world dataset containing event data from the Belgian railway network[3], and corresponds to the train monitoring use case explained in Example 1. Each tuple registers the stop of a train at a particular station (attribute `name_of_stop`), together with information concerning its delay (if any), the track in the station where the train stopped, and an identification of the endpoints of the journey that the train is making (attribute `relationship`). The dataset contains only records of normal weekdays (i.e., weekends and holidays are omitted). In all related experiments, we set $\ell = 70$ and $t_{\text{supp}} = 7$. In particular, $t_{\text{supp}} = 10\%\ell$, so a key needs to occur in 10% of the historical ingestion batches to be considered as a candidate for unit testing. In the scenarios described next, we use the following parameters: $K \in \{1, 5, 10\}$ and $p \in \{0.5, 0.75, 1\}$, except for $N2$ where we use $p \in \{0.5, 0.75\}$. For every choice of parameter p and K we perform 5 experiments. In total, we ran 120 experiments.

- *Scenario N1: setting delay to zero.* This error scenario corresponds to that of Example 1. Specifically, we pick K train numbers at random from those occurring in the test batch, and set $p\%$ of their stops to report zero delay for the *delay_at_arrival* attribute.
- *Scenario N2: deleting trains.* We wish to gauge how well conditional metrics can detect that certain (expected) tuples are missing. Specifically, we select K trains at random from those occurring in the test batch, and delete $p\%$ of their records (i.e. stops). We do not consider the case $p = 100\%$ where we remove all the records.
- *Scenario N3: changing stops.* We pick K train numbers at random from those occurring in the test batch, and for each such train n, set $p\%$ of their stops

[3] http://www.nmbs.be

to occur in train station s_n, where s_n is drawn randomly from the stations occurring in the test batch.

In scenarios N1 and N3, we define MT(B_{mod}) to consist of the tuples corresponding to the selected stops. The scenario N2 is different in that all modified tuples are removed. Because those tuples are no longer present in B_{mod} and because SET(B_{mod}) is always a subset of B_{mod}, we necessarily always have an F1-score and $\mathrm{F1}^{\max}$of zero. To evaluate this scenario, we therefore report precision, recall and F1 on the level of entities rather than tuples. That is, we compute these metrics with respect to the *suspected erroneous entities* of batch B_{mod}, SEE(B_{mod}) $:= \{e \mid e \in gap(f, C), C$ a connected component of $G\}$, and the set of modified entities in N2 consisting of the selected trains.

SCMRF(IQR, cd). We discuss the effectiveness of the best method SCMRF(IQR,*cd*)when $t_{\text{stable}} = 0.05$. Figure 3 displays the per-scenario precision, recall and F1-score, reported as an average over all experiments in the considered scenario. SCMRF(IQR,*cd*)attains an average F1 $\geq$ 0.79 in scenarios N1and N2, and attains F1 $\geq$ 0.64 in N3. The ratio between the F1-score attained by SCMRF(IQR,*cd*)and $\mathrm{F1}^{\max}$is 95% (for N1), respectively 99% (N2), 82% (N3).

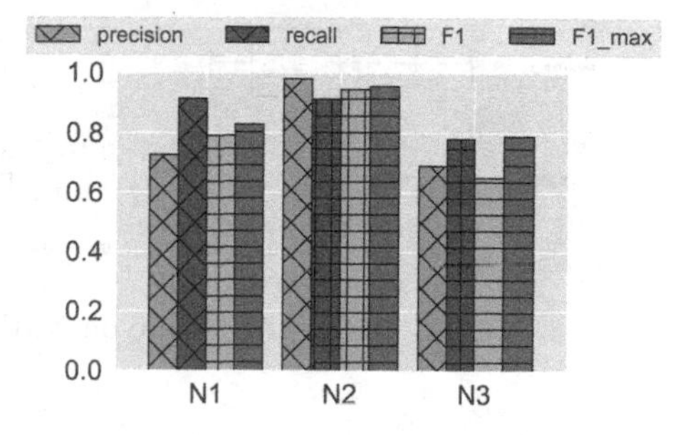

Fig. 3. Performance of SCMRF(IQR,*cd*).

The raw F1-scores demonstrate that SCMRF(IQR,*cd*)is capable of reliably detecting a significant amount of errors in the studied scenario's. The ratio w.r.t. $\mathrm{F1}^{\max}$shows that within the proposed method SCMRF, the choice of IQRand *cd* is almost optimal.

We have also experimentally verified that IQR performs better than avg-kNN: IQR retains the same recall as avg-kNN but has a much higher precision. Furthermore, IQR returns less entities which makes it easier for *cd* to identify the correct entities, resulting in a higher overall F1-score.

Monitoring and Error Identification Phase. We evaluate the effectiveness of several ranking metrics and take into account measurements that are often used to assess the importance of nodes in a graph in addition to the measure *cd* as defined in Sect. 4. Specific measures we consider are PageRank [5], HITS [10], betweenness centrality [8], closeness centrality [9], (bipartite) degree centrality (the fraction of nodes a node is connected to), and the degree of a node. For HITS we display the authority score, which in our tests we found to be marginally superior to the hub score. Fig. 4 displays the F1-score on the various scenario's for SCMRF(IQR,*f*) under different ranking measures *f* with $t_{\text{stable}} = 0.05$. We see that for N1and N2, all ranking measures behave rather similarly while *cd* performs slightly better. For N3*cd* is clearly superior. We can therefore conclude that *cd* is the superior ranking measure.

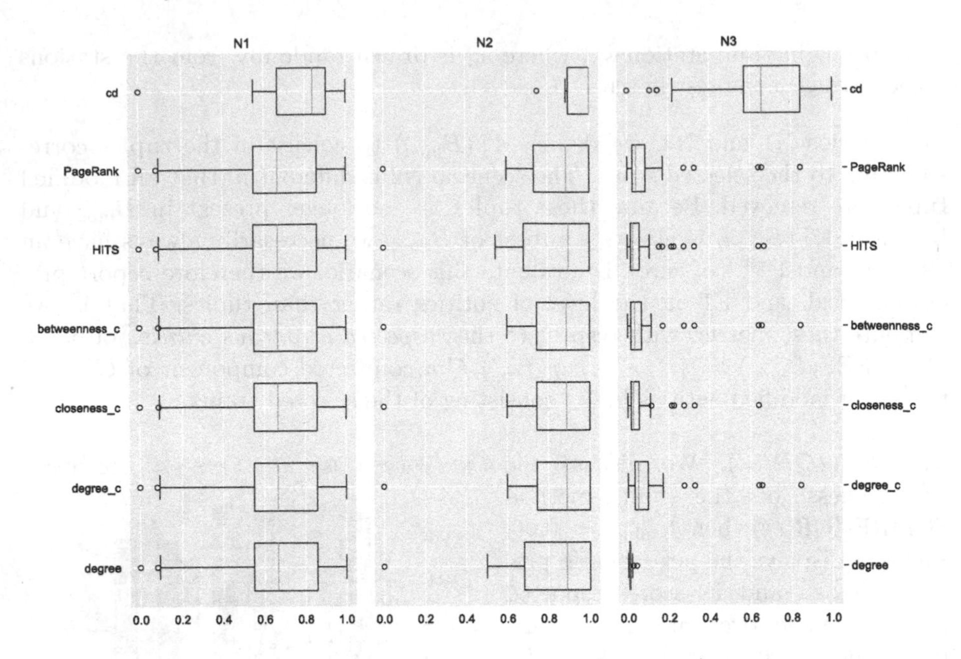

Fig. 4. F1-score on NMBS per ranking measure

Comparison with Global Metrics. Recall from Sect. 2, that global metrics differ from conditional ones in that they do not take a subset but a whole column into account. In this section, we evaluate the ability of global metrics to identify errors in the different scenarios. Each relation R thus induces a d-dimensional feature vector $v(R)$ from the d global metrics (in this paper, $d = 17$). The data quality problem then reduces to deciding whether for an ingestion batch B, $v(B)$ is an outlier w.r.t. $v(\overline{R})$ for a given historical sequence $\overline{R}$. If so, then B is said to be erroneous and therefore rejected. It should be noted that this approach operates on the granularity of the batch which is accepted or rejected as a whole and does not supply a direct method for identifying suspected erroneous tuples within a rejected batch.

Redyuk et al. [11] propose to use avg-kNN with the Euclidean distance metric, $k = 5$, and a contamination parameter of $\rho = 0.01$, as an outlier detection method. We implemented this method and tested it on the NMBSexperiments. We found that for only 3 of the 165 experiments, the modified batch was rejected. This hence confirms that global metrics are less-suited for identifying fine-grained errors that occur in a specific (potentially small) part of the batch.

6 Conclusion

We presented the SCMRF framework based on stable conditional metrics as an approach for fine-grained data quality validation that in addition aids in identifying suspected erroneous tuples. We showed that correlation density, a novel ranking measure, in combination with IQR as an anomaly detection method results in a high F1-score for erroneous tuple detection in the various scenario's. Furthermore, we argued that this combination is close to optimal within the proposed framework: no other ranking measure can improve much over *cd*. Finally, we showed that errors over ingestion sequences can be detected that are missed by global metrics. The latter underscores that conditional metrics can be a valuable addition to the more traditional but coarser global metrics.

We view this work as an initial exploration of the potential of conditional metrics as a tool for error detection in ingestion sequences. While the initial results are most promising, various challenges remain that can be addressed in follow-up work: (1) How to improve the filtering step? What methods improve over the employed 'gap method' to choose an optimal cut off point in the ranking of entities? (2) Rather than naive enumeration, what are principled ways to more efficiently discover stable conditional metrics? (3) Minimize the number of data unit tests for validation while minimally sacrificing accuracy.

Acknowledgements.. We thank Kris Luyten for many helpful discussions on this paper. S. Vansummeren was supported by the Bijzonder Onderzoeksfonds (BOF) of Hasselt University under Grant No. BOF20ZAP02. This work is partially funded by the Research Foundation - Flanders (FWO-grant G055219N). The computing resources and services used in this work were provided by the VSC (Flemish Supercomputer Center), funded by the Research Foundation - Flanders (FWO) and the Flemish Government.

References

1. Abiteboul, S., Hull, R., Vianu, V.: Foundations of databases. AW (1995)
2. Baylor, D., et al.: TFX: A tensorflow-based production-scale machine learning platform. In: SIGKDD (2017)
3. Boese, J., et al.: Probabilistic demand forecasting at scale. In: VLDB (2017)
4. Breck, E., et al.: Data validation for machine learning. In: MLSys (2019)
5. Brin, S., Page, L.: The anatomy of a large scale hypertextual web search engine. Comput. Netw. **30**(1–7), 107–117 (1998)
6. Caveness, E., et al.: Tensorflow data validation: Data analysis and validation in continuous ML pipelines. In: SIGMOD (2020)
7. Chandola, V., Banerjee, A., Kumar, V.: Anomaly detection: a survey. ACM Comput. Surv. **41**(3), 1–58 (2009)
8. Freeman, L.C.: A set of measures of centrality based on betweenness. Sociometry **40**(1), 35–41 (1977)
9. Freeman, L.C.: Centrality in networks conceptual clarification. Social Netw. **1**(3), 215–239 (1979)
10. Kleinberg, J.M.: Authoritative sources in a hyperlinked environment. J. ACM **46**(5), 604–632 (1999)
11. Redyuk, S., Kaoudi, Z., Markl, V., Schelter, S.: Automating data quality validation for dynamic data ingestion. In: EDBT (2021)
12. Schelter, S., et al.: Unit testing data with deequ. In: SIGMOD (2019)
13. Schelter, S., Lange, D., Schmidt, P., Celikel, M., Bießmann, F., Grafberger, A.: Automating large-scale data quality verification. In: VLDB (2018)

An Ontology for Representing and Querying Semantic Trajectories in the Maritime Domain

Georgios M. Santipantakis, Christos Doulkeridis(✉), and George A. Vouros

Department of Digital Systems University of Piraeus, Karaoli & Dimitriou 80, 18534 Piraeus, Greece
{gsant,cdoulk,georgev}@unipi.gr

Abstract. This paper presents the design of an ontology for the representation of enriched semantic trajectories in the maritime domain. The ontology supports vessel trajectories, at different levels of detail, enriched with vessel characteristics, weather information and events, as well as topological and proximity relations to geographical areas of interest (such as ports and protected areas). The paper describes how raw data from diverse data sources are integrated in order to produce enriched semantic trajectories, represented as linked data in RDF, in compliance with the proposed ontology. Moreover, the added-value of the ontology is demonstrated by means of semantic queries that retrieve information about moving vessels and their trajectories using complex criteria. The linked data and our ontology are publicly available as five star linked open data, offering a valuable resource to the research community.

Keywords: Ontology · data integration · semantic trajectories · maritime domain

1 Introduction

In the maritime domain, a wide range of heterogeneous data sources need to be integrated, enriched and exchanged between different endpoints to enable data analytics and machine learning tasks [1]. A maritime domain ontology provides the conceptualizations and descriptions in the domain, and enables the interplay between different environments. In this work we propose an ontology which supports various data sources on the Web, including text, sensor data, semi-structured and binary files, in both archival and streaming nature. For example, vessels report through a live streaming endpoint their positions using AIS messages (Automatic Identification System) which typically include the vessel's current position, its transcoder's identification number (Maritime Mobile Service Identity – MMSI), and optionally, a general vessel type, e.g., cargo. However, positional data can become useful for further analysis, when combined with data provided by other sources, reporting departure/destination seaports, weather conditions, type of lading, protected areas, coastline, etc. [20].

A. Abelló et al. (Eds.): ADBIS 2023, LNCS 13985, pp. 224–237, 2023.
https://doi.org/10.1007/978-3-031-42914-9_16

Although a few attempts towards semantic modeling in the maritime domain exist (such as MarineTLO [18], EUCISE-OWL [12], RMSAS Movement Ontology [3], GeoLink [9]), to the best of our knowledge, they provide only a limited view of trajectories and do not explicitly represent concepts (possibly stemming from different data sources) that are critical for mobility, as will be explained in the related work section. For instance, a trajectory of a vessel needs to be associated with the prevailing weather conditions, in order to perform accurate route forecasting. As another example, a trajectory needs to be annotated with the regions crossed (such as protected areas), in order to detect illegal activity. Finally, it is important to annotate trajectories with complex events (e.g., overspeeding near ports) to enforce regulations by maritime authorities, but also for the timely detection of real-time incidents that affect safe navigation.

Motivated by these real-world requirements and based on our experience from previous research projects related to maritime data, we propose a refined ontology for the representation of semantic trajectories of moving vessels. The proposed ontology was developed by group consensus with domain experts and maritime data producers and consumers, over a period of 12 months, following a data-driven approach according to the HCOME methodology [8].

Specific contributions that this work makes are as follows:

- We propose an ontology for the representation of maritime data, focusing on mobility aspects by means of semantic trajectories enriched with various types of information.
- We show how the ontology can be used in practice, by populating the ontology with data from various data sources and by demonstrating complex queries over integrated maritime data.
- We improve the accessibility to both the surveillance, sensor and contextual data, as they are made available online through a single SPARQL endpoint.

Apart from the variety of concepts that can be represented (trajectories, weather, events, context, geographical objects), notable features of our ontology include: (a) flexible trajectory representations at different levels of detail (raw positions, sequence of segments, trips between ports), (b) time-evolving characterization of trajectories (e.g., a trajectory is considered "dangerous" based on the cargo of the vessel for the specific trip), (c) re-using (importing) concepts from existing, well-established ontologies (DUL, OGC, TIME), (d) the representation of geometries in compliance with GeoSPARQL.

The rest of this paper is structured as follows: Sect. 2 describes the diverse data sources supported, while Sect. 3, presents the ontology and its constituent modules in more detail. In Sect. 4, we showcase the added value of our approach and Sect. 5 briefly reviews existing related efforts. Finally, in Sect. 6, we conclude the paper and describe our plans for future work.

2 Maritime Data Sources

In the maritime domain, a wide set of diverse data sources [7,11], need to be combined in order to derive an enriched data set, suitable for downstream analysis tasks [1]. We classify existing raw data sources in dynamic and static data. The former category includes time-evolving (streaming) data, mainly the positions of moving vessels as well as the respective weather conditions. The latter category contains archival (batch) data sets with information about the maritime domain: vessel characteristics, ports, protected areas, coastlines, bathymetry data, etc.

Positional Data (AIS): The maritime Automatic Identification System (AIS) data is obtained from many different terrestrial and satellite sources which collect messages transmitted from vessels [2]. Typically, each vessel reports its information by sending an AIS message every few seconds. Each message contains the coordinates of the vessel, a timestamp, an identifier for the vessel, its speed and heading. This is a dynamic data set of streaming surveillance data which reports positions of vessels, provided for past dates as CSV files.

Weather Data: Weather data typically are provided as forecasts of certain variables, such as u-wind, v-wind speed, precipitation, air temperature, etc. The Copernicus Climate Change Service (C3S) is a commonly used weather information data source, which provides the data in the binary GRIB2 format.

Places of Interest: Places of interest in the maritime domain may include seaports, protected areas, fishing areas, endangered species habitat areas, exclusive economic zones (EEZ), etc. Any geometry type can be used for the representation of a place of interest, e.g. a seaport usually is represented by a point geometry, while a protected area can be represented by a polygon. A commonly used source for seaports around the globe can be considered the United Nations Code for Trade and Transport Locations (UN/LOCODE), which is a geographic coding scheme developed and maintained by United Nations Economic Commission for Europe. Features and facilities of the seaport such as lift and cranes, railway terminal, etc. are also useful when available. Data from this source can be also combined with attributes from the World Port Index (WPI), a database about ports, to extract additional seaport characteristics and facilities. The World Database on Protected Areas (WDPA) can be considered as a typical data source for protected areas. WDPA is arguably one of the largest global database of marine and terrestrial protected areas that is publicly available.

Vessel Characteristics: Finally, data sources for the vessel characteristics need to be consulted in combination with the surveillance data. Since vessel transcoders can be transferred from one vessel to the other over time, the vessel characteristics data source needs to be temporally aligned with the data source providing the surveillance data, especially if archival data are being processed. The vessel characteristics data usually are collected via static AIS messages, reporting the type of vessel, dimensions, type of lading, and some times the departure and destination ports. For this reason, data sources providing surveillance data can also provide the vessel characteristics of most of the vessels reported in the surveillance data.

The challenge is to integrate these disparate data sources under a common semantic model that is able to represent the main entities along with associations. The significance of having such a conceptualization of the maritime domain is immense for future maritime applications, as it fosters efforts for standardized data representation, exchange and interoperability. This motivates the design and implementation of our ontology.

3 The VesselAI Ontology

The developed ontology was initially guided from specific use case scenarios and data sources in maritime domain, all provided in the context of the VesselAI project[1]. International standards and publicly available documentation for maritime[2] data were taken into consideration towards standardizing terminology across domains, complying with standard specifications. The ontology imports and reuses concepts from the upper ontologies DUL[3], which provides generic concepts and properties, OGC[4] ontology for representing spatial information, TIME[5] ontology, which provides the vocabulary for representing temporal information, and the World Meteorological Organization (WMO)[6] for the description of weather conditions.

3.1 Methodology and Basic Concepts

The views of stakeholders including domain experts in the maritime domain and data scientists, have been considered towards providing conceptualizations that are valid and coherent in the domain. However, not all the stakeholders share the same skills on ontology engineering and a collaborative ontology engineering methodology was required. A shared space should be also available where the ontology would be accessible by anyone, as well as a protocol which allows message exchange for information sharing and argumentation, ontology updates and multiple ontology versions.

We selected a methodology that best fits the requirements for this task, namely the HCOME methodology [8]. HCOME supports collaborative ontology engineering, with the direct involvement of domain experts in conjunction to other stakeholders. HCOME distincts the steps of the ontology evolution with respect to the work space of participants: Phases specify clear requirements for accessing the shared ontology versions, allowing intermediate versions of the ontology before an agreed commit to the next version. HCOME also specifies a generic protocol to be followed for the argumentation and discussion of changes that need to be taken. The current version of the ontology features 508 classes,

1 http://www.vessel-AI.eu/.
2 https://help.marinetraffic.com/hc/en-us/articles/205579997-What-is-the-significance-of-the-AIS-SHIPTYPE-number, http://www.fao.org/fishery/cwp/handbook/h/en.
3 http://www.ontologydesignpatterns.org/ont/dul/DUL.owl.
4 http://www.opengis.net/ont/geosparql.
5 https://www.w3.org/2006/time.
6 https://codes.wmo.int/.

206 object properties and 68 data properties. In the following paragraphs we present the ontology modules and the core concepts and relations of the ontology.

Any entity that can change its position over time is considered as a *moving object*. therefore, a position of a moving object is always given as a pair of temporal and spatial values, namely a spatio-temporal position. Formally, a moving object in this work is an entity defined as a tuple:
`<ID, name, physical attr: (width, height, length, weight)>`
A vessel is a direct subclass of moving object, which in addition to the inherited description of moving object is further specialized by the tuple:
`<IMO, MMSI transcoder, LISTOF trajectory>`
where LISTOF trajectory is a possibly empty list of trajectories as described below.

Trajectories are defined as the segments of a moving object's track, that are important for an application. For the purposes of maritime domain applications, we consider trajectories of vessels that are important for marine safety awareness tasks and marine automation operations. In general, a trajectory of a vessel is expected to connect seaports, i.e., it departures from one seaport and arrives to another (or the same) seaport. The intermediate set of positions can also define meaningful segments of the trajectory. For example, a fishing vessel can depart from a seaport and move to a fishing region, where it applies a fishing movement pattern (with respect to the vessel type and activity) and then move to another fishing region or return to a seaport. Each of the actions described in this example, are represented in the trajectory as trajectory segments defined by initial and final spatio-temporal positions. Formally, a trajectory is defined as a sequence of trajectory parts, i.e., meaningful segments in the trajectory trace, defined by one or more positions and the time interval needed for the moving object to cover the distance in between, where each trajectory part can be enhanced with annotations. The trajectory is defined as a tuple:
`<trajID, trace: LISTOF trajectory part: (LISTOF (time, position), annotation)>`
where the list of trajectory parts is temporally sorted.

This definition of a trajectory allows representing and converting a trajectory at varying levels of spatio-temporal analysis, from sequences of positions over time, to sequences of meaningful segments (comprising one or more positions). Semantic information concerning trajectories and their segments is provided by means of annotations, i.e., descriptions of movement-related aspects, activities, or anything relevant to segmenting the trajectory (e.g., regions crossed, arriving to or departing from seaport).

Trajectory parts can be associated with events. An event is a set of spatio-temporal positions (e.g., a segment, or region) that satisfy a given predicate. The event is defined as a tuple:
`<eventID, annotation, time interval, geometry: LISTOF position>`
A predicate can be an abstract pattern of spatio-temporal movement features of the moving object and its context. It must be noted that it is not within the scope of the ontology to specify these patterns, but rather to represent the classes of

events to be detected and their association to objects trajectories. Events are further distinguished to environmental events affecting moving objects' behaviour such as weather conditions reported for associated regions, and low level events, i.e., a sub-class of event that can be identified directly from the sensor's data.

Finally, trajectory parts can be also associated with regions or points of interest. A region of interest is a place defined by one or more positions that satisfy a given predicate. A region of interest as a tuple:
`<regionID, annotation, geometry: LISTOF position>`
Predicates in the case of regions of interest, can be any combination of geometrical (e.g., the region defined by a given distance from the centre of a seaport) and domain-specific physical (e.g., coastline, island) or artificial (e.g., seaport, lighthouse) structures.

3.2 Main Modules of the Ontology

The proposed ontology provides a generic conceptual framework for the representation of semantic information related to the trajectory, the moving object and the contextual information in the maritime domain. We justify the objectives of this ontology by concrete examples for its use.

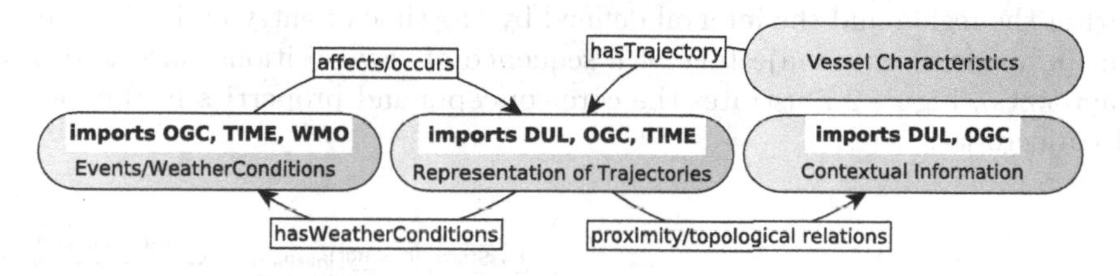

Fig. 1. The connections between the modules of the ontology.

Figure 1 illustrates the interconnections between the main modules of our ontology. The module describing *Trajectories* of moving objects is at the core of the ontology, since the objective is to represent vessel trajectories as "first-class citizens". Each trajectory is associated with a moving object (a vessel) which is described in the *Vessel Characteristics* module. Also, trajectories (and trajectory parts) can be related to *Contextual Information*, that is physical places or regions of interest, by topological relations, such as crosses, within, touches, etc. Also, trajectories are associated with complex events and weather attributes (wind, temperature, wave length and period, etc.) via the *Events/Weather Conditions* module. Finally, both the modules of *Contextual Information* and *Trajectories* of moving objects import GeoSPARQL for the spatial representations of entities.

The ontology has been implemented in OWL2 as a set of modules. Each module is a self-contained ontology, describing a certain part of the domain. The modular structure of the ontology enables the arbitrary combination of modules with respect to the data available or the task applied. The following paragraphs briefly present the contents of each module.

Trajectories of Moving Objects: This module describes the trajectories of moving objects according to the reported surveillance data. It enables various levels of abstraction, to satisfy a wide range of analytics tasks. The core concepts in this module are the `Trajectory` and the `TrajectoryPart`. A trajectory resource can be associated with any number of resources of type `TrajectoryPart`, via the property `hasPart` (inherited from dul:Entity). Each `TrajectoryPart` can be also associated with another `TrajectoryPart` resource, via the properties `hasPart` and `hasNext`, to model either its components or the sequence of trajectory parts forming the trajectory.

Both trajectories as well as trajectory parts inherit properties for the specification of temporal (i.e., a time interval or instance) and spatial constituents (the geometry of the trajectory, trajectory part, or semantic node). Special cases of trajectory parts are: (a) the `RawPosition` (i.e., a reported position of the moving object), b) a `Node` (i.e., a reported position annotated with contextual data), c) a `Segment` (i.e., includes at least two raw positions or nodes). For example, reported positions of a vessel that is anchored in a port can be represented by a single node resource, where the temporal constituent will specify the time interval that the vessel is anchored. In this way, compressed trajectories [4] can also be represented. Similarly, the trajectory part of a vessel crossing a region of interest can be represented by a `Segment`, comprised of the trajectory parts within the region and the interval defined by the time of entry and exit. A route can be modeled as a trajectory of a sequence of raw positions (also known as waypoints). Figure 2 illustrates the core concepts and properties in the module of trajectories.

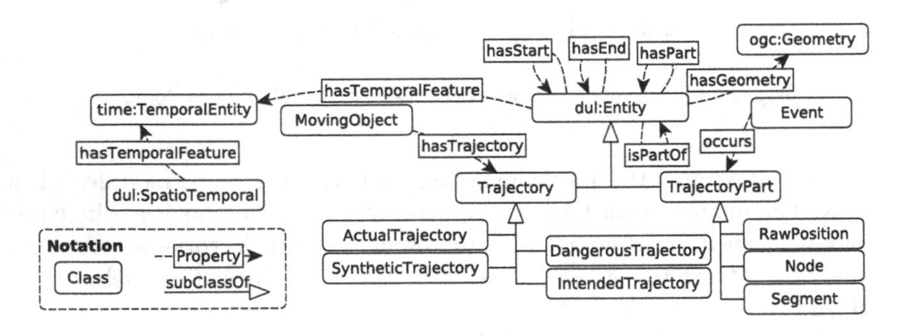

Fig. 2. The main concepts and relations in the trajectories of moving objects module.

Vessels Characteristics: This module describes the vessel types and their characteristics, such as dimensions of vessels (length, width, height, depth), identification information (name, callsign, International Maritime Organization (IMO) code, Maritime Mobile Service Identity (MMSI) code, etc.), depictions if available, Deadweight, draft and vessel type. A taxonomy of 256 vessel types is also provided, to enable subsumption on types when needed. Figure 3 illustrates the core concepts and some of the properties in the module of vessel characteristics.

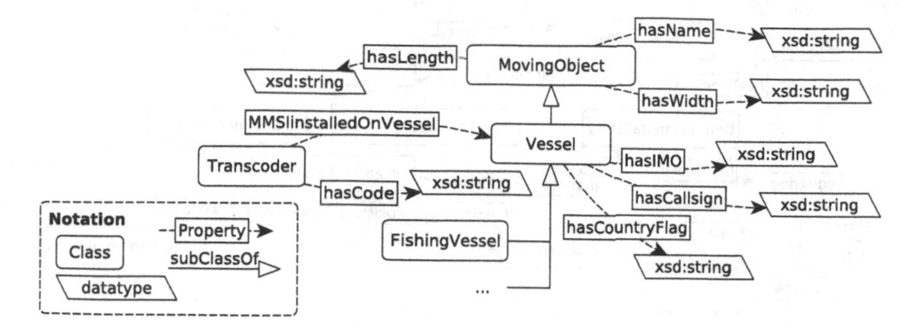

Fig. 3. The main concepts and relations in the module of vessel characteristics.

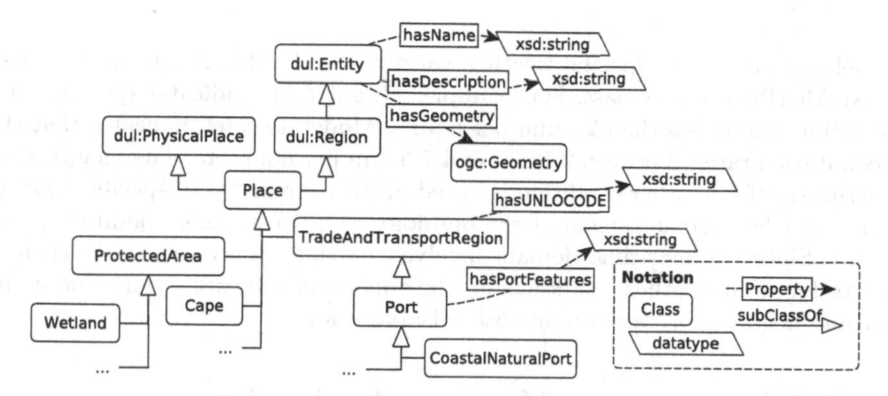

Fig. 4. The main taxonomies of concepts and their relations in the module for contextual information.

Contextual Information: This module describes regions of interest such as fishing regions, protected areas (e.g., Natura2000), as well as physical places that are important for the VesselAI domain, such as ports, lighthouses, coastlines, etc. Physical places are always related to some geometry, also described in this module. Regions of interest in this module are often connected to the module describing trajectories of moving objects, via the origin and destination properties, as well as any proximity and topological relations detected between trajectory parts and the geometries of regions of interest. Figure 4 illustrates a sample of the taxonomies described in the module for contextual information.

Events and Weather Conditions: This module, depicted in Fig. 5, imports the World Meteorogical Organization (WMO) conceptualizations and properties. The purpose of this import is to enable interlinking of weather condition resources with third party external sources. The goal of this module is to describe simple (e.g., moving, turn, stop) and complex events in the domain and their association with trajectory parts via the affects/affectedBy properties. This module also enable the description of weather conditions as events, both using qualitative and quantitative descriptions. The qualitative description employs 400

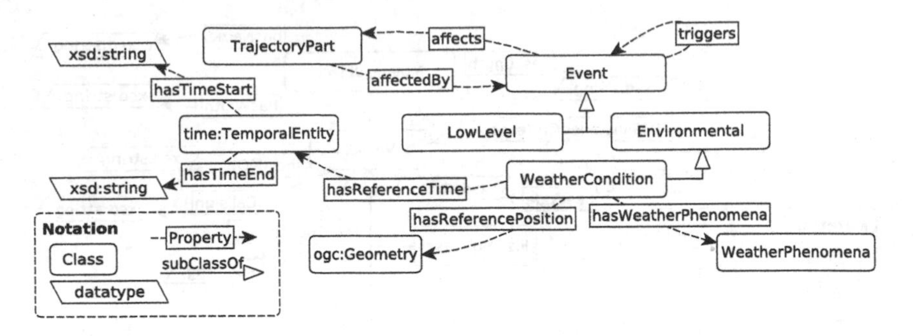

Fig. 5. The main concepts and their relations in the module of events.

predefined presets on specific weather variables and value ranges as instances of WeatherPhenomena class. For example, "Light rain" indicates that the precipitation rate is less than 2.5 mm per hour, "Moderate rain" indicates that the precipitation rate is between 2.5 mm and 7.6 mm per hour, etc. The quantitative description of a weather condition is based on measurements of specific weather variables (the current version of the ontology uses 10 weather condition properties). Since the maritime domain involves moving objects at sea level, it is safe to focus on weather conditions on the surface of the earth, thus ignore any measurements or forecasts on any other isobaric level.

4 Ontology Usage in Maritime Applications

In this section, we describe how the ontology can be used in practice by maritime applications. First, we briefly describe the process of populating the ontology with data from the various data sources (Sect. 4.1), and then we demonstrate the added-value of our semantic modeling approach by means of examples of expressive queries over integrated maritime data (Sect. 4.2).

4.1 Populating the Ontology

Since the maritime data sources come in various input formats (CSV, XML, JSON, Shapefiles, GRIB2) and support different data types (including spatial data), we need a flexible mechanism for data transformation to RDF triples that supports different data sources. Moreover, apart from the variety dimension, maritime data also affects the other two main dimensions of big data: volume and velocity. Historical batch data sets of positional information and weather can easily take several TBs just for a timespan of a couple of years. On the other hand, AIS data is provided in real-time as a bursty data stream of high rate, due to the peculiarities of data collection through a combined network of terrestrial and satelite receivers.

Consequently, these domain characteristics specify the requirements for a highly-efficient solution for data transformation to RDF triples. To this end, we employ RDF-Gen [15], a state-of-the-art tool for data transformation to RDF triples that supports diverse data sources. RDF-Gen has been selected because of its excellent performance and scalability compared to competitive tools on both archival and streaming data sources [14]. The applied method for data transformation to RDF triples is equally scalable on both archival and streaming data sources.

In brief, the architecture of RDF-Gen comprises two main software modules: the *Data Connector* and the *Triple Generator*. Specifically, the Data Connector is responsible for fetching and filtering data from the source, based on user-defined and source-dependent constraints, converting the input data into a sequence of records. It is important to note that input data is accessed in a record-by-record fashion, where a record is formed by the Data Connector. In our implementation, we provide data connectors for all maritime data sources.

Thereafter, each record is provided as a vector of variables to the Triple Generator which is responsible for producing the RDF triples. This is achieved by means of *triple templates*, which are similar to SPARQL graph patterns. A triple template prescribes exactly how a record will be represented as a set of triples. Additionally, each input record's values can be accessed in the triple template by template variables. Then, at execution time, a binding is established between the record's values and the template variables, and the corresponding RDF triples are produced. Any kind of data conversion, filtering, cleaning or processing can be applied via custom functions that are implemented in RDF-Gen and are applicable to the template variables. The transformation process eliminates any outliers (using spatio-temporal criteria), and links resources (e.g., positional data to weather conditions). This approach minimizes the cost of adapting a new data source to the description of a new triple template, while custom functions or triple patterns used for other sources can be reused in each new template, if needed.

4.2 Semantic Queries

A query that features a simple graph pattern and the topological function "within" can request the positions of vessels transporting chemical products within any protected area, at any time. This query combines information stored in trajectory, vessel characteristics and contextual information modules. Since the topological function "within", is not supported by all the triple stores available, precomputed assertions for topological relations can be also supported.

```
SELECT DISTINCT ?g ?gLabel ?p ?pLabel WHERE {
  [] :MMSIinstalledOnVessel ?v . ?v a :Chemical_productsTanker ;
    :hasName ?gLabel ; :hasTrajectory ?tr. ?tr :hasPart ?node .
 ?node :hasGeometry ?geom/ogc:asWKT ?g .
   ?pLabel a :ProtectedArea ; :hasName ?name ;
      :hasGeometry/ogc:asWKT ?p . FILTER(ex:within(?g, ?p)) . }
```

A query that retrieves the nearest port for all fishing vessels being under certain weather conditions (e.g., :10_metre_U_wind-_component_surface and :10_metre_V_wind_component_surface ranging from 3 to 7 m/sec), can provide useful information about behaviour of vessels under these weather conditions, w.r.t. their distance from the nearest port:

```
SELECT DISTINCT ?vessel ?wkt ?wkt2 ?dist ?t WHERE {
   ?vessel a :FishingVessel .
   { SELECT ?wkt ?wkt2 ?dist ?t WHERE {
      ?vessel :hasTrajectory/:hasPart ?node .
      ?node :hasGeometry/ogc:asWKT ?wkt2 .
      ?node :hasWeatherCondition ?c .
      ?c :10_metre_U_wind_component_surface ?Ucomp ;
         :10_metre_V_wind_component_surface ?Vcomp .
      ?port a :Port ; :hasGeometry/ogc:asWKT ?wkt .
      BIND(s:distance(?wkt,?wkt2) as ?dist) .
      FILTER((?Ucomp>"3") && (?Ucomp<"7") &&
         (?Vcomp>"3") && (?Vcomp<"7") ) . }ORDER BY ?dist
   LIMIT 1 }
}
```

A sample data set is available online through a SPARQL endpoint[7], and it can be accessed by third party query engines via federated queries. For example, the following SPARQL query can be submitted at the OpenStreetMap endpoint[8] and combine information about the position of a specific vessel at a given time (provided by our SPARQL endpoint), with information about regions (provided by OpenStreetMap) to compute the distance between the vessel and the region of interest (by wikibase:around service):

```
prefix ogc: <http://www.opengis.net/ont/geosparql#>
prefix : <http://www.vesselAI-project.eu/ontology#>
prefix d: <http://www.datatypes.org/>
SELECT ?area ?distance ?p WHERE {
    ?area osmt:leisure "nature_reserve" ; osmt:name
                                     "V\"{a}der\"{o}arnas naturreservat" .
  SERVICE<http://83.212.101.70:9999/bigdata/namespace/kb/sparql>{
  SELECT (d:toDatatype(strafter(str(?pos),
  "<http://www.opengis.net/def/crs/EPSG/0/4326> "),
  "http://www.opengis.net/ont/geosparql#wktLiteral") as ?p) WHERE {
        ?vessel a :Chemical_productsTanker ; :hasName "TERNVAG" ;
 :hasTrajectory/:hasPart ?node .
    ?node :hasTemporalFeature/:hasTimeStart "2021-01-15 12:53:37" .
        ?node :hasGeometry/ogc:asWKT ?pos .
  } } SERVICE wikibase:around {
    ?area osmm:loc ?coordinates. bd:serviceParam wikibase:center ?p.
    bd:serviceParam wikibase:radius "300". # kilometers
    bd:serviceParam wikibase:distance ?distance. } }
```

[7] Accessible at: http://83.212.101.70/demo.html.
[8] https://sophox.org/.

5 Related Work

Our work is tightly related to semantic modeling of trajectories [10,17] as well as semantic ontologies for the representation of maritime information. The datAcron ontology [13,16,19] has been previously proposed for representing semantic trajectories of vessels and aircraft at multiple, interlinked levels of detail. Trajectories can be enriched with miscellaneous data, such as prevailing weather conditions, vessel characteristics, spatial and spatio-temporal relations with geospatial data, as well as complex events. The VesselAI ontology adopts the design of the datAcron ontology, but it focuses on and extends the part that targets the maritime domain. That is we have refined the conceptual model to cover more concepts from the maritime domain, including bathymetry data, more detailed representation of information about ports, as well as a wider coverage of geographical regions (including WDPA).

In previous work on modeling semantic trajectories, the Baquara[2] ontology [5] provides a rich set of constructs for the representation of semantic trajectories, specified as sequences of episodes, each associated with raw trajectory data, and optionally, with a spatio-temporal model of movement. Beyond representing trajectories only as sequences of episodes, there is no fine association between abstract models of movement and raw data, thus limiting its applicability for analysis tasks that explicitly need this association.

Yan et al. [21] also present their approach for semantic querying of trajectories, by integrating three ontologies into one. In more detail, the Geometric Trajectory Ontology describes the spatio-temporal features of a trajectory, the Geographic Ontology describes the geographic objects, while the Domain Application Ontology describes the thematic objects of the application.

EUCISE-OWL [12] is an ontology representation for information sharing in the maritime domain. However, the main focus is on standardized information exchange, rather than modeling of semantic trajectories. Similarly, MarineTLO [18], the Marine Top Level Ontology, is focused on biodiversity data, rather than mobility data and its associated concepts (events, routes, weather, etc.).

GeoLink [9] is a modular ontology, which consists of an interlinked collection of ontology design patterns, aiming at reuse of existing geoscience repositories. One of the content patterns, namely the Cruise pattern, supports the representation of trajectories, however there are two important distinctions with our work, stemming from the fact that GeoLink does not explicitly focus on mobility. First, trajectories (and parts of trajectories) in GeoLink cannot be associated with critical information for understanding mobility (notable examples include complex events and prevailing weather conditions). This is due to the fact that GeoLink reuses the Semantic Trajectory pattern proposed in [6], which offers a generic way to model trajectories as sequences of *fixes* (i.e., spatio-temporal points) that can form *segments* and be associated with attributes via a generic *hasAttribute* property. In contrast, the VesselAI ontology has higher expressiveness by means of the content-rich and (maritime) domain-specific representation that it offers. Second, GeoLink cannot explicitly support different levels of detail for the same trajectory, thus providing a "single" view of a trajectory, instead of multiple views at different levels of detail.

6 Conclusions and Future Work

In this paper, we presented an ontology for representing semantic trajectories in the maritime domain interlinked with various data, including weather, events, static vessel characteristics, as well as geographical areas and points of interest. The ontology has been populated with data from miscellaneous sources of the maritime domain, in different formats and types. Moreover, we demonstrated the added value of our conceptualization, by formulating queries with diverse constraints for flexible data retrieval. We make the ontology (https://zenodo.org/record/7450963) and the linked data (https://zenodo.org/record/7102043) publicly available as five star linked open data.

In technical terms, we have adopted the HCOME methodology for the design of the ontology, and we follow best practice by reusing existing top-level ontologies (including DUL, OGC, TIME, WMO). We have provided public access to our resources, both the ontology and integrated maritime data produced using our tools, using the Zenodo repository to ensure availability. Regarding compliance with the FAIR principles, the ontology has an open license, it is available in a permanent URI and all the resources are completely available online and archived in Zenodo (with a corresponding DOI).

The potential impact of our ontology is arguably very high, as it targets the maritime domain and it is widely accepted that the "blue economy" is a major contributor for the global economy. Companies and organizations working with maritime data have to struggle with data heterogeneity and lack of standardized data exchange and interoperability. Typically, this is confronted using ad-hoc solutions. Meanwhile, shipping generates extremely large amount of data in every minute, which potential, however, is still not fully exploited. Our resource is a step forward towards enabling standardization and interoperability.

Regarding future work, we intend to extend the ontology so as to accommodate more data sources and make integrated data at larger scale publicly available. Furthermore, we intend to complement the data set with a set of queries, in order to prepare a benchmark that can be used for evaluation.

Acknowledgment. This work was supported by the Horizon 2020 Framework Programme of the European Union under grant agreement No 957237 (project VesselAI).

References

1. Artikis, A., Zissis, D. (eds.): Guide to Maritime Informatics. Springer, Cham (2021). https://doi.org/10.1007/978-3-030-61852-0
2. Batty, E.: Data analytics enables advanced AIS applications. In: Doulkeridis, C., Vouros, G.A., Qu, Q., Wang, S. (eds.) MATES 2017. LNCS, vol. 10731, pp. 22–35. Springer, Cham (2018). https://doi.org/10.1007/978-3-319-73521-4_2
3. Brüggemann, S., Bereta, K., Xiao, G., Koubarakis, M.: Ontology-based data access for maritime security. In: Sack, H., Blomqvist, E., d'Aquin, M., Ghidini, C., Ponzetto, S.P., Lange, C. (eds.) ESWC 2016. LNCS, vol. 9678, pp. 741–757. Springer, Cham (2016). https://doi.org/10.1007/978-3-319-34129-3_45

4. Fikioris, G., et al.: Fine-tuned compressed representations of vessel trajectories. In Proceedings of CIKM, pp. 2429–2436 (2020)
5. Fileto, R., et al.: The Baquara^2 knowledge-based framework for semantic enrichment and analysis of movement data. Data Knowl. Eng. **98**, 104–122 (2015)
6. Hu, Y., et al.: A geo-ontology design pattern for semantic trajectories. In: Tenbrink, T., Stell, J., Galton, A., Wood, Z. (eds.) COSIT 2013. LNCS, vol. 8116, pp. 438–456. Springer, Cham (2013). https://doi.org/10.1007/978-3-319-01790-7_24
7. Kalyvas, C., Kokkos, A., Tzouramanis, T.: A survey of official online sources of high-quality free-of-charge geospatial data for maritime geographic information systems applications. Inf. Syst. **65**, 36–51 (2017)
8. Kotis, K., Vouros, G.A.: Human-centered ontology engineering: the HCOME methodology. Knowl. Inf. Syst. **10**(1), 109–131 (2006)
9. Krisnadhi, A., et al.: The geolink modular oceanography ontology. In: Arenas, M., et al. (eds.) ISWC 2015. LNCS, vol. 9367, pp. 301–309. Springer, Cham (2015). https://doi.org/10.1007/978-3-319-25010-6_19
10. Parent, C., et al.: Semantic trajectories modeling and analysis. ACM Comput. Surv. **45**(4):42:1–42:32, 2013
11. Ray, C., et al.: Mobility data: a perspective from the maritime domain. In: Big Data Analytics for Time-Critical Mobility Forecasting, pp. 3–31. Springer, Cham (2020). https://doi.org/10.1007/978-3-030-45164-6_1
12. Riga, M., et al.: EUCISE-OWL: an ontology-based representation of the common information sharing environment (CISE) for the maritime domain. Semantic Web **12**(4), 603–615 (2021)
13. Santipantakis, G.M., et al.: Specification of semantic trajectories supporting data transformations for analytics: the datAcron ontology. In Proceedings of SEMANTiCS, pp. 17–24. ACM (2017)
14. Santipantakis, G.M., et al.: RDF-Gen: Generating RDF triples from big data sources. Knowl. Inf. Syst. **64**(11), 2985–3015 (2022)
15. Santipantakis, G.M., Kotis, K.I., Vouros, G.A., Doulkeridis, C.: RDF-Gen: Generating RDF from streaming and archival data. In Proceedings of WIMS, New York, USA (2018)
16. Santipantakis, G.M., Vouros, G.A., Glenis, A., Doulkeridis, C., Vlachou, A. .: The datAcron ontology for semantic trajectories. In Proceedings of ESWC, vol. 10577, pp. 26–30. Springer (2017). https://doi.org/10.5281/zenodo.570885
17. Spaccapietra, S., et al.: A conceptual view on trajectories. Data Knowl. Eng. **65**(1), 126–146 (2008)
18. Tzitzikas, Y.: Integrating heterogeneous and distributed information about marine species through a top level ontology. In: Garoufallou, E., Greenberg, J. (eds.) MTSR 2013. CCIS, vol. 390, pp. 289–301. Springer, Cham (2013). https://doi.org/10.1007/978-3-319-03437-9_29
19. Vouros, G.A., et al.: The datAcron ontology for the specification of semantic trajectories - specification of semantic trajectories for data transformations supporting visual analytics. J. Data Semant. **8**(4), 235–262 (2019)
20. Vouros, G.A., et al. (eds.). Big Data Analytics for Time-Critical Mobility Forecasting, From Raw Data to Trajectory-Oriented Mobility Analytics in the Aviation and Maritime Domains. Springer (2020). https://doi.org/10.1007/978-3-030-45164-6
21. Yan, Z., Macedo, J., Parent, C., Spaccapietra, S.: Trajectory ontologies and queries. Trans. GIS **12**(s1), 75–91 (2008)

Extracting Provenance of Machine Learning Experiment Pipeline Artifacts

Marius Schlegel(✉) and Kai-Uwe Sattler

TU Ilmenau, Ilmenau, Germany
{marius.schlegel,kus}@tu-ilmenau.de

Abstract. Experiment management systems (EMSs), such as MLflow, are increasingly used to streamline the collection and management of machine learning (ML) artifacts in iterative and exploratory ML experiment workflows. However, EMSs typically suffer from limited provenance capabilities rendering it hard to analyze the provenance of ML artifacts and gain knowledge for improving experiment pipelines. In this paper, we propose a comprehensive provenance model compliant with the W3C PROV standard, which captures the provenance of ML experiment pipelines and their artifacts related to Git and MLflow activities. Moreover, we present the tool MLFLOW2PROV that extracts provenance graphs according to our model from existing projects enabling collected pipeline provenance information to be queried, analyzed, and further processed.

Keywords: Machine Learning · Model Development · ML Experiment · ML Artifact · Provenance · MLflow · W3C PROV

1 Introduction

The development of machine learning (ML) models typically requires highly exploratory and iterative approaches [2,3]: From data ingestion, dataset processing and feature engineering, over model design and algorithm selection to model training and optimization, it is often a long journey to produce a model that maximizes a given target metric. Each step, iteration, and evolution of an experiment pipeline produces new artifacts, such as data, model, metadata and software artifacts. As a result, achieving traceability and reproducibility of ML pipeline artifacts is often quite challenging [19].

Since the manual management of ML experiment pipelines and their artifacts is simply inefficient and error-prone, ML experiment management systems (EMSs) have been established [1,10,14,16,23], which enable the systematic collection and management of ML experiment pipelines and their artifacts. A popular open-source EMS representative is MLflow [5,24]. Its experiment tracking component captures ML pipeline data, model, metadata and software artifacts based on intrusive log statements that call the MLflow Python API.

A. Abelló et al. (Eds.): ADBIS 2023, LNCS 13985, pp. 238–251, 2023.
https://doi.org/10.1007/978-3-031-42914-9_17

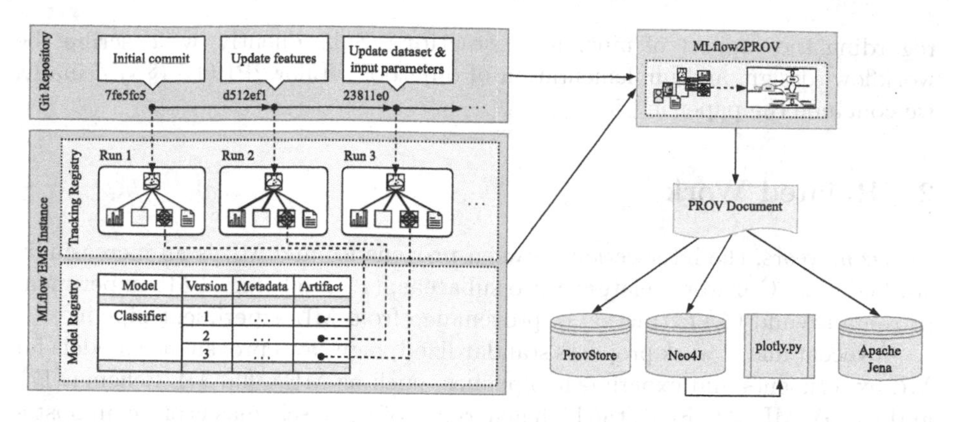

Fig. 1. The evolution and execution of an ML experiment pipeline utilizing the Git VCS and the MLflow EMS as input for the provenance extraction and provenance graph generation with MLFLOW2PROV.

Unfortunately, EMSs, including MLflow, generally offer limited provenance capabilities: First, they do not capture an experiment pipeline's complete provenance, including all development and experiment activities, which originate from the pipeline's source code repository as well as EMS instance managing experiment pipeline runs and corresponding artifacts produced (see Fig. 1). Second, the provenance documentation cannot be exported in an interoperable format such as W3C PROV [8] and, thus, cannot be consumed by PROV-compatible tools enabling further processing, analysis, and visualization.

To address these issues, we adapt the approach pursued in previous work [20]: Based on the extraction of provenance information from both experiment code repositories and EMS instances, we generate provenance graphs according to our PROV-compliant model. By importing such a provenance graph into a database (e.g., Neo4J), queries can be executed that require knowledge of EMS and version control system (VCS) activities (see Fig. 1). However, the approach is not capable of expressing and processing ML experiment pipeline abstractions. Therefore, our contributions are the following:

1. We introduce an extended version of our PROV-compliant provenance model capturing the provenance of ML experiment pipelines and their artifacts related to activities in the corresponding Git repositories and MLflow instances.
2. We present an extended version of our tool MLflow2PROV[1], which, given a Git repository and an MLflow instance, extracts a provenance graph based on our introduced provenance model (see Fig. 1).

This paper is structured as follows: First, we give an overview of related work (§ 2). In § 3, we present our PROV-compliant model for the provenance of collaborative ML model development and experiment projects and the extensions

[1] https://github.com/mariusschlegel/mlflow2prov/.

regarding the support of pipeline abstractions. Subsequently, we describe the workflow, design, and implementation of our tool MLflow2PROV (§ 4). Finally, we conclude the paper in § 5.

2 Related Work

In recent years, the intersection between provenance and ML is an increasingly studied area. Our work relates to two subareas: (1.) modeling of ML experiment provenance and (2.) extraction of provenance from ML experiment pipelines.

A recent line of work proposes standardized, high-level provenance models for ML experiments and experiment pipelines, such as MEX [6], ML Schema [17], and PROV-ML [22]. First, the high genericity of these schemas is often an obstacle in practice, underlined by their low adoption. Second, these schemas address only a part of the provenance-relevant activities: The evolution of pipeline artifacts residing in version-controlled repositories, such as source code, software, and, partly, data and metadata, is not considered.

Specific systems for tracking provenance in ML experiment pipelines typically use their own provenance model, which fits specific use cases and environments: ExperimentTracker [18] uses a schema geared towards SparkML and scikit-learn, which, for example, also considers feature transformation flows. DPDS [4] captures fine-grained provenance of data processing pipelines based on basic data preparation operators. Vamsa [13] is a fine-grained, provenance-based analysis approach for data science scripts in Python, which does not require changes to user code and uses a knowledge base about different ML libraries. mlinspect [7] enables fine-grained, provenance-based inspection of ML preprocessing pipelines based on a directed acyclic graph representation of the dataflow (e.g., for detecting data distribution bugs). However, none of the approaches considers activities within VCS repositories related to the evolutions of experiment pipelines.

In contrast, MLflow2PROV and its underlying provenance model combine high-level provenance (also referred to as lineage) extracted from EMSs, such as MLflow, with provenance information from VCSs, such as Git. That enables dependencies, interrelationships and evolutions of ML experiment pipelines in collaborative environments to be captured, queried and analyzed. We argue that MLflow2PROV's approach is complementary to more fine-grained, record-level provenance tracking approaches.

3 Provenance Model

This section introduces our W3C-PROV-compliant provenance model capturing the provenance of ML experiment pipeline artifacts based on Git-related and MLflow-related activities. First, § 3.1 gives an overview of the PROV standard and relevant parts of its data model. Then, § 3.2 presents the provenance model, which is composed of submodels for Git-related and MLflow-related activities.

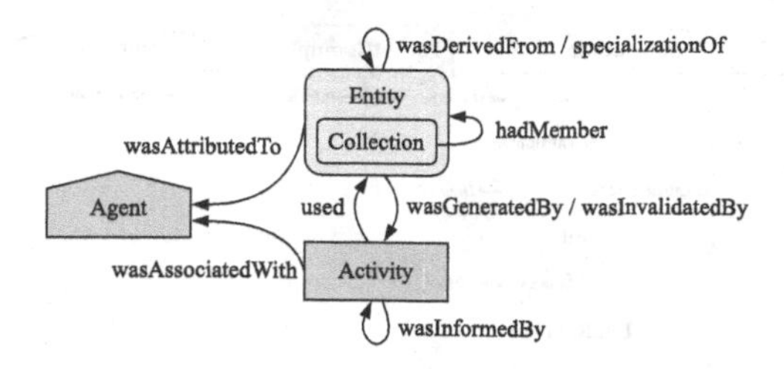

Fig. 2. Overview of relevant PROV concepts.

3.1 W3C PROV

The W3C PROV standard [8] defines PROV-DM [12] – a generic provenance data model that allows domain- and application-specific representations of provenance to be expressed and interoperably exchanged. PROV-DM defines the following structures relevant to this work (see Fig. 2): (1.) types *entity*, *activity*, and *agent* that are involved in producing a piece of data or artifact, and (2.) relations that relate entities, activities, and agents, such as

- *wasGeneratedBy*: the production of a new entity by an activity,
- *wasInvalidatedBy*: the destruction or expiry of an existing entity by an activity,
- *used*: the utilization of an entity by an activity,
- *wasInformedBy*: the exchange of some entity by two activities, where one activity uses some entity generated by the other,
- *wasDerivedFrom*: the transformation of an entity into another one, update of an entity resulting in a new one, or creation of a new entity based on an existing entity,
- *wasAttributedTo*: the attribution of an entity to an agent,
- *wasAssociatedWith*: the assignment of responsibility for an activity to an agent,
- *specializationOf*: the specialization of a general entity by a more specific one, and
- *hadMember*: the belonging of an entity to a collection entity.

Capturing the provenance of the entities, such as files, experiment runs, or any other type of artifact, as well as their interrelations produces a directed acyclic graph (DAG). All nodes and edges of this DAG have defined semantics, thus, resulting in a specific knowledge graph. Consequently, for specific uses of PROV-DM, each type (i.e., entity, activity, and agent) has at least one specialization with application-specific semantics.

3.2 Model Overview

The goal is to capture Git-related activities as well as MLflow-related activities using the provenance model and thus to enable different types of questions:

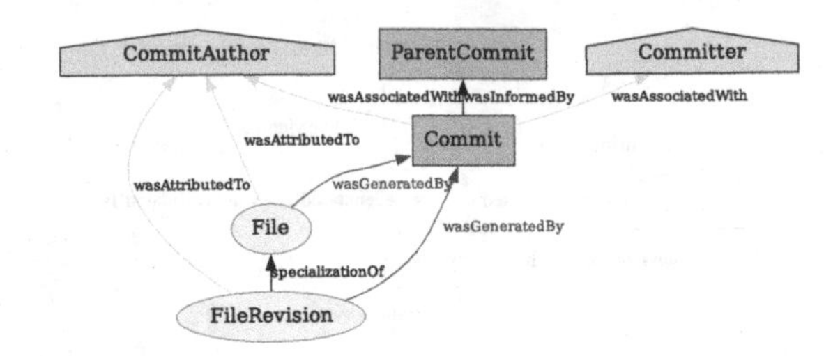

Fig. 3. Submodel representing the creation of a file.

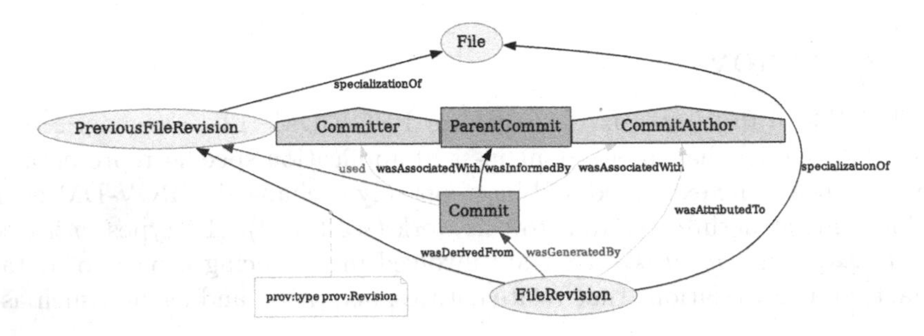

Fig. 4. Submodel representing the modification of a file.

from simple questions, e.g., *"What hyperparameters were used to train the most current version of model x?"* to questions that require knowledge about experiments, runs, etc. as well as source code history, e.g., *"Who created the best-performing model and at what time, and who contributed most to that model?"*. Thus, our provenance model is composed of several submodels for Git-related and MLflow-related activities.

Types and relations of the submodels include various attributes which are essential for later querying. Some examples are:

- predefined attributes, such as `prov:role` and `prov:type`,
- entity attributes, such as `name`, `value`, `timestamp`, `step` of the `Metric` (Entity),
- activity attributes, such as `prov:startTime` and `prov:endTime` of the `*Creation` [Activity], and
- agent attributes, such as `name`, `email`, and `username` of the `*Author` [Agent].

In the following figures, the attributes are omitted for clarity.

Git-Related Provenance Submodels. Inspired by Schreiber et al. [21], three submodels capture the effects that Git commits have on files' status and contents: the creation of a new file (see Fig. 3), the change of a file (see Fig. 4), and the deletion of a file (see Fig. 5). A commit, that adds a file, results in the following provenance information being captured:

- The file and the file revision at the point of creation (current file version) are represented as `File` (Entity) and `FileRevision` (Entity). `FileRevision` is a specialization of `File`.

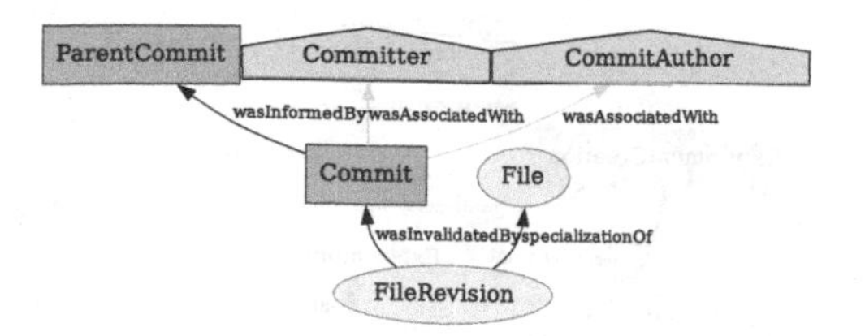

Fig. 5. Submodel representing the deletion of a file.

- The commit is represented as `Commit`, which generated both `File` and `FileRevision`. The previous commit (`ParentCommit`) is also directly linked (wasInformedBy).
- The commit author and committer are represented as `CommitAuthor` and `Committer`. `File` and `FileRevision` are attributed to `Author` representing the responsibility for her content. Both agents, `CommitAuthor` and `Committer`, are responsible for that commit and, thus, are associated with `Commit`.

A commit modifying a file has the following differences regarding the captured provenance information:

- Since `Commit` used `PreviousFileRevision` of `File`, the newly generated `FileRevision` is derived from `PreviousFileRevision`. Thus, `FileRevision` is also a specialization of `File`.

The deletion of a file by a commit leads to the following provenance information:

- The current `FileRevision` is invalidated by `Commit`.

MLflow-Related Provenance Submodels. MLflow's tracking component is organized around the concept of runs, which are executions of some piece of data science code [24]. Runs can be organized into experiments, that group runs belonging to the same task. Four submodels capture the effects of the creation of a new experiment (Fig. 6), the creation of a new run (Fig. 7), the deletion of an experiment (Fig. 8), and the deletion of a run (Fig. 9). While the provenance information collected for the creation of an experiment are structured straightforward and comparable to the file creation model (centered around the `ExperimentCreation`), the run creation model is much more complex and has some noticeable characteristics:

- `RunCreation` uses `FileRevision` (e.g., model training code) to execute a run and, thus, was informed by the corresponding `Commit`. `RunCreation` generates `Run`, which is a collection entity, and `Metric`, `Parameter`, `RunTag`, `Artifact`, and `ModelArtifact`, which are members of `Run`. In addition, the membership relation between `Run` and, if specified, the corresponding `Experiment` is captured. The remaining relations are analogous to the previous submodels.

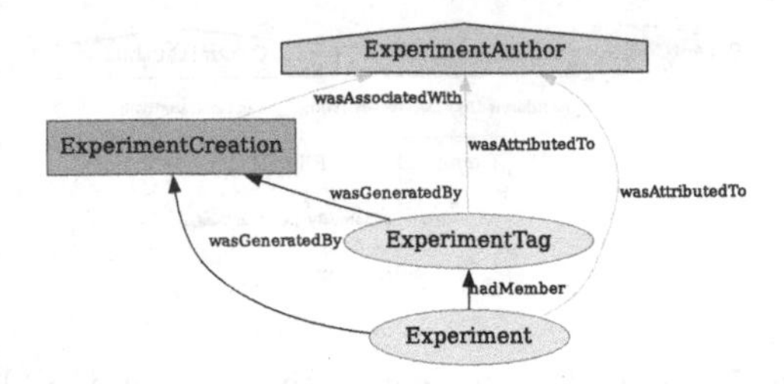

Fig. 6. Submodel representing the creation of an experiment.

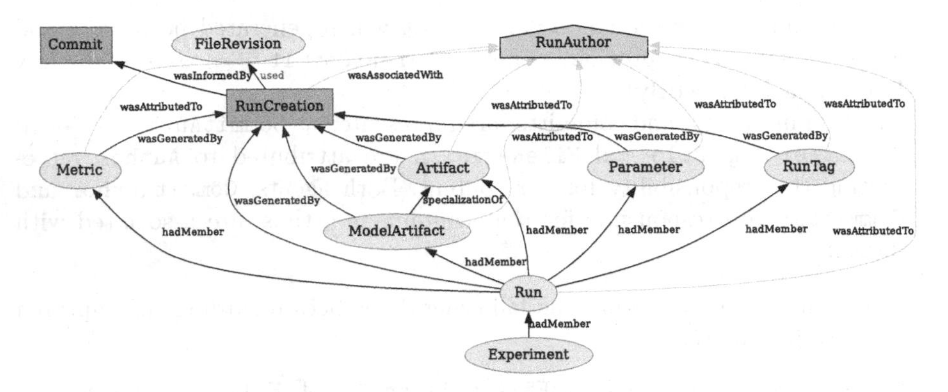

Fig. 7. Submodel representing the creation of a run.

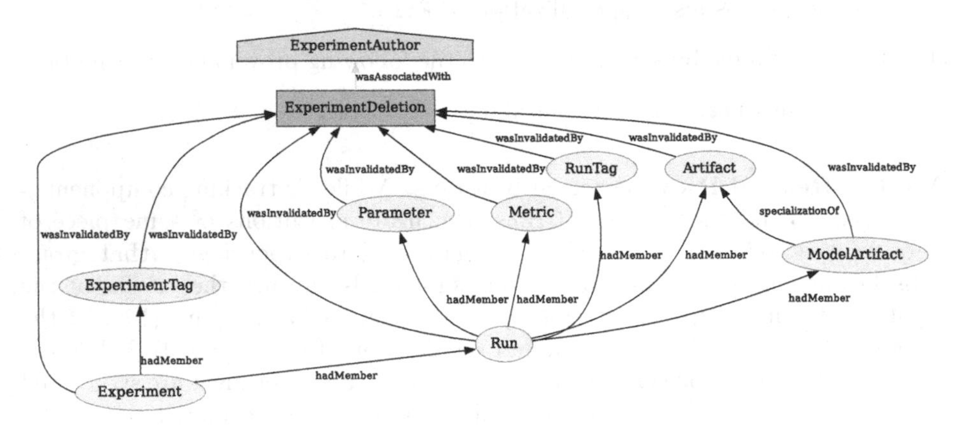

Fig. 8. Submodel representing the deletion of an experiment.

Moreover, deleting experiments (see Fig. 8) and runs (see Fig. 9) is similar to the file deletion model. The only difference is that all members of the two collection entities are deleted as well (whereas `ExperimentDeletion` also includes all associated runs and their members).

Besides tracking, MLflow provides a model registry, which is a centralized model store for selected model artifacts (registered models) with support for

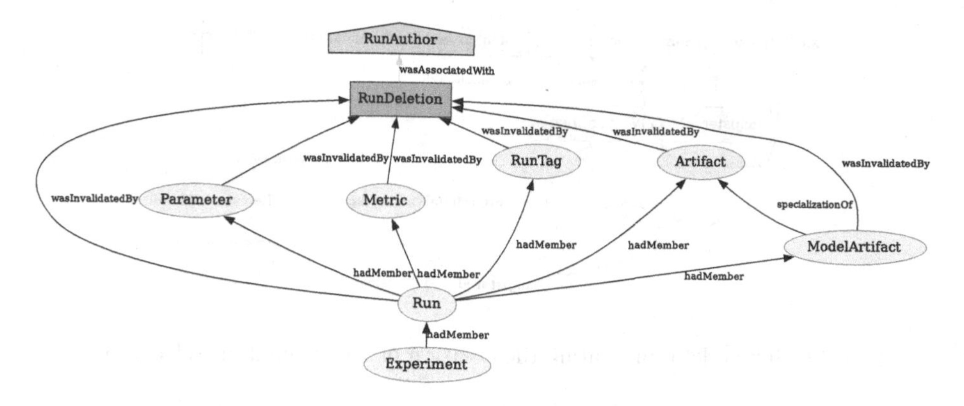

Fig. 9. Submodel representing the deletion of a run.

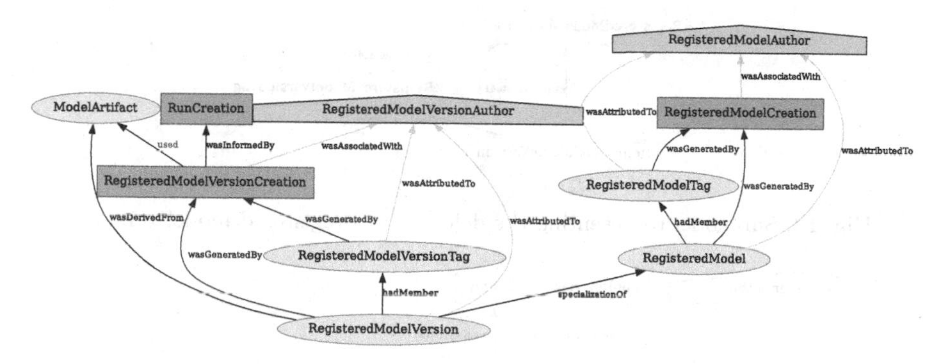

Fig. 10. Submodel representing the creation of a registered model.

model lineage, model versioning, stage transitions (e.g., from staging to production), and annotations. The provenance submodels for registered model creation (see Fig. 10), registered model version creation (see Fig. 11) and registered model version deletion (see Fig. 12) are analogous to the models already discussed. However, a provenance submodel capturing registered model deletion does not and cannot exist, because the deletion of a `RegisteredModel` cannot be traced (e.g., using the file store backend, the entire model directory, including all model versions, is irrevocably deleted).

With MLflow 2.0, the MLflow Recipes framework was integrated. This framework simplifies the implementation of ML experiment pipelines through predefined templates for common ML tasks, conventions (e.g., folder and file structure), and tool support (e.g., command line tool, data profiling, and hyperparameter tuning). These predefined conventions have the additional advantage that provenance information can be extracted from ML experiment pipelines in a more detailed way than was previously possible, further enriching generated provenance graphs.

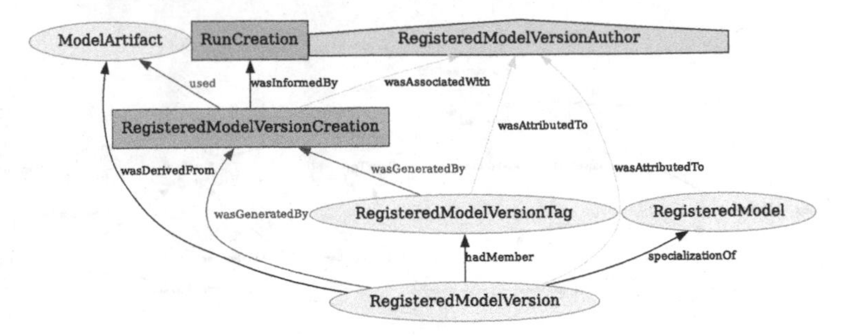

Fig. 11. Submodel representing the creation of a registered model version.

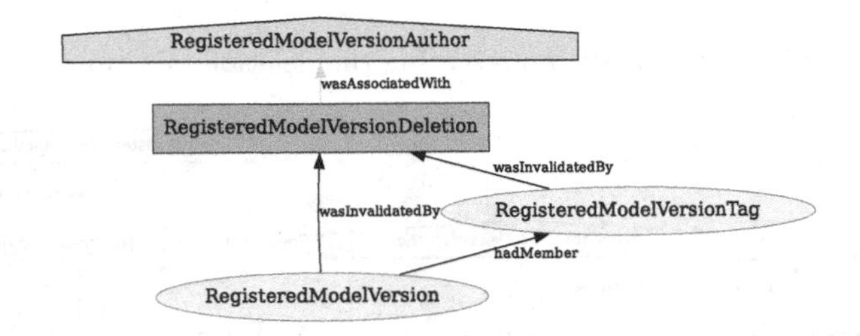

Fig. 12. Submodel representing the deletion of a registered model version.

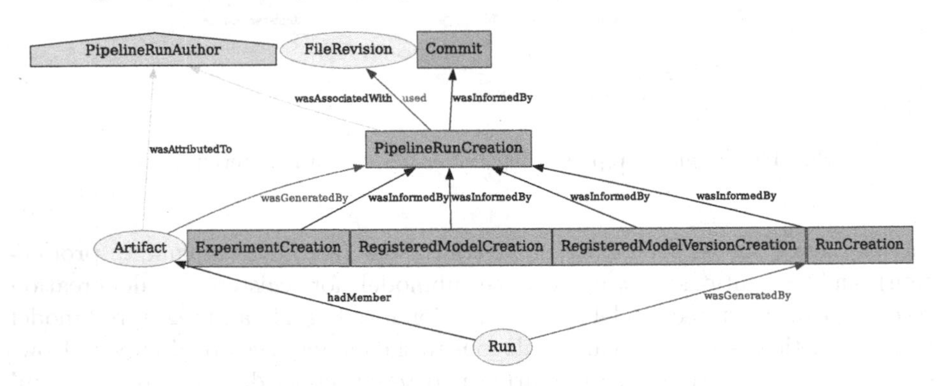

Fig. 13. Submodel representing the creation of a pipeline run.

MLflow Recipes comprises the following relevant concepts: A step represents an individual ML operation, such as ingesting data, fitting an estimator, or evaluating a model against test data. Each step accepts a collection of well-defined inputs and produces well-defined outputs according to user-defined configurations and code. A recipe is an ordered composition of steps used to solve an ML problem.

From the perspective of the tracking and model registry components of MLflow, there are no abstractions for steps or recipes. The artifacts produced in each step of a pipeline are captured in a structured manner so that they can be attributed to the steps. In addition, models and model versions are also registered. A pipeline run (i.e., execution of a recipe) results in the following provenance information being captured (see Fig. 13):

- `PipelineRunCreation` Activity uses `FileRevision` to execute a pipeline run and, thus, was informed by the corresponding `Commit`. Moreover, `PipelineRunCreation` informed `ExperimentCreation`, `RunCreation`, `RegisteredModelCreation`, and `RegisteredModelVersionCreation`, since running the pipeline triggered these creations.
- Each pipeline step produces a step card `Artifact` Entity which is generated by `PipelineRunCreation` and attributed to the `PipelineRunAuthor` Agent.

Since `PipelineRunCreation` does not create a separate pipeline entity, it is not necessary to delete it. Any actions for deleting experiments, runs, and registered model versions are already covered by the submodels explained above (see Figs. 8, 9, and 12).

4 MLFLOW2PROV

This section introduces our tool MLFLOW2PROV that extracts provenance information from the VCS and EMS resources associated with collaborative ML experiment projects. First, § 4.1 explains the underlying workflow of MLFLOW2-PROV. Subsequently, § 4.2 describes the tool's design and implementation.

4.1 Workflow

MLFLOW2PROV provides a command line interface (CLI) that takes a chain of CLI commands and corresponding options (comparable with piped shell commands). Alternatively and functionally equivalent, a `.yaml` file specifying the commands and their options can be used for configuration. The following commands are available:

- `extract`: This command extracts provenance information from the resources of ML experiment projects (using paths and URLs of the associated Git repositories and MLflow instances) and produces a corresponding provenance graph for each.
- `load`: This command loads PROV files (e.g., `.rdf`, `.json`, or `.xml`) containing provenance graphs from previous extractions (e.g., for updating them).
- `save`: This command saves each extracted provenance graph in a separate file according to the specified format (e.g., `.rdf`, `.json`, `.xml`, `.provn`, or `.dot`).
- `merge`: This command merges multiple provenance graphs originating from the extraction of multiple ML experiment projects into a common one.
- `transform`: This command applies various transformations, such as the merging of multiple agents belonging to the same person based on user-defined mappings, the elimination of duplicates, and the replacement of user-related information (e.g., usernames and email addresses) with pseudonyms in the given provenance graphs.

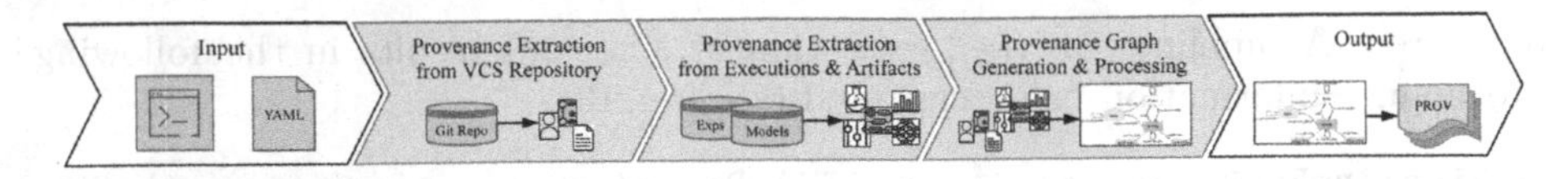

Fig. 14. The workflow underlying MLFLOW2PROV. The major steps for provenance extraction and graph creation are highlighted.

- `statistics`: This command prints statistics related to the given provenance graphs (e.g., number of PROV elements and PROV relations for each type).

By calling MLFLOW2PROV with the `extract` command, a provenance graph is generated for each project based on the given Git repositories and MLflow instances. For each project, this is accomplished in three major steps (see Fig. 14):

1. First, each given Git repository is cloned if the user specified a URL instead of a local path. Following this, a `git log` command is invoked per repository, the output is parsed, and converted to intermediate data structures that represent Git-related entities such as commits, files, users, etc.
2. Next, the given MLflow servers are queried via the MLflow APIs retrieving stored metadata of pipelines, experiments, runs, and artifacts (models, parameters, metrics, etc.) and translated into intermediate data structures[2].
3. Finally, the intermediate data structures are converted into PROV-compliant data structures, which are used to generate the provenance graph associated with each project. In addition, post-processing steps are performed according to further commands (e.g., `transform` and `merge`) as well as the output of the PROV document in the selected formats (`save`).

The generated provenance graph in the output PROV document can be imported for further processing into specific provenance storage systems (e.g., ProvStore [9]) or graph database systems (e.g., Neo4J which can be queried with the graph query language Cypher), or visualized (e.g., with Plotly). We provide a selection of sample integrations in the GitHub repository of MLflow2PROV.

For example, we assume that an extracted provenance graph was imported into Neo4J. Then the question *"Who created the best-performing model and at what time, and who contributed most to that model?"* can be answered with the result of the Cypher query from Fig. 15.

4.2 Design and Implementation

MLFLOW2PROV is implemented in the Python programming language using libraries for MLflow, Git, and W3C PROV. For the design and implementation of MLFLOW2PROV, we adapt architecture design patterns that follow domain-driven design principles, allow managing application complexity, and

[2] Further EMSs can be supported by adapting or extending intermediate data structures and implementing additional procedures for extracting from EMS APIs.

```
CALL {
  MATCH
  (run:Entity)-[:hadMember]-(metric:Entity),
  (run:Entity)-[:hadMember]-(runTagUser:Entity),
  (run:Entity)-[:hadMember]-
    (runTagSourceName:Entity),
  (run:Entity)-[:wasGeneratedBy]-
    (runCreation:Activity)

  WHERE
  metric.`prov:type`="Metric" AND
  metric.name="accuracy" AND
  runTagUser.`prov:type`="RunTag" AND
  runTagUser.name="mlflow.user" AND
  runTagSourceName.`prov:type`="RunTag" AND
  runTagSourceName.name="mlflow.source.name"

  RETURN
  run.run_id AS model_run_id,
  metric.value AS accuracy,
  runTagUser.value AS created_by,
  runCreation.`prov:startTime` AS created_at,
  runTagSourceName.value AS run_source
  ORDER BY
  metric.value DESC LIMIT 1
}
```

```
MATCH
(commit:Activity)-[:wasGeneratedBy]-
  (fileRevision:Entity),
(user:Agent)-[:wasAssociatedWith]-
  (commit:Activity)

WHERE
commit.`prov:type`="Commit" AND
fileRevision.name=run_source AND
fileRevision.`prov:type` = "FileRevision"

WITH
model_run_id,
accuracy,
created_by,
created_at,
run_source,
user.name AS contributed_most,
COUNT(DISTINCT commit) AS commit_count

RETURN
model_run_id,
accuracy,
created_by,
created_at,
contributed_most

ORDER BY
commit_count DESC

LIMIT 1
```

Fig. 15. Cypher query that determines the model that maximizes the metric *accuracy*, the creator and the creation date (left column, subquery), and the user who contributed most to the pipeline's training step (right column).

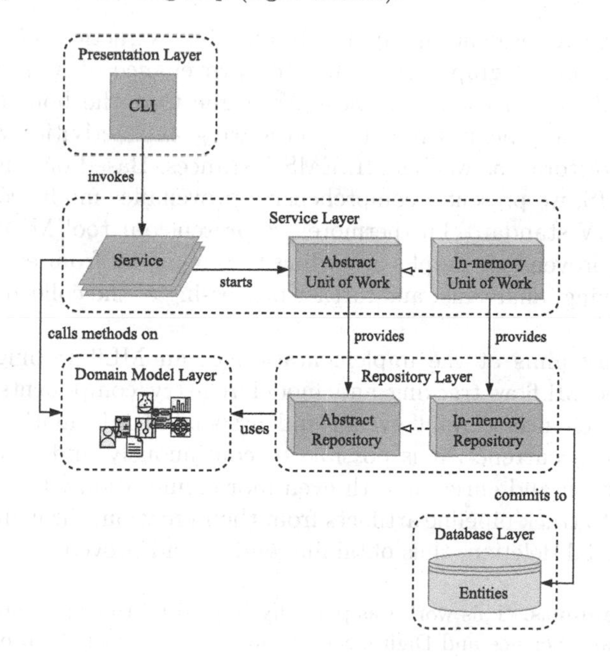

Fig. 16. The software architecture of MLFLOW2PROV.

have been established for Python software [15]. Figure 16 provides an overview of MLFLOW2PROV's software architecture.

According to the *Domain Model Pattern*, our provenance model (see § 3) forms our domain model. It is implemented in a separate subpackage and independent of the other layers of MLFLOW2PROV, which eases testing and refactoring, since our domain model is not tangled up with the other layers. By applying the dependency inversion principle [11], everything else is built around the domain model.

The *Repository Pattern* provides an abstraction over data storage decoupling the Domain Model Layer from the Database Layer, which stores the extracted provenance information. Moreover, we apply the *Service Layer Pattern* to take care of orchestrating our workflows and defining MLFLOW2PROV services for fetching and extraction, processing, as well as provenance graph creation and serialization. This way, the Service Layer is the entry point to our domain model and the CLI as the Presentation Layer exclusively talks to the Service Layer.

For tieing together the Repository and Service Layers, we apply the *Unit of Work (UoW) Pattern*. This pattern provides an abstraction over data access by representing atomic operations. It enables decoupling the services within the Service Layer from the Repository and Database Layers. Each service runs in a single UoW (as part of the Service Layer) that is responsible for accessing the Repository and Database Layers and succeeds or fails as a block.

5 Conclusion

In this paper, we present an approach for the extraction of W3C-PROV-compliant provenance graphs capturing the provenance of ML artifacts originating from ML experiment pipelines. We argue that the holistic provenance of ML experiment pipelines requires considering any activities within associated VCS repositories as well as ML EMS instances. Based on Git as VCS and MLflow as EMS, we present a comprehensive provenance model compliant with the W3C PROV standard. Furthermore, we present our tool MLFLOW2PROV that extracts provenance graphs according to our model from existing projects enabling querying, analyzing, and further processing of the collected provenance information.

Future work aims at the implementation of an MLflow plugin anchoring hooks into the MLflow tracking and model registry components. By directly capturing provenance-relevant events and associated information at the time of their actual occurrence, it is possible to continuously update the generated provenance graph and enrich it with even more annotations. Furthermore, it is then possible to track pipeline artifacts from their creation, through any changes, to their potential deletion, thus obtaining end-to-end provenance.

Acknowledgements. This work was partially funded by the Thuringian Ministry of Economic Affairs, Science and Digital Society in the context of the project "Learning Products" (grant 5575/10-3).

References

1. Allegro AI. ClearML - MLOps for Data Scientists, ML Engineers, and DevOps (2023). https://clear.ml
2. Amershi, S., Begel, A., Bird, C., et al.: Software engineering for machine learning: a case study. In: SEIP@ICSE '19, pp. 291–300 (2019)
3. Chaoji, V., Rastogi, R., Roy, G.: Machine learning in the real world. PVLDB **9**(13), 1597–1600 (2016)
4. Chapman, A., Lauro, L., Missier, P., Torlone, R.: DPDS: assisting data science with data provenance. PVLDB **15**(12), 3614–3617 (2022)
5. Chen, A., Chow, A., Davidson, A., et al.: Developments in MLflow: a system to accelerate the machine learning lifecycle. In: DEEM@SIGMOD '20, pp. 5:1–5:4 (2020)
6. Esteves, D., Moussallem, D., Neto, C.B., et al.: MEX vocabulary: a lightweight interchange format for machine learning experiments. In: SEMANTiCS '15, pp. 169–176 (2015)
7. Grafberger, S., Guha, S., Stoyanovich, J., Schelter, S., et al.: MLINSPECT: a data distribution debugger for machine learning pipelines. In: SIGMOD '21, pp. 2736–2739 (2021)
8. Paul Groth and Luc Moreau. PROV-Overview: An Overview of the PROV Family of Documents. Tech. rep., W3C, 2013
9. Huynh, T.D., Moreau, L.: ProvStore: a public provenance repository. In: IPAW '14, volume 8628 of LNCS, pp. 275–277 (2014)
10. LF Projects. MLflow - A platform for the machine learning lifecycle (2023). https://mlflow.org
11. Martin, R.C.: Agile Software Development. Patterns, and Practices. Prentice Hall, Principles (2003)
12. Moreau, L., Missier, P., Belhajjame, K., et al.: PROV-DM: The PROV data model. Tech. rep., W3C (2013). www.w3.org/TR/prov-dm/
13. Namaki, M.H., Floratou, A., et al.: Vamsa: automated provenance tracking in data science scripts. In: KDD '20, pp. 1542–1551 (2020)
14. Neptune.ai. Neptune.ai - ML Metadata Store (2023). https://neptune.ai
15. Percival, H., Gregory, B.: Architecture Patterns with Python. O'Reill (2020)
16. Polyaxon. Polyaxon - MLOps Tools For Managing & Orchestrating The Machine Learning Lifecycle (2023). https://polyaxon.com
17. Publio, G.C., et al.: ML-Schema: Exposing the Semantics of Machine Learning with Schemas and Ontologies. arXiv:abs/1807.05351 (2018)
18. Schelter, S., Boese, J.H., Kirschnick, et al.: Automatically tracking metadata and provenance of machine learning experiments. In: MLSys@NIPS '17 (2017)
19. Schlegel, M., Sattler, K.-U.: Management of machine learning lifecycle artifacts: a survey. ACM SIGMOD Rec. **51**(4), 18–35 (2022)
20. Schlegel, M., Sattler, K.-U.: MLflow2PROV: extracting provenance from machine learning experiments. In: DEEM@SIGMOD '23, pp. 9:1–9:4, (2023)
21. Schreiber, A., de Boer, C., von Kurnatowski, L.: GitLab2PROV - provenance of software projects hosted on GitLab. In: TaPP '21 (2021)
22. Souza, R., Azevedo, L.G., Lourenço, V., et al.: Workflow provenance in the lifecycle of scientific machine learning. Concurr. Comput. Pract. Exp. **34**(14) (2022)
23. Weights & Biases. Weights & Biases - Developer Tools for ML (2023). https://wandb.ai
24. Zaharia, M., Chen, A., Davidson, A., et al.: Accelerating the machine learning lifecycle with MLflow. IEEE Data Eng. Bull. **41**(4), 39–45 (2018)

References

Author Index

A. AbellxF3 et al. (Eds.): ADBIS 2023, LNCS 13985, pp. 253–254, 2023.
https://doi.org/10.1007/978-3-031-42914-9

Printed in the United States
by Baker & Taylor Publisher Services

Printed in the United States
by Baker & Taylor Publisher Services